# Sustainable Transport in Eastern Europe

Exploring the crucial shift towards sustainable transport, this book examines how we can ensure safe and efficient movement of people and goods without compromising future mobility.

Transport, while vital to humanity, currently generates negative externalities that can offset its benefits. This necessitates a paradigm shift in how we approach transport systems. This book provides a comprehensive analysis of sustainable transport, from its conceptual foundations and current achievements to future perspectives. Grounded in original research, the book introduces concepts such as 'Sustainability 2.0' and 'sustainable infrastructure', offering new pathways for achieving this paradigm shift. It also examines diverse global transport developments, focusing particularly on the European Union and Romania and utilising Romania's recent EU integration and its experience building integrated transport systems to offer valuable comparative insights.

Presenting arguments and providing solutions for steering transport towards sustainable development at both micro- and macro-levels, this book will be invaluable for students, academics, researchers, policymakers, transport professionals and anyone with an interest in the current state and future of sustainable transport.

**Frantz Daniel Fistung** is an engineer with a degree in transportation since 1986 and a PhD in transportation economics since 2001. His main scientific research activity is focused on the field of transport economy, especially the development of sustainable transport. Employed at a prestigious research institute of the Romanian Academy since 1989 to the present, throughout his entire career, most of his research has been focused on the issue of sustainable transport development, whose definition he imposed on the Romanian market since 1999. For over three decades, he has been leading a scientific research group in transport economics within the Romanian Academy research institute called The Institute of Industrial Economics, recently renamed the Centre for Industrial and Services Economy. He has been a member of the Scientific Council for over 25 years. He has received awards from the Romanian Academy (2022) and the Romanian Ministry of Research (2008).

## Routledge Research in Sustainable Urbanism

This series offers a forum for original and innovative research that engages with key debates and concepts in the field. Titles within the series range from empirical investigations to theoretical engagements, offering international perspectives and multidisciplinary dialogues across the social sciences.

**Regenerative Urban Design and Ecosystem Biomimicry**
*Maibritt Pedersen Zari*

**The Politics of Urban Sustainability Transitions**
Knowledge, Power and Governance
*Edited by Jens Stissing Jensen, Philipp Späth and Matthew Cashmore*

**Ecologies Design**
Transforming Architecture, Landscape, and Urbanism
*Edited by Maibritt Pedersen Zari, Peter Connolly and Mark Southcombe*

**Planning Wild Cities**
Human-Nature Relationships in the Urban Age
*Wendy Steele*

**Games and Play in the Creative, Smart and Ecological City**
*Edited by Dale Leorke and Marcus Owens*

**Geographies of Post-Industrial Place, Memory, and Heritage**
*Edited by Mark Alan Rhodes II, William Price and Amy Walker*

**Land Use Management to Support Sustainable Settlements in South Africa**
*Verna Nel and Stuart Paul Denoon-Stevens*

**Sustainable Transport in Eastern Europe**
Foundational Concepts and Perspectives
*Frantz Daniel Fistung*

https://www.routledge.com/Routledge-Research-in-Sustainable-Urbanism/book-series/RRSU

# Sustainable Transport in Eastern Europe

## Foundational Concepts and Perspectives

**Frantz Daniel Fistung**

Routledge
Taylor & Francis Group
LONDON AND NEW YORK

First published 2026
by Routledge
4 Park Square, Milton Park, Abingdon, Oxon OX14 4RN

and by Routledge
605 Third Avenue, New York, NY 10158

*Routledge is an imprint of the Taylor & Francis Group, an informa business*



*British Library Cataloguing-in-Publication Data*
A catalogue record for this book is available from the British Library

ISBN: 978-1-041-24669-5 (hbk)
ISBN: 978-1-041-24668-8 (pbk)
ISBN: 978-1-003-74306-4 (ebk)

DOI: 10.4324/9781003743064

Typeset in Times New Roman
by SPi Technologies India Pvt Ltd (Straive)

**To my wife Lucia and my daughter Bianca**

# Contents

# Introduction

Why do we need to consider a paradigm shift in transportation?

Climate change risks to cities, settlements and key infrastructure will increase rapidly in the medium to long term with further global warming, particularly in places already exposed to high temperatures, along coasts or with high vulnerabilities. Globally, the population in cities and low-lying settlements (about one billion people) is expected to be at risk from coastal climate hazards. The population potentially exposed to coastal inundation is projected to increase by about 20% if global mean sea level rises by 0.15 m from 2020 levels; this exposed population doubles at a 0.75 m rise in mean sea level and triples at 1.4 m (Rossman, 2023). Sea level rise poses an existential threat to some small islands and some low-lying coasts.

The costs of maintaining and rebuilding urban infrastructure, including buildings, transportation and energy systems, will increase with the level of global warming and the associated functional disruptions are expected to be substantial, particularly for cities, settlements and infrastructure located on permafrost in cold regions and on the coasts (IPCC, 2022).

In all this, most experts have concluded that one of the major causes is greenhouse gases (GHGs), which have been produced in increasing amounts in recent years.

Industrial activities, uncontrolled burning of forests and, in particular, transportation, are the main factors producing GHGs and obviously accelerating global warming.

The UN Secretary-General recently stated that: ‘The era of global warming is over; the era of *global boiling* is here. The air is unbreathable. The heat is unbearable. And the level of fossil fuel profits and climate inaction are unacceptable’ (UN, 2023).

Under these conditions, the need for a paradigmatic shift in the processes of human development has become evident, with biodiversity becoming crucial for the development of such a new way of managing the relationship between the human species and the environment.

More specifically, for half a century, since the Stockholm Conference on the Environment held in 1972, mankind has accepted that environmental problems

DOI: 10.4324/9781003743064-1

are inseparable from those of well-being and economic processes in general. The UN's World Commission on Environment and Development was set up with the task of defining a new concept of the overall development of human society for the last decades of the second millennium and beyond. By 1987, the Commission had identified over 60 definitions of the new concept. However, after debates, not infrequently with partisan overtones, the Brundtland report, titled 'Our Common Future', was adopted. It called for a reconciliation between the economy and the environment, a new development path that would sustain human progress not just in a few places and for a few years, but for the whole planet and for a long future.

The concept of sustainable development seeks a permanent reassessment of the links between man and nature and advocates solidarity between generations as the only viable option for long-term development.

If we are at all clear about why we need to intervene in the current trend in human activities and adapt them to the new conditions, we must also understand why transport can no longer continue to operate in the current framework and why it needs a paradigm shift.

It is generally recognised that well-developed economies benefit from an efficient transport system.

Transport, in turn, influences and is influenced by the characteristics of economic development. It contributes substantially to the formation of GDP, creates opportunities for employment and provides indirect benefits oriented towards regional development and globalisation.

Road and rail networks have laid the foundations for the concentration and specialisation of production processes, and the modernisation of transport infrastructure has facilitated the expansion of trade and increased competition between countries and economic regions. Such developments have usually taken place in parallel with economic growth, and the development of transport has contributed to increased employment and income growth.

An efficient transport network is, therefore, a basic condition for economic growth, competitiveness and employment.

For these reasons, we can see transport as an economic power, a link and an influencing factor for other sectors of activity.

At the same time, it is becoming increasingly clear that transport policy is at a crossroads.

While it is recognised that modern and efficient transport systems are of particular importance in the new economic structures, other aspects that are less well understood and taken into account, such as increased congestion, noise and chemical pollution and accidents, must also be taken into account. From this perspective, the old orientation of transport policies towards a permanent increase in the volume of activity and hence in the profits obtained, to the detriment of environmental factors, can no longer continue, as this may otherwise increase the cost of transport activities significantly without a corresponding improvement in the economic effects (satisfying demand, increasing efficiency, etc.) or other effects (environmental, social, etc.).

It is, therefore, becoming increasingly worrying to note that in recent decades the transport system has not been functioning as it should. In many cases, due to negative effects such as chemical pollution, congestion and other such effects, the demand for jobs in outdoor activities, especially in urban areas, has fallen sharply, and the number of cases of people falling ill in areas with heavy traffic has increased.

Transport, at the beginning of the third millennium, is developing in an environmentally unfriendly way. The continued growth in the use of motorised vehicles may soon lead to major disruptions in the world's ability to supply resources such as oil for consumption. At the same time, both the specific emissions from vehicle engines and the increasing spatial occupation or artificial fragmentation of ecosystems are causing major damage to the natural environment and hence to human existence.

From this perspective, new transport development policies (activities and infrastructure) can be built on current trends or take into account the requirements of sustainable development. In the first case, the construction of policies based on past statistical trends leads to the adoption of the solutions and measures needed to sustain or modify these trends in the future. In the second case, policies can be developed on the basis of clear, imposed objectives, with the aim of defining plans for achieving them.

In other words, in the first type of policy, every effort is made to avoid undesirable situations, while in the second type of policy, the necessary measures are taken to achieve the proposed objectives. In practice, transport development policies often use both methods, unfortunately with a greater emphasis on the present situation to the detriment of future objectives. From this perspective, specific policies present the unconditional development of transport as the cornerstone of sustaining overall development, of increasing general welfare. Unfortunately, this is only conjunctural, since most of the actions undertaken in the transport sector involve discretionary use of non-renewable or hardly renewable resources, with serious long-term effects on the environment and human health.

It is, therefore, inevitable that a new approach to transport policies is needed in order to achieve a balance between transport and the environment. One such approach could be that of sustainable transport, a definition based on the parent concept of sustainable development of society.

Looking only at the European continent, the EU transport system offers its citizens safe and secure mobility. The transport sector in the EU-27 employs over 10 million workers and encompasses around 1.1 million enterprises, providing services to people and businesses in the European Union and worldwide (EC, 2020a).

The development of the European transport network over the last decades has been a huge success, succeeding in improving links between communities, economic, social and cultural areas while substantially reducing travel times. At the same time, the negative environmental and health impacts of transport have increased, and this is a situation that should not be tolerated.

Therefore, in its Communication of 11 December 2019, the European Commission presented a European Green Deal (EGD) for the European Union and its citizens (EC, 2020b), in which the transport sector is a major pillar.

The EGD primarily calls for accelerating the transition towards sustainable and smart mobility, given that the transport sector is responsible for a quarter of the Union's GHG emissions and that this share continues to grow. Road transport has had and continues to have an overwhelming influence on this situation. In 2018 alone, for example, road transport accounted for 71% of total GHG emissions from transport. Air transport was responsible for 14.4% of the total (over 4.5 billion tons of $CO_2$ equivalent), shipping for 13.5%, with rail transport having the smallest share at 0.5% (EC, 2020c).

To achieve climate neutrality, transport emissions need to be reduced by 90% by 2050 (EC, 2019).

The shift to sustainable transport means putting users first and providing more efficient, affordable, environmentally and health-friendly mobility alternatives for those living in the European Community area. This will allow the development of smart applications and 'mobility as a service' solutions. Automated mobility and intelligent traffic management systems will give transport the chance to become more efficient and cleaner.

A key objective of the EGD is to significantly boost the use of clean vehicles and alternative fuels.

Achieving ambitious targets from an environmental perspective also requires a shift to more environmentally friendly modes of transport, such as rail and waterways. For this to happen, the EGD strategy aims to expand and improve the capacity and management of both modes. The aim is to increase the amount of freight transported by rail and inland waterways. The Single European Sky should also significantly reduce aviation emissions (by around 10%) without imposing additional costs on consumers and businesses.

Multimodal transport can also support the increased use of environmentally friendly modes, but it needs strong support in the development process. The Energy Efficiency Directive (EED) on Combined Transport is important in this respect, as it is designed to support combined freight transport operations involving mainly rail and waterborne modes, including short-distance transport.

Transport is the only economic sector where GHG emissions are higher than in 1990 and have started to increase again since 2013, despite mitigation efforts.

This is why the EGD identifies a number of concrete initiatives in ten key areas, such as increasing the number of zero-emission vehicles; making sustainable alternatives available to the public and businesses; supporting digitisation and automation; and improving connectivity and access, with the aim of reaching the climate neutrality target for the transport sector by 2050 (EC, 2020b). The EGD calls for a 90% reduction in GHG emissions from transport to make the European Union a climate-neutral economy by 2050, while working towards the ambitious goal of zero-emission transport.

Every year, trillions of dollars are invested in transport infrastructure. The most important question is in which direction the investments should be focused from now on: *do we develop the same types of infrastructure or do we change the paradigm and orient infrastructure development to support/accelerate sustainable transport development*?

An extensive and efficient transport infrastructure is essential for the smooth functioning of economies and the development of regions and cities. When designed efficiently, transport networks can be a driver for productivity and improved quality of life for citizens.

Currently, the design, construction and operation of transport infrastructure systems have a substantial negative impact on natural resources and ecological systems. If allowed to continue, this excessive use of natural resources will have devastating consequences.

Current mega-trends – demographic transition, climate change, digitalisation and artificial intelligence – create additional pressures on transport networks. From this perspective, infrastructure demand is expected to grow in response to increasing urbanisation rates. The provision of transport infrastructure will have to keep pace if urbanisation is to expand without a slowdown in economic growth or a loss in living standards. Even during the recent COVID-19 pandemic, it has been shown that, in any event, the growing pressures on transport links are unlikely to diminish.

Public transport networks play a crucial role in helping cities adapt to these trends. As bottlenecks in OECD countries eased in the wake of the health crisis, travel resumed rapidly, although public transport use remained below pre-crisis levels for a while.

For example, in London, there were less than half the average number of daily tube journeys between April and October 2020 compared to the same periods in previous years (OECD, 2024). The increase in teleworking accounts for part of this decline, but at the same time, there has also been a significant shift towards other individual modes of transport. One example is the increasing use of individual vehicles, especially in urban areas, such as scooters, bicycles, electric scooters, etc.

But the challenge facing developed countries around the world is how to reduce the net environmental footprint without *sacrificing* quality of life.

In other words, how to make our lives better, more resilient and sustainable, without over-consuming resources. The actual transition to sustainability is not and will not be a smooth and easy process. Achieving sustainability implies a complete overhaul of infrastructure, replacing old components with more efficient ones that are better adapted to current and future needs.

In this context, future investments in transport infrastructure need to address environmental sustainability concerns. As countries start to commit to net-zero carbon emissions, an increasing number of *green* and innovative mobility *solutions* are becoming available and need to be integrated into transport planning. It may even be necessary to completely rethink and redesign mobility systems around *sustainable accessibility*.

Globally, according to the Global Infrastructure Outlook, it is estimated that more than two trillion USD of investment in transportation infrastructure will be needed annually by 2040 to support economic growth (UBS, 2023). According to the same study, globally, the infrastructure sector continues to benefit from recent impulses such as the four Ds – decarbonisation, digitalisation, globalisation and demographic shifts. As the inflation and interest rate shocks of 2022–2023 have been overcome, market expectations are for increased trading activity, especially as long-term investments have remained and remain intact (UBS, 2023).

Addressing sustainable transport infrastructure in isolation, on its own, is, however, totally wrong. What is right is to adopt a strategy that encourages the reduction of transport demand for non-essential trips, while at the same time increasing the attractiveness of environmentally friendly modes of transport. *Smart* development, with its emphasis on putting housing close to shops, workplaces and public transport, responds to this challenge.

Rapid urbanisation, increased demand for freight services and plans to spur response to COVID-19 have put pressure on stakeholders to accelerate the pace of infrastructure development. Also, although the pandemic temporarily *pushed* people out of the super-crowded cities of Europe and the United States and into less populated areas, current trends indicate that people are returning to urban areas.

'Urban area' can refer not only to cities but also to towns and suburbs. An urban area includes the city itself as well as the surrounding areas.

Today, about 56% of the world's population – 4.4 billion people – live in cities. This trend is expected to continue, with the urban population expected to almost double its current size by 2050, when almost seven out of ten people will live in cities (World Bank, 2022).

This is why the concept of the 'green city', combining human and natural living together at the urban level, was launched not long ago.

Looking at Europe, around 70% of the EU population lives in urban areas. The state of a city's environment affects everyone who lives there or visits the city. Today, cities across the European Union face a host of environmental challenges that pose risks to the health and well-being of their citizens. In many cities, air pollution continues to exceed EU standards. Rivers, lakes and coastal waters remain under significant pressure. Noise pollution is increasing. Urban sprawl is affecting the availability of green spaces, and waste generation has a major impact on the local environment.

As a result, in January 2021, the European Commission launched the *Green Cities Agreement*, an initiative of European cities to protect the environment. By signing the Green Cities Agreement, city leaders agreed to take further steps to make their cities cleaner and healthier places (EC, 2021). The Green Cities Agreement also considers the interconnection with other EU initiatives, such as the European Green Pact or the UN Sustainable Development Goals.

Clearly, within the utility grid of urban areas, communication infrastructure, and in particular transport infrastructure, is an essential 'pawn'. In strict coexistence with urban development plans, the transport network must support the development strategy of each urban area. In this context, one of the main challenges facing major European cities and beyond is sustainable urban mobility. This issue is a matter of concern for a considerable number of EU citizens. Economic growth, reduced environmental pollution, and personal health are just some of the areas that would benefit from sustainable urban mobility.

But what are the real effects of projects to support sustainable transport in line with the desire for sustainable cities? Is it possible to achieve a 'green city' by developing a sustainable transport network?

Some sources state that no radical change in sustainable urban mobility has been realised(Idu, 2023). For example, air quality has improved in many urban nodes, but pollution continues to exceed safe thresholds.

However, since 2013, the European Commission has allocated more and more funds to sustainable urban transport. In the same year, it published the 'Urban Mobility Package'. Some €13 billion was allocated to this area for the period 2014–2020. Despite this financial progress, managing urban mobility remains an important challenge for urban areas.

The challenge is current and needs to be considered for all urban agglomerations (cities with more than 20,000 inhabitants).

Romania must not remain outside this initiative. That is why 2022 was the year in which three Romanian cities became part of the 'Mission 100 Climate-Neutral Cities by 2030' (Bojneagu, 2023). Bucharest (Sector 2), Cluj-Napoca and Suceava are committed to reducing GHG emissions.

Having made this introduction, we must specify that in the present book, the concept of sustainable transport (derived from the general concept of sustainable development) and throughout the chapters aims to present an overview of the concept of sustainable transport (history and structures, defining elements, etc.), followed by a detailed description of some of the fundamental elements, in our opinion, of this concept. In this context, Chapters 2 and 3 present the two major components of the sector: infrastructure and the activity itself, introducing a new element related to sustainable transport infrastructure. Chapter 4 highlights the connections between sustainable transport and overall economic development, where it is increasingly evident that the external costs caused by transport negatively affect both economic growth and living conditions. Public transport is sustainable, even with less environmentally friendly vehicles, because when used under normal parameters, it replaces a large part of motorised transport in urban areas, which are most affected by pollution (chemical, noise and space occupation). Continuing the theoretical development of the concept of sustainable transport, Chapter 5 presents its prospects for evolution, while Chapters 6, 7 and 8 provide a detailed presentation of solutions, with examples, to support sustainable mobility. The aim was to present economic solutions (e.g., the evolution of markets for different types of eco-friendly vehicles), administrative and strategic planning solutions, as well as a rethinking of overcrowded urban spaces, advocating a new concept of the 'green city'.

Of course, analyses of the interactions between transport and general economic development are not neglected either, with a greater focus on the EU Member States and Romania, nor are the strategies, policies and measures already adopted in the process of developing sustainable transport worldwide.

## References


Bojneagu A., 2023, https://urbanizehub.ro/cum-se-transforma-sectorul-2-bucuresti-si-cluj-napoca-in-orase-verzi/

EC, Sustainable mobility. *The European green deal*, SBN 978-92-76-13916-4. doi:10.2775/588442 NA-02-19-958-RO-N, 2019, https://ec.europa.eu/commission/presscorner/detail/ro/fs_19_6726

EC, *Sustainable and smart mobility strategy-putting European transport on track for the future*, (COM (2020)789 final), 2020a, https://www.eumonitor.eu/9353000/1/j9vvik7m1c3gyxp/vlehreuol0wv

EC, *Transport and the European green pact*, 2020b, https://ec.europa.eu/info/strategy/priorities-2019-2024/european-green-deal/transport-and-green-deal_ro

EC, *EU energy and transport in figures, statistical pocketbook 2020*, ISBN 978-92-76-17565-0, 2020c.

EC, *Exchange rates (Infor Euro)*, 2021, https://ec.europa.eu/info/funding-tenders/how-eu-funding-works/information-contractors-and-beneficiaries/exchange-rate-inforeuro_ro

Idu M., *New standards in urban mobility*, 2023, https://urbanizehub.ro/noi-standarde-in-mobilitatea-urbana/

IPCC (Intergovernmental Panel on Climate Change), Climate Change 2022: Impacts, Adaptation and Vulnerability, 2022, https://www.ipcc.ch/report/sixth-assessment-report-working-group-ii/

OECD, *Transport infrastructure trends and regional development*, 2024, https://www.oecd-ilibrary.org/sites/512e786f-en/index.html?itemId=/content/component/512e786f-en, accessed January 2024.

Rossmann D., *The ninth tentacle of the octopus*, Ed. Rao, ISBN 978-606-006-839-6, 2023

UBS, *Infrastructure outlook 2024*, 2023, https://www.ubs.com/content/dam/assets/asset-management-reimagined/global/insights/asset-class-perspectives/equities/docs/infrastructure-outlook-2024.pdf

UN, Secretary-General's press conference - on climate, 2023, https://www.un.org/sg/en/content/sg/press-encounter/2023-07-27/secretary-generals-press-conference-climate

World Bank, *Urban development*, 2022, https://www.worldbank.org/en/topic/urban development/overview

# 1 Is sustainable transport development necessary?

## 1.1 General considerations

Even though sustainability started as an ideology (a concept) and has had a normative influence on public administration, it can also be seen as a tool to improve decision-making. Some communities use this principle as a major argument in reconciling complex situations in particular areas, overlapping with systems-based solutions. These holistic approaches change the role of transportation from a pathway to an intermediate goal to part of a comprehensive solution. In this context, transportation sustainability may be less about engineering (technical) or economic solutions and more about social, financial and environmental solutions to existing situations in the communities under consideration (Leuenberger et al., 2014).

The need to introduce the principles of the new concept into transportation activities has become increasingly necessary for various reasons.

Surface transportation plays an essential role in people's everyday lives and is a decisive factor for economic competitiveness and employment. From an environmental and resource point of view, for surface transportation (road, rail, water, etc.), the use of 'clean energy' is the main issue (Gray and Frost, 1998). From this perspective, there are several reasons for concern, as carbon emissions need to be reduced at the same time as transportation activity increases, and energy needs to be conserved.

The world energy market is still dominated by fossil fuels, but innovation and adoption of new energy technologies can have a significant influence on efforts to achieve sustainability (Goldemberg, 2006). Renewable energy sources are essential to achieve this goal (Shafiee and Topal, 2009).

Creta and Miller highlighted in their research that the use of batteries and fuel cells can provide sustainable solutions for the existing transportation sector (Dubey and Gunasekaran, 2015). This will certainly help to reduce or eliminate the dependence on the current transportation system based on the consumption of petroleum products.

In another study, Sorrentino emphasised the need for electric vehicles (EVs) and examined the impact of full electrification on greenhouse gas (GHG) emissions was examined (Sorrentino et al., 2014).

DOI: 10.4324/9781003743064-2

Moreover, in recent decades, the analysis of the competition between vehicle infrastructure and agricultural land has intensified. So far, roads have gained ground over agricultural land in industrialised countries, where four-fifths of all motor vehicles were registered. In 2015, for example, statistics showed a worldwide total of almost 1.3 billion motor vehicles (STATISTA, 2019), of which only 26.1% were commercial (goods transportation, special, etc.).

According to forecasts, there will be two billion cars in 2035. Only 2.5% of these will be electric, and 8% will be hybrids. By 2050, the figure will reach 2.5 billion, and the world population will reach 9.731 billion. There will be almost one car for every four people (Auto Bild, 2024).

Incidentally, the global population grew by 12% between 2010 and 2020, to about 7.7 billion people, and the urban population grew by almost 20% during this period.[1]

As the population grows, more and more people around the world need reliable transportation services to access socio-economic activities and opportunities (SLOCAT, 2022).

Global air travel grew by 4.2% from 2018 to 2019. At the end of July 2019, approximately 225,000 airplanes were active in a single day, the highest daily airplane movements ever recorded. For all the major constraints suffered during the health pandemic period between 2019 and 2022, global air travel has grown rapidly in recent years and could prospectively double to 8.2 billion per year by 2037, if globalisation is not affected by protectionism (CBM, 2024)

Global demand for public transportation grew by 4% per year between 2012 and 2017. In 2023, the average ratio of freight supply to cargo space was about 60:40, making it somewhat more balanced than in previous years. Even in 2024, the ratio is expected to remain unchanged from the end of the COVID pandemic, when many transport capacities were utilised above the maximum level. Transporters, who have increasingly entered into long-term contracts in recent years as demand has risen sharply, will again have to rely more on ad hoc deals (TIMOCOM, 2024).

Urban public transport by bus, tram, trolleybus, metro and urban rail has expanded, but in different ways in almost all regions, with bus systems gaining momentum in Europe and urban rail becoming more widespread in Oceania.

At the same time as the number of vehicles has grown rapidly, the infrastructure for moving and parking them has taken up ever larger areas of land.

Millions of hectares of agricultural land, especially in heavily industrialised areas, have been largely redistributed to road transport. Every car in the United Sates, for example, has ended up with about 0.07 ha of land (part of the road infrastructure in that area). Thus, for every five cars that added to the national car fleet, a football-sized piece of land was paved over, redirected from agricultural sectors to transportation infrastructure.

Moreover, well-drained land, useful for farms, was asphalted as a priority, as it was ideal for road construction. Past experience also confirms that land, once asphalted, cannot easily be reclaimed and returned to its former activity (forests, agricultural crops, farms, etc.). As Rupert Cutler put it[2] 'asphalt is the last harvest of the land'.

It has become evident that areas with large vehicle (especially car) fleets have significantly reduced areas of land that could be used for agriculture or housing but, unfortunately, are occupied by roads, railways or buildings ancillary to these infrastructures. In less industrialised countries with smaller vehicle fleets, agricultural land 'has a slight advantage'.

The United States of America (USA), with 214 million motor vehicles at the beginning of the third millennium, had 6.3 million kilometres of *paved* roads, enough to encircle the Earth 157 times around the equator. This means that *the conflict between* cars and farmland is being paid for, for example, by wetlands and rice paddies in countries where famine is everywhere. The outcome of this conflict in China and India, for example, two countries that together account for 38% of the planet's population, will affect the security of everyone's food supply.

If every second person in China were allocated a car, as is the current situation in Japan, the total number of cars in China, relative to the country's surface area, would be 640 million cars, compared to the 13 million that exist today. When farmers in southern China lose a hectare of land planted with rice crops twice a year to the supremacy of the car, building the necessary road transportation infrastructure will be a big shock to rice production.

The situation in India is similar. With the rapid growth of villages and cities, agricultural land has already been greatly reduced. If we add to this the paving of roads for cars, we can see that large tracts of land have also been lost in India. In a country that expects its population to increase by 515 million people by 2050, it is not possible to cover agricultural land needed for food production with asphalt for roads and parking lots.

India, China and other overpopulated countries such as Bangladesh, Pakistan, Iran, Egypt, and Mexico do not have enough land to support a powerful transportation system and simultaneously provide food for their populations.

The competition between cars and farmland becomes a competition between poverty and wealth, between those who can afford cars and those who struggle to buy the food they need.

However, as the pandemic disrupted the supply chain, global vehicle production fell by 14.5% in 2020, with 13 million fewer vehicles produced during the year (SLOCAT, 2022). The decline was mainly in passenger car sales, while global commercial vehicle sales fell by about 2.3 million. Only EVs showed strong growth, reaching a total of 11.2 million electric cars on the world's roads in 2020 (exceeding 2019 estimates by about 1.9 million cars).

By comparison, sales of bicycles in the United States increased by 62% between January and October 2020 (compared to the same period in 2019), and sales of electric bicycles increased by 144%. Eleven European countries averaged an 8% increase in cycling in 2020.

It is time to re-evaluate the future of the car, and of all transportation systems that enable mobility for the entire population, not just a minority, and this must be done without jeopardising food security. Solutions must be found quickly to reduce global demand for transport, while at the same time adopting transport policies in line with the needs of a livelihood and the planet's sustainability, which has been severely tested by the unsustainable development of most of the activities undertaken by mankind over the last century, in particular.

This is why transport infrastructure should not be developed just anywhere and everywhere.

However, adopting policies for the sustainable development of transport systems does not mean an immediate and total change of lifestyle. It is not necessarily necessary to reduce mobility or limit the purchase of vehicles (especially cars). The most important objective is that any desire for mobility, which generates negative environmental and health impacts, must be met, while at the same time measures to mitigate such impacts must be implemented.

In this context, in recent years, the European Union's transport policy has begun to increasingly include environmental issues and, in particular, the concept of sustainable transport. In the framework of the Common Transport Policy, the European Community points out that maintaining the current upward trends in road and air transport activity, both civil and commercial, is not in line with the principles of sustainable development in terms of environmental impact at local, regional and global levels.

Together with other wishes related to the promotion of a sustainable transport system (STS), the European Transport Commission considers that in the long term, in the development of the future transport policy, environmental issues related to climate change, biodiversity loss, the use of conventional energy resources, environmental degradation and quality of life including health will play as important a role as economic and social factors in the development of the future transport policy. Under these circumstances, economic growth must be achieved without the need for a concomitant increase in traffic and the automatic negative environmental and health effects that transport produces. At the same time, the smooth functioning of the entire transport system and fair access to transport services must be ensured.

In this context, the European Union's Sustainable Development Strategy highlights the following main objectives: decoupling transport from overall economic growth; increasing the development of less polluting modes of transport; and a balanced development of regional infrastructure, which will reduce the current regional disparities.

That is why the EU decision was taken in 1980 to create a sustainable common transport market and to develop a Trans-European Network (TEN) as a strategic element to accelerate the achievement of social and economic cohesion. This project was thus designed to ensure interconnection, interoperability and access to national infrastructures. Achieving this goal has required and still requires the use of huge funds with very long payback periods, which has often made them unattractive for private investors. Unfortunately, for a long time, urban transport infrastructure was not included in this policy. However, in recent years, initiatives such as the European Union's CIVITAS or CUTE programmes have tried to remedy this situation and to draw attention to the issue of urban traffic and its better management by promoting sustainable transport. These initiatives are still in their initial phase, with the first projects having started at the end of the 20th century.

The concept of sustainable transport has thus become a reaction to the harmful consequences of its own activity on the economic performance achieved. In fact, unsustainable transport can be characterised by excessive energy consumption, major environmental pollution, significant contributions to the deterioration of public health and often unsatisfactory performance for the various economic and social groups it serves.

The European Commission's Green Paper on Security of Supply of November 2000 noted the important role of transport in increasing energy demand and $CO_2$ emissions. At that time, $CO_2$ emissions from transport accounted for 26% of total emissions from activities in the EU. By 2010, it was estimated that $CO_2$ emissions from transport would increase by 40% compared to 1990.

But the reality was different.

Between 1990 and 2022, GHG emissions from transport in the European Union increased by more than 25%, reaching a total of 1.044 million tonnes of $CO_{2e}$, according to the European Environment Agency. During the same period, the United States has been the biggest contributor to global transportation $CO_{2e}$ emissions for more than three decades. In 2023, transportation emissions in the United States totalled 1.71 billion metric tons ($GtCO_{2e}$). In the US, increased demand for travel, particularly from light-duty and heavy-duty trucks, was the primary driver of its rising transportation emissions. US transportation emissions from fossil fuel combustion rose by approximately 19% between 1990 and 2022. Up until 2014, the European Union was the second-largest contributor to global transportation emissions, but in 2014, it dropped below China. After 1990, China's transport greenhouse gas (GHG) emissions increased dramatically, experiencing a tenfold rise, while the United States saw a more modest, approximately 19% increase in its transport sector's GHG emissions by 2022. The growth in China was fuelled by rapid economic expansion and industrialisation, leading to increased demand for travel and infrastructure. In Romania, between 1990 and 2022, GHG emissions caused by transport increased by over 60%, reaching 21.5 million tonnes of $CO_{2e}$. In the Czech Republic and Hungary, the increase was almost 70% during the same period, while Poland recorded an increase of more than 3.2 times, reaching 73.1 million tonnes of $CO_{2e}$ in 2022 (EUROSTAT, 2019).

Clearly, the development of a STS requires adequate infrastructure. This can be translated, at the urban level, by prioritising routes for less polluting modes of transport (public transport, transport by non-polluting means of transport or walking) and by providing appropriate multi-purpose terminals.

Currently, there are several initiatives to promote sustainable urban transport, most of them aiming at the integration and development of public transport (as in Dublin), the use of green fuels (as in Budapest, the capital of Hungary), restricting access to means of transport in central areas, the development of walking and cycling infrastructure and the development of park-and-ride at public transport terminals.

In North America, the 20th century saw dramatic changes in the specific evolution of each mode of transportation. With the westward expansion, the focus of transportation began to shift towards accessibility, mobility and safety

(Leuenberger et al., 2014). The predominant use of automobiles has led to numerous problems in the context of sustainable development, such as increased levels of GHG emissions, air pollution, traffic congestion, traffic-related deaths and urban sprawl. Air pollution from vehicles has created smog, and the growing awareness of pollution led to the enactment of the first federal legislation to control automobile emissions in 1966 and was also a factor in the creation of the US Environmental Protection Agency in 1970 to regulate motor vehicle pollution (EPA, 1996).

The federal government began subsidising public transit in 1964 and played a larger role in transit funding in the 1960s and 1970s. Public transit also provides an alternative mode of urban transportation and brings many environmental and social equity benefits. However, despite substantial investment, the transition from excessive use of personal cars to public transit has not been significant.

Sustainable transport (sustainable mobility) does not have a universally accepted definition. As this sector has, since 1988, been part of the sustainable development mainstream, it is broadly characterised by the ability to meet the demands of today's mobility without compromising the ability of future generations to meet their needs.

One of the best-known definitions is that given by the Organisation for Economic Cooperation and Development (OECD) in 1994, namely, 'Transport that does not endanger public health or ecosystems and meets the requirements of mobility, under conditions of (a) use of renewable resources below their rates of regeneration and (b) use of non-renewable resources below the rates of development of renewable substitutes'.

In our view, sustainable transport is '**that complex system which is designed to meet the mobility needs of current generations without damaging environmental factors and health, and which must make its energy consumption so efficient that it can meet the mobility needs of future generations'** (Fistung, 1999).

Recently, there has been a trend towards a more frequent use of the term 'sustainable mobility', with the view that sustainable mobility broadens the scope of the original concept by considering the social dimension of the activity, whereas sustainable transport is more focused on the operational aspects of sustainability. According to this approach, the World Business Council for Sustainable Development defines sustainable mobility as 'the ability to meet society's needs to move freely, access more freely and communicate, relate more quickly, without sacrificing present and future human or environmental values' (WBCSD, 2004).

This statement encompasses a wide range of sustainable mobility issues:

- Accessibility.
- Financial resources to be spent by users.
- Time needed to make the journey.
- Safety.
- Security.
- GHG emissions.

- Impact on the environment and human well-being.
- Resource use.
- Equity implications.
- Impact on public revenue and expenditure.
- Expected rate of return on private capital.

The above definitions lead to the idea that the overall objective of sustainable development is to find an optimum of the interaction of the four systems – economic, human, environmental and ecological – in a dynamic and flexible way. The optimal level corresponds to that long-term development that can be sustained by the four systems.

In relation to the transport sector, there is a broad consensus that it is not sustainable, often causing negative social, economic and environmental damage. There is also consensus on the fundamental principles underpinning sustainable transport, the most frequently invoked being the need to internalise negative externalities, the application of the 'user pays' principle or the support of fair competition between different modes of transport.

A STS is designed to contribute to economic development and social well-being without harming the natural environment or human health. Major areas of concern related to this include increasing traffic volume and congestion, noise and air pollution (including climate change phenomena). Actions should focus on encouraging the use of environmentally friendly modes of transport, internalising the external costs caused by transport (right-sizing the total social costs of transport) and reducing transport demand.

The relationship between economic growth and growth in the transport sector needs to be reconsidered so that economic growth does not go hand in hand with growth in transport activities, but is based on making them more efficient and improving their quality. In other words, the growth of industrial activities should not, for example, go hand in hand with an increase in road transport related to these activities, but rather by converting a large part of transport from road transport, which is the most polluting mode, to much less polluting modes. In this context, multimodal, rail or non-conventional transport (e.g. cableways in mountainous areas) are the best options. Similarly, collective transport (public, private) must always be encouraged at the expense of individual transport.

The relationship between the development of transport and the economy in general is difficult to establish in concrete terms and has been debated for many years. There are several sides of the business that transport can capitalise on, starting from an appropriate location that supports the development of its accessibility, to investment in infrastructure to make better use of existing transport resources. This is further nuanced by the nature, scale and scope of possible impacts:

- The timing of development varies as transport effects may precede, exist during or after the economic process of development. Delays, consequences and main effects make it difficult to separate the specific contributions of

transportation from development, per se. Each case study appears to be specific to a set of situational or timing circumstances that are difficult to replicate elsewhere.
- The types of effects vary considerably. The spectrum of effects ranges from positive to permissive to negative. In some cases, transportation effects may promote development, in others, they may impede economic development in a region. In many cases, these connections can be clearly established.

Economic development cycles provide a revealing conceptual perspective on how transportation systems evolve over time and space because they include the timing and nature of transportation impacts on economic development. This perspective emphasises that, after a phase of economic initiation and growth, a transport system will eventually reach a phase of maturity through geographic and market saturation. There is also the risk of over-investment when economic growth is credit-driven, which can lead to significant misallocation of capital, including in the transport sector. In periods of recession, which typically follow periods of expansion, transport activities may regress, mainly due to falling demand and a shortage of capital investment.

In line with the above, European transport policy has increasingly included specific emphases on sustainable transport development. A case in point is the OECD Conference on Sustainable Transport, held in Vienna in 2000, where the necessary guidelines for the development of a STS were established:

- The development, over a long period of time, of a scenario that supports a STS with positive environmental and health effects and that is advantageous for mobility and access requirements.
- Assessing the long-term evolution of transport, taking into account all influencing aspects, economic, environmental and social.
- The definition of specific qualitative objectives based on environmental and health protection criteria and standards, and thus respecting the principles of sustainable development.
- Setting specific targets derived from the specific qualitative environmental and health objectives, including target dates for their achievement.
- Identifying the strategies needed to achieve sustainable transport development and the measures needed to ensure changes in transport activity and the introduction of new technologies.
- Assessment of the socio-economic implications of the scenario and their place in the evolution based on the principles of sustainable development;
- The development of packages containing measures and instruments to support the development of a STS. Policies will be adopted targeting the side, infrastructural investments, charging, transport demand and traffic management, as well as improving public transport and encouraging walking and cycling.
- An implementation plan will be drawn up, including the steps for the use of packages of measures and instruments aimed at achieving a STS, taking

into account local, regional and national characteristics and circumstances. A clear timeframe will be established, including the timeframe and responsibilities for the implementation of each step.
- Implementation of measures to monitor the realisation of the scenario and public presentation of the strategy for developing a STS. Well-defined sustainable transport indicators will be used to communicate the results achieved. Continuity in the implementation of the scenario actions will be ensured by adapting to possible new scientific requirements.
- Building international support and cooperation in support of the scenario. Political parties and civil society will be involved in order to obtain their support. Specific training and knowledge-building programmes will be implemented.

Moreover, sustainable development has been a political objective of the European Union since 1997, when it was included in the Maastricht Treaty. On the basis of the identification of the marked regional disparities, in particular as a result of EU enlargement, the Council of the European Union adopted the renewed Sustainable Development Strategy for an enlarged Europe on 9 June 2006.

In September 2001, the European Commission adopted a new White Paper on Community Transport Policy. The document proposed measures to revise transport policy to make it more sustainable and avoid the huge economic losses caused by congestion, pollution and accidents.

In October 2004, 'A Constitution for Europe' was adopted, and in 2011, a new White Charter 'Towards competitiveness and resource efficiency of transport systems' was launched. The main elements of this latest European transport strategy simultaneously aim at:

- Sustainable development.
- Integrated development of all transport modes.
- Increasing accessibility.
- Increasing the level of technology used.

These objectives aim to achieve, simultaneously with:

- Economic progress.
- Competitive growth (within the single market).
- Resource efficiency.

Together with other wishes related to the promotion of a STS, the European Transport Commission considered that, in the long term, in the development of future transport policy, environmental issues related to climate change, the reduction of biodiversity, the use of conventional energy resources and the degradation of the environment and quality of life including health, will play as important a role as economic and social factors in the development of future transport policy. However, the conditions must be provided for economic development to take place without the need for a concomitant increase in traffic and the automatic negative environmental and health effects that transport causes. At the same time, the smooth functioning of the entire transport system and fair access to transport services must be ensured.

In the framework of the Common Transport Policy, the European Community points out that maintaining the current upward trends in road and air transport activity, whether civil or commercial in nature, is not in line with the principles of sustainable development, relating to the environmental impeccability of the local, regional and global levels.

This is why the aim of the European Transport Commission has been to promote mobility that is efficient, safe, environmentally friendly and can meet the mobility needs of European goods and people.

Transport is about connectivity and mobility. Europe's competitiveness on the global stage will increasingly depend on its level of connectivity and efficiency both internally and with the rest of the world. Investment, especially in infrastructure, will always be a major factor in strengthening prosperity and cohesion in the Union, as well as in the fight against climate change and externalities from transport activities. The right infrastructure, innovative Intelligent Transport Systems (ITS), measures to improve safety, security and the performance of the transport sector are essential.

The shift to low-carbon mobility was already an objective in the 2011 Transport White Paper and has been supported by various initiatives.

Through its Low Emission Mobility Strategy (adopted in July 2016) and subsequently in its 'Europe on the Move' Communication of 31 May 2017, the European Commission aims to address the shortcomings, in particular in the market integration of road transport, through a set of initiatives for a socially just transition to clean, competitive and connected mobility (EC, 2018a). The Communication accompanies a series of legislative proposals (the so-called 'mobility package') aiming at revising the rules for access to the road freight market, improving social legislation for road transport and a review of road charging rules.

On 8 November 2017, the Commission adopted a second set of proposals (the 'second wave of the Mobility Package'). They respond to the challenge of making mobility clean, competitive and connected through a combination of supply and demand-side measures on low-emission mobility. They combine a supportive regulatory framework with actions to mobilise finance where necessary, while ensuring consumer acceptance and safeguarding the employment and social protection dimension. In concrete terms, the proposals include a number of measures to enable the transition to low and even zero-emission mobility, such as a reform of the Clean Vehicles Directive or a continuation of the Action Plan on the Alternative Fuels Directive.

In December 2019, at the EU level, the *European Green Deal (The Green Deal)* was adopted, a document that sets the guidelines for various European policies for the next five years. The document is in connection with a number of legislative and non-legislative initiatives in multiple areas, such as environment, climate change, energy, industry, transport, agriculture, digitisation and the financial sector (MFA, 2022).

A first set of proposals for the implementation of the European Environmental Pact (EEP) was launched on 14 January 2020 with the European Commission's Communication on the Sustainable Europe Investment Plan and the establishment of the Just Transition Facility. The Facility will focus on the regions and sectors most affected by the transition, as they are dependent on fossil fuels or carbon-intensive processes. It will receive funding from both the EU budget and the European Investment Bank Group to mobilise the necessary private and public resources. The Facility complements the substantial contribution from the EU budget, including through the European Regional Development Fund and the European Social Fund Plus.

On 4 March 2020, the European Commission published its proposal for a Regulation establishing the framework for achieving the objective of climate neutrality and amending Regulation (EU) 2018/1999 – *Climate Law*. The Regulation – the main element deriving from the EEP – is, in this context, the European climate legislative act enshrining the objective of achieving climate neutrality by 2050. Thus, for the first time at the Community level in the field of climate change, it was proposed to adopt a long-term objective in the form of a legislative act, which will provide the framework for Community measures to be taken at the European Union and national level to achieve the objective of climate neutrality.

The EU strategy on sustainable mobility, presented in December 2020, emphasised that the current paradigm of incremental change needs to be reoriented towards a fundamental transformation (EC, 2020a). Thus, the strategy set out ten flagship areas with an action plan to guide work in the coming years. The scenarios that underpinned the development of the strategy showed that, with the right level of ambition, the combination of policy measures set out in the strategy could lead to a 90% reduction in transport emissions by 2050.

The act, suggestively called 'A strategy for sustainable and intelligent mobility – putting European transport on the road to the future', set out the EU's vision for transport in the near future, based on nine major influencing elements:

1 The free movement of people and goods across internal borders is one of the fundamental freedoms of the European Union and its single market. As the second largest area of expenditure for European households, the transport sector contributes 5% to European GDP and directly employs around 10 million workers.
2 Policy efforts and measures to date have not yet sufficiently addressed the external costs caused by transport. GHG emissions from the transport sector, for example, have increased over time and now account for no less than a quarter of the EU total.
3 Given its high share of total EU GHG emissions, the EU's objective of at least a 55% reduction in GHG emissions by 2030 and climate neutrality by 2050 will only be achieved by introducing, without delay and in synergy with efforts

to zero-clear pollution, more ambitious policies to reduce transport's dependence on fossil fuels. The success of the European Green Deal (EGD) depends on the EU's ability to make the transport system sustainable as a whole.

4 Maintaining supply chains and a coordinated European approach to connectivity and transport activity are essential to overcome any crisis, but also to strengthen the EU's strategic autonomy and resilience.
5 The COVID-19 pandemic has had a significant impact on mobility. As we recover from this severe crisis, public support must help mobility to 'rebuild more thoroughly' and make important progress towards a sustainable and smarter future. In this context, the completion of the Single European Transport Area, as envisaged in the 2011 White Paper, remains a cornerstone of European transport policy.
6 Mobility must be based on an efficient and interconnected multimodal transport system for both passengers and freight, reinforced by an affordable high-speed rail network, numerous recharging and refuelling infrastructures for zero-emission vehicles and the provision of renewable, low-carbon fuels, cleaner and more active mobility in greener cities that contribute to the health and well-being of citizens.
7 Digitisation and automation processes need to be accelerated to further enhance levels of safety, security, reliability and comfort, thus maintaining the EU's leading position in equipment manufacturing and transport services and improving our global competitiveness through efficient and resilient logistics chains.
8 Mobility must be accessible and affordable for all, so that even rural and remote regions are better connected, accessible also for people with reduced mobility and people with disabilities, and the economic sector offers good social conditions, retraining opportunities and attractive jobs.
9 The current paradigm needs to be reoriented from a focus on incremental change towards fundamental transformation.

On the basis of this strategy, the milestones for the future development of the transport sector in the EU area have also been mapped out:

- By 2030:
  - At least 30 million zero-emission vehicles will be in operation on European roads.
  - One hundred European cities to be declared climate neutral.
  - Double high-speed rail traffic.
  - Within the European Union, collective travel of less than 500 km should be carbon neutral.
  - Widespread implementation of sustainable transport automation.
  - Bring zero-emission ships to market.
- By 2035, zero-emission large-capacity zero-emission ships must be brought to market.

- By 2050:
  - Nearly all new cars, vans, buses and heavy-duty vehicles must be zero-emission.
  - Doubling rail freight traffic.
  - Tripling high-speed rail traffic.
  - Full operationalisation of the trans-European multimodal transport network equipped for sustainable and intelligent transport with high-speed connectivity.

On 24 February 2021, the Commission published the new EU Strategy on Adapting to the Impacts of Climate Change, which set out the way forward to deal with the inevitable consequences of climate change. The strategy encouraged Member States to make further progress to boost adaptive capacity, strengthen resilience and reduce vulnerability to climate change. The main objectives of the Strategy were as follows:

- Improving the level of knowledge on climate change impacts and uncertainty management.
- Strengthening planning and risk assessment measures/actions.
- Accelerating adaptation to the impacts of climate change.
- Increasing global resilience to the impacts of climate change.

In order to achieve the climate objectives in the period 2030–2050, sustainable mobility became a major element in the community development strategy

Reducing traffic-related pollution in cities was, in this context, one of the objectives pursued by the new strategic vision, with a view to proposing more ambitious standards for pollutant emissions (other than $CO_2$) from vehicles powered by internal combustion engines. At the same time, the adopted legislation also envisaged the revision of $CO_2$ emission standards for passenger cars and light commercial vehicles (June 2021) to cover the post-2025 period and to set a clear trajectory towards electromobility. In parallel, the inclusion of road transport in the EU ETS[3] as a complementary measure to emission standards will also be considered. Under the European Union Emissions Trading Scheme (EU ETS), one European Union Allowance (EUA) corresponds to the right to emit one tonne of carbon dioxide equivalent ($CO_{2e}$) over a specified period.[4] These allowances can be used by operators in the EU ETS to cover their verified emissions, or they can be traded with other operators who need them.

On 14 December 2021, the European Commission communicated new transport proposals aimed at greater efficiency and more sustainable travel. The document stated that,

> By increasing connectivity and shifting more passengers and freight to rail and inland waterways, supporting the introduction of charging points, alternative refuelling infrastructure and new digital technologies,

> putting a stronger focus on sustainable urban mobility and facilitating the choice of different transport options in an efficient multimodal transport system, the proposals will put the transport sector on track to reduce its emissions by 90%.

In broad terms, the proposals outlined in the paper aimed to:

- Modernising and completing missing links in the TEN-T, which:
  - Requires major TEN-T passenger rail lines to allow trains to travel at 160 km/h or faster by 2040, creating competitive high-speed rail connections across the Union. Canals and rivers must provide good navigable conditions, undisturbed, for example, by water levels, for a minimum number of days per year.
  - Calls for more transhipment terminals, improved handling capacity at freight terminals, reduced waiting times at rail border crossings, longer trains to shift more freight to more environmentally friendly modes and the option for rail freight wagons to be carried on the rail network as a whole.
  - Introduces a new interim deadline of 2040 to bring forward the completion of important parts of the network ahead of the deadline (2050) that applies to the wider and more comprehensive network. So new high-speed rail connections between Porto and Vigo and Budapest and Bucharest – among others – must be completed by 2040.
  - It calls on all 424 major cities along the TEN-T network to develop sustainable urban mobility plans to promote zero-emission mobility and to increase and improve public transport and walking and cycling infrastructure.
- Increase cross-border and long-distance rail traffic. Although the number of people travelling by train has increased in recent years, only 7% of rail kilometres travelled between 2001 and 2018 involved cross-border travel. To encourage more people to consider rail travel for trips abroad, the proposed plan sets out concrete actions to remove barriers to cross-border and long-distance travel and to make rail travel more attractive for passengers. The proposal is accompanied by an action plan on long-distance and cross-border rail transport, which sets out a roadmap of further actions to help the EU meet its target of doubling high-speed rail traffic by 2030 and tripling it by 2050. The actions include:
  - A multimodal legislative proposal in 2022 to incentivise user-friendly multimodal ticketing.
  - Developing measures to enable passengers to find the best tickets at the most attractive price and to better support passengers facing rail disruptions, as well as a commitment to investigate an EU-wide VAT exemption for train tickets.
  - Repealing redundant national technical and operational rules.

   - Proposals for 2022 on timetabling and capacity management, which will boost faster and more frequent cross-border rail services.
   - Development of guidelines on rail access pricing in 2023, which will facilitate rail operators' access to infrastructure, increasing competition and allowing more attractive ticket prices for passengers.
   - The Commission supporting the launch of at least 15 cross-border pilot projects by 2030 to test the Action Plan approach before the new TEN-T requirements enter into force.

- Providing intelligent transport services for drivers. The Commission proposes to update the 2010 ITS Directive, adapting to the emergence of new road mobility options, mobility apps and connected and automated mobility (EC, 2010). The proposal aims at faster deployment of new, intelligent services by proposing that certain key road, travel and traffic data, such as speed limits, traffic circulation plans or road works, should be made available in digital format along the TEN-T network and eventually cover the entire road network. It will also ensure that drivers along the TEN-T network are provided with essential safety-related services.
- Developing a cleaner, greener and more accessible urban mobility system. The Urban Mobility Framework sets European guidelines on how cities can reduce emissions and improve mobility, including through sustainable urban mobility plans. The main focus will be on public transport, walking and cycling. The proposal also prioritises zero-emission solutions for urban fleets, including taxis and ride-sharing services and the construction and upgrading of multimodal nodes, as well as new digital solutions and services. The proposal also outlines financing options for local and regional authorities to implement these priorities. In 2022, the Commission is considering proposing to EU Member States a recommendation for the development of national plans to support cities in developing their mobility plans.

As we mentioned, in order to accelerate work on sustainable development, the European Parliament adopted the EGD at the end of 2020.

Complementing this strategy, in October 2023, a plan called 'Fit for 55' was adopted, which presents a set of legislative proposals and amendments to existing European legislation designed to support the European Union in taking action to reduce net GHG emissions by 55% by 2035 and to achieve climate neutrality (by 2050).

With the new $CO_2$ standards, all new cars and vans registered in Europe will have zero emissions by 2035 (EC, 2024a). As an intermediate step towards zero emissions, average emissions from new cars will have to fall by 55% and new vans by 50% by 2030. This will put road transport on a firm path towards zero-emission mobility in 2050. In addition, from 2027, road transport will have an emissions trading scheme in place, putting a price on pollution, stimulating the use of cleaner fuels and reinvesting in clean technologies.

## 1.2 How to develop sustainable mobility

In order to be able to outline the prospects for the sustainable development of transport, we need to start from an analysis of the current situation of the sector. We can therefore briefly highlight some of the characteristic features that make the transport sector unsustainable.

Take the European Union, for example:

- In 2016, 25,600 lives were lost and more than 1.4 million people were injured in road accidents (EC, 2018b), fortunately reaching a decrease of more than 4% in 2019 (EC, 2021). Moreover, in 2022, the number of lives lost in road crashes fell to 20,653, a reduction of almost 20% compared to 2016 (EC, 2024b)
- Transport activity generated about 24% of total GHG emissions, which amounted to more than 1 billion tons of $CO_2$ 'dumped' into the atmosphere (more precisely 1.26 billion tons in 2019, according to – EC, 2021, p. 141). Fortunately, in 2024, there was a significant decrease to 1.04 billion tons (EC, 2024b)
- The total external costs, caused by transport in the European Union, at the level of 2016 amounted to around €987 billion (including congestion costs), equivalent to 6.6% of the GDP of the 28 Member States (EC, 2019).
- Of the total external costs, the most important category is accident costs, accounting for about 29% of total costs, followed by congestion costs (27%). Climate change specific costs as well as air pollution costs contribute 14% of total costs, noise costs 7% and costs associated with habitat damage 4%.
- Road transport generates by far the highest external costs (83%). Maritime transport is responsible for 10% of the total external costs, aviation for 5%, rail for 1.8% and inland waterways for 0.3% of the costs. By category of transport, 69% of total costs are due to passenger transport (with a large share due to passenger car transport) and only 31% to freight transport.

In the United States, at the beginning of the third millennium, the situation was as follows (Delucchi, 2010):

- For both freight and passenger travel, the road mode generally has higher external costs than the other modes.
- The share in total external costs specific to road transport was about 83.8% for passenger transport and 86.7% for freight.
- In passenger transport, rail accounted for 8.1% of total external costs, air 5.3% and sea 2.8%. For freight, the situation was as follows: rail 3.1%, air 6.0% and ship 5.2%.
- External costs caused by passenger transport were almost 16% higher than those caused by freight transport.
- Among the multiple component categories of total external costs, accidents and congestion were generally the largest, followed by air pollution and climate change.

- Road congestion has increased rapidly and dramatically over the last two decades (Schrank and Lomax, 2007), regardless of the remedial measures taken. The dramatic increase in congestion has been due in large part to the extremely rapid growth in the active vehicle fleet, often leading to increased travel distances and reduced travel speeds.
- The growth in the number of motorised vehicles, oil consumption and the level of GHG emissions have increased continuously since the 1970s, as vehicle travel has steadily increased. Over the same time period, fuel consumption per kilometre has fallen only modestly. Since the mid-1990s, oil use has risen in line with distance travelled, meaning that fuel use per kilometre has remained constant.

In line with adopted policies, the downward trend in transport emissions in the European Union (since 2005) is expected to continue until 2030 (–12% for the period 2005–2030). For the period 2019–2022 alone, the data presented above show a decrease of about 17.5% in total GHG emissions from transport, which means that Europe can go beyond its initial commitments in this area.

The main factor supporting this development is the increase in fuel efficiency (EC, 2018a). This is supported by the adoption of $CO_2$ emission standards for light-duty vehicles, the increase in fossil fuel prices over time and the use of lower and/or zero-$CO_2$-emitting fuels.

However, major efforts will still be needed to maintain low or reducing GHG emissions. The cost of air pollution from road transport is expected to remain high, due to traffic congestion and increasing transport demand. This was also evident during the health pandemic and during the energy crisis caused by the Russian–Ukrainian military conflict that started in February 2022.

Therefore, it is evident once again that the current transportation system is not sustainable. The level of $CO_2$ emissions and air pollution generated by transport are currently the main concerns related to the transport–environment binomial.

However, $CO_2$ emission levels are difficult to 'attribute' to specific countries. They are calculated on the basis of fuels sold and do not correspond to transport activity within countries' borders, leading to biased conclusions, especially for transit countries.

In the case of maritime or air transport, there are additional problems related to the attribution of territoriality of emissions for the parts of the journeys that take place within a given territory.

In contrast to other sectors, aviation emissions are expected to increase dramatically as air traffic increases worldwide. $CO_2$ emissions have increased by about 2.3 times between 1990 and 2019 (EC, 2021, p. 150) and are projected to increase by a further 44% between 2014 and 2035 (EC, 2018a).

The switch to alternative fuel vehicles varies substantially from country to country, although there is an overall positive trend. The share of plug-in electric vehicles (PEVs) in new car registrations indicates progress in the deployment of electric cars.

In 2016, according to the European Alternative Fuels Observatory, the Netherlands was the leader, mainly due to a large number of newly registered plug-in hybrid vehicles (PHEVs), which accounted for almost 5% of new passenger car registrations (EC, 2018a). The Netherlands was followed by Sweden and Belgium. France and Austria have the highest share of battery EVs in new registrations. The lowest shares are held by Greece, Bulgaria, Malta and Slovakia, with a 0.1% share of EVs in new car registrations.

The global market share of EVs has, however, made a tremendous leap over the last decade, and we expect the trend to accelerate exponentially.

The year 2020 did not see a significant increase in total registrations of new EVs. The global market for all types of cars has been negatively impacted by the COVID-19 pandemic and the ensuing economic slowdown. Against the backdrop of the pandemic, the outlook for global sales of EVs was quite unpredictable at the beginning of the year.

However, 2020 turned out to be a surprisingly good year, with global sales of EVs up 43% from the previous year, and the global electric car industry's market share of the global electric car industry growing to a record 4.6% (VIRTA, 2022).

Electric vehicle sales more than doubled in the following year, 2021, reaching around 6.75 million units sold globally. The volume of EVs sold in a week in 2021 was more than the number of EVs sold in a full year in 2012.

EVs will have a major role to play in reaching the EU's ambitious goal of zero transport emissions by 2050, and the industry is clearly preparing for this goal.

Based on this challenge, most developed nations are taking plans to renew their vehicle fleets with EVs more and more seriously, which is why sales grew rapidly in 2022. Analysing the first five months of 2022, over 3.2 million new cars were registered worldwide (VIRTA, 2022).

The increase in electric car registrations in Europe since 2019, which has been much faster than in other parts of the world, can be attributed to the incentives introduced by many EU governments. Thus, within the specific market, sales of EVs in the European Union grew by 17% in 2019 compared to the previous year. The most important markets from this perspective were in Germany, Norway, Iceland and Sweden.

In total, EVs consumed about 55 terawatt-hours of electricity in 2021, of which 10% can be attributed to two-wheelers in China.

Under these circumstances, during 2021, the use of EVs globally reduced air pollution by more than 40 million tons of $CO_{2e}$ during 2021. Forecasts for the further development of this process indicate that by 2030 the use of EVs could contribute to a reduction of more than 460 million tons of $CO_{2e}$ in the atmosphere.

With all this in mind, European sustainable transport policy has increasingly become an issue on the agenda of the European Parliament, and is being brought in line with energy and environmental policies. It is constantly being complemented by recommendations from the Committee on the Environment, Public Health and Food Safety and the Committee on Industry, Research and Energy.

The approach is complex and is driven by a number of current aspects of social and economic life, including:

- Approximately one third of the total energy consumption of the countries of the European Union is due to the transport sector, with the exception of maritime and pipeline transport, and road transport is the largest energy consumer, with a share of 83%, of the total of the whole sector.
- Transport is responsible for 70% of the EU's oil demand, 97% of which depends on fossil fuels, with electricity accounting for the remaining 2% and biofuels for 1%.
- The energy efficiency of the different modes of transport has improved significantly in recent years, which has led to a significant reduction in the corresponding $CO_2$ emissions per kilometre.
- NOx emissions from road transport have decreased, based on technological innovations and intensive efforts by the automotive industry. This has been achieved in particular through the introduction of catalysts, fine particle filters and other technologies, which have contributed to a reduction of NOx and particulate pollutant emissions by 30%–40% over the last 15 years.
- However, the improvements achieved and mentioned above have been neutralised and outweighed by the effects of the steady increases in demand and volume of transport sector activities respectively. As a result, emissions for the sector as a whole have maintained a steady upward trend.
- Urban traffic accounts for 40% of $CO_2$ emissions and 70% of other pollutant emissions from motor vehicles, and road congestion, concentrated mainly in metropolitan areas, causes economic and health losses in the European Union estimated at around 1% of GDP per year;
- Seventy per cent of the priority trans-European transport projects approved recently have been oriented towards rail and ship transport, which are known to be the least polluting modes.

Important changes in European transport policy have led to an increasing emphasis on sustainable transport development. Thus, as early as the OECD Conference on Sustainable Transport in Vienna in October 2000, the necessary guidelines for the development of a STS were established:

- The development, over a long period of time, of a scenario that supports a STS with positive environmental and health effects and that is advantageous to mobility and access requirements.
- Assessing the long-term evolution of transport, taking into account all influencing aspects, economic, environmental and social.
- The definition of specific qualitative objectives based on environmental and health protection criteria and standards, and thus respecting the principles of sustainable development.
- The setting of specific targets derived from the specific qualitative environmental and health objectives, including target dates for their achievement.

- Identification of the strategies needed to achieve sustainable transport development and the measures needed to ensure changes in transport activity and the introduction of new technologies.
- Assessment of the socio-economic implications of the scenario and their place in the evolution based on sustainable development principles.
- The development of packages containing measures and instruments to support the development of a STS. Policies will be adopted on technology, infrastructure investment, charging, transport demand and traffic management, improving public transport and encouraging walking and cycling.
- Developing an implementation plan that includes the steps for using the packages of measures and instruments to achieve a STS, taking into account local, regional and national characteristics and circumstances. The establishment of a clear time-frame and responsibilities for the implementation of each step.
- Implementation of measures to monitor the realisation of the scenario and public presentation of the strategy for developing a STS. Well-defined sustainable transport indicators will be used to communicate the results achieved. Continuity in the implementation of the scenario actions will be ensured by adapting to possible new scientific requirements.
- Building international support and cooperation in support of the scenario. Political parties and civil society will be involved in order to obtain their support. Specific training and knowledge-building programmes will be implemented.

A set of specific indicators has been agreed internationally to assess the state of implementation of such measures.

EU transport policy is based on the current modal split (road 44%, air 41%, rail 8%, inland waterway 4%), which is considered alarming, particularly because it conceals a worsening trend, with road transport increasing by 50% between 1998 and 2010, without measures to divert freight flows to other modes. In this context, and with a view to reducing the number of accidents (40,000 road accidents per year in recent years) and environmental degradation, the European Union is encouraging the development of intermodal transport and the redirection of freight flows from road to inland waterway and rail transport, which are less harmful to the environment and human health.

EU policy is increasingly emphasising that all modes of transport are indispensable to the EU transport system and therefore all must become more sustainable.

As a priority action to this end, the uptake of low and zero-emission vehicles, as well as renewable and low carbon fuels, for road, ship, air and rail transport must be stimulated without delay. Research and innovation on competitive, sustainable and circular products and services should also be supported.

Over the last five years, significant efforts have been made internationally, but especially within the European Union, to reduce air pollutant emissions from vehicles. Cars on sale in 2020 emit significantly less pollutants than in 2015. Although growing rapidly, the proportion of low or zero-emission vehicles in the fleet is far too low today. $CO_2$ and air pollutant emission standards, as well as public procurement rules such as those under the Clean Vehicles Directive, must continue to be key policy drivers in the transition to zero-emission mobility in road transport, and by increasing the supply of zero-emission vehicles, sustainable mobility can be made more affordable for all. Therefore, in order to meet the objectives proposed in the 2030 Climate Target Plan and to ensure a clear trajectory towards zero-emission mobility from 2025, the revision of the $CO_2$ emission standards for passenger cars and light vans (Euro 7) should be adopted.

From a modal perspective, at the EU level, support for the implementation of the transport system had the following characteristics in 2020 (EC, 2020a):

- In terms of road transportation, solutions for zero-emission vehicle travel had been implemented. Manufacturers had started to invest heavily in battery EVs. Market penetration was already growing, especially in cars, vans and buses used in cities, while trucks and coaches were just starting. Manufacturers also started to invest in vehicles with hydrogen-powered engines, especially for use in commercial fleets, buses and heavy transport. These promising options were supported in the context of the EU's hydrogen and energy system integration strategies, as well as in the context of the Strategic Battery Action Plan. Energy efficiency was also a criterion for prioritising the choice of appropriate technologies in the future, taking into account the whole life cycle. Transitional technological solutions had to fully comply with $CO_2$ emission and pollution standards.
- On rail transport, efforts supported its electrification, and where this has not been possible, the use of hydrogen-powered trains has been encouraged.
- Unfortunately, aviation and shipping have faced greater decarbonisation challenges in recent decades, and look set to continue to do so for some time, due to the current lack of zero-emission engine technologies, the long development and life cycles of aircraft and ships, the significant investments required in refuelling equipment and infrastructure and international competition in these sectors. EU international emissions from shipping and aviation have increased by more than 50% since 1990. Urgent action is, therefore, needed in these sectors, including in the recovery from the current crises. These modes must be given priority access to other types of liquid and gaseous fuels from renewable and low-carbon sources because in the short term there is a lack of suitable alternative propulsion systems. Significant efforts are needed to develop disruptive technologies to bring zero-emission ships and aircraft to the market.
- Ports and airports have been identified as essential for international connectivity, the European economy and the regions concerned. Therefore, as part of their transition to zero-emission hubs, the best practices highlighted by the most sustainable airports and ports must become the new norm and enable the development of more sustainable connectivity infrastructures. Ports and airports must become multimodal mobility and transport hubs, linking all relevant modes. This will improve local air quality, thus

contributing to better health for local residents. Inland ports and seaports have great potential to become new clean energy hubs for integrated power systems, hydrogen and other low-carbon fuels, as well as laboratories for waste reuse and the circular economy.

- Sustainable alternatives must now become widely available as part of a fully integrated and fluid multimodal mobility system. It has been quite clear that the European Union cannot rely solely on technological solutions, and therefore immediate action was and is needed to adapt the mobility system to combat climate change and reduce pollution. Multimodality builds on the strengths of different modes of transport such as convenience, speed, cost, reliability, predictability and by combining them, can provide more efficient transport solutions for people and goods. The COVID-19 pandemic has demonstrated the importance of increased multimodality for improving the resilience of our transport system and has shown how ready citizens have been to adopt alternative, sustainable modes of transport.
- Important actions have also been taken and must be continued to transform the transport sector into a truly multimodal system of sustainable and intelligent mobility services. To achieve this, Europe needs to build a high-quality transport network with high-speed, short-distance rail and 'clean' aviation services that will improve coverage of long-distance routes and enable them to offer travellers carbon-neutral choices for journeys of less than 500 km within the European Union by 2030. Subject to competition law, airlines must sell an increasing number of multimodal tickets. Investment should be directed towards upgrading the necessary TEN-T infrastructure to allow a shift towards more sustainable links. Steps must also be taken to build a comprehensive transport system where EU investment, state aid, capacity allocation rules and Public Service Obligations (PSOs) are geared towards meeting mobility needs and stimulating different multimodal options.
- It has been realised that fluid multimodality, facilitated by digital solutions, is vital in urban and suburban areas. The increasing pressure on transport systems of travellers has increased the demand for new and innovative solutions, with various transport services being integrated into an on-demand accessible service, according to the Mobility as a Service (MaaS) concept. At the same time, many cities have reoriented towards shared and collaborative mobility services (car sharing, bikes, on-demand vehicles and other forms of micro-mobility) facilitated by the emergence of intermediate platforms, thus allowing for a reduction in the number of vehicles in daily traffic.

Sustainable development in general, however, is constantly changing due to factors such as population growth (SOCIAL), climate change (ECOLOGICAL) and resource constraints (ECONOMIC).

Therefore, the recalibration and increased adaptability of these three factors to real-world conditions must be added as a fourth pillar of sustainability which we define as resilience, despite the fact that the economic meaning of the term refers to the ability to return to normal after suffering a shock (Dexonline, 2024).

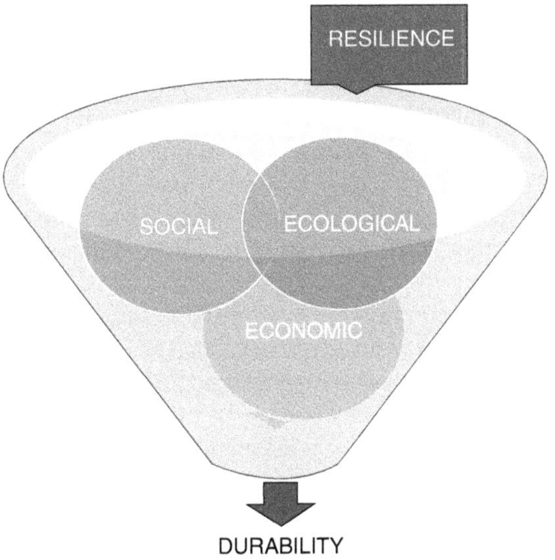


*Figure 1.1* The four components of the new concept of sustainable development – Sustainability 2.0. (Figure by author.)

Structurally, the four pillars represent premises (social, economic and ecological) and a guiding-adaptation mechanism (resilience). The mechanism (see Figure 1.1) is the binder of the three premises, and the result is the continuous process of sustainability (sustainability).

We define the result of this continuous adaptive process of sustainability as *neo-sustainability* (or Sustainability 2.0).

In fact, the concept of sustainable development involves establishing, over time, a balance between the three major pillars: environmental, social and economic. It is evident that when one of these pillars changes its state (e.g. when the population increases), the other two must reorient themselves to another level of balance in order for the entire system to evolve sustainably, which can only be achieved through interactions that lead to this state. This, I believe, is the fourth pillar of Sustainability 2.0, namely, resilience, the ability of the system to rebalance itself when one or more of its component dimensions change. The process is sustainable because, through resilience, the three initial pillars (environmental, social and economic) will always be guided so as to respect the conditions for the balanced development of society in its various stages of evolution without reaching irreversible imbalances.

## 1.3 Specific policies to develop an efficient sustainable mobility system

The development of sustainable transport policies implies the simultaneous adoption of environmental, social and economic objectives and requires additional approaches in different areas specific to transport within the European Union.

According to the conclusions of the European Conference of Ministers of Transport (ECMT), held in Prague (Czech Republic) in May 2000, death and injury from transport accidents are the most important issues from the perspective of sustainable transport and not environmental damage, as may have been emphasised in other documents. Current rates of death and injury caused mainly by road crashes are considered far from acceptable by governments, even in countries at the forefront of road safety performance (ECMT, 2000). Accident rates for other modes of transport, although much lower, are still not considered acceptable.

The challenge in developing transport policies for sustainable development is to steer the sector towards a compromise that maximises the economic and social aspects, the benefits of transport and minimises the environmental, social and economic aspects associated with the costs. Many of the measures needed to achieve this balance are not new; the difficulty lies in their effective implementation.

From this perspective, the development of a STS requires the simultaneous consideration of issues such as:

- Socio-ecological:
  - Increasing accessibility to Sustainable Transportation Development (STD) by adopting the most viable options that fully meet society's mobility requirements.
  - Ensuring transparency of all measures and making them explicit, while encouraging public participation in the debate on the usefulness, purpose and costs of these actions.
  - Assessing social and environmental impacts before any action is implemented.
  - Ensuring permanent information and education in the spirit of promoting STD.
  - Prioritising the development of transport systems in favour of environmentally friendly ones with minimal or no negative impact.
  - Reorienting and reorganising all modes of transport towards STD.
  - Minimising the generation and disposal of waste, for each phase of the life cycle of vehicles, while at the same time recycling them.
  - Halting habitat degradation, ecosystem fragmentation and loss of agricultural and social land.
  - Reducing or eliminating sources of pollution (fuels, vehicle components, etc.).
  - Covering all transportation needs for community members, while increasing quality of life and safety.
- In economic and environmental terms:
  - Internalisation of external costs, including those with long-term effects.
  - Promotion of research and development programmes focused on strategies, plans and measures necessary for the development of STD.

- The establishment of specific STD indicators, which will have to be constantly monitored, and the value of which needs to be in line with that of the general indicators of the socio-economic development of society.
- Ensuring a higher utilisation rate for renewable resources than for non-renewable resources, while using new types of technically and environmentally efficient materials.
- Reducing overall fuel consumption by making systems and operations more efficient (mainly by adopting both minimum travel routes and appropriate driving patterns).
- The adoption of administrative and fiscal measures, with a view to encouraging environmentally friendly transport (public transport should be a priority) at the expense of polluting transport.
- Setting up special funds for the development of STD.
- The use of specific forecasting models, necessary for the long-term adoption of all actions contributing to the implementation of STD, taking into account environmental, economic, social, financial, urban planning aspects.

Clearly, a just policy on the development of a STS must contain specific elements at both national and local level.

The implementation of measures to achieve the above policy objectives is sometimes slow or politically difficult. It is important to better understand how implementation can best be achieved and how obstacles to the introduction of appropriate measures can be overcome.

Consultation of the public and interest groups is extremely important not only for the preparation of infrastructure project proposals but also for the introduction of traffic management, pricing and other policy initiatives. Effective consultation takes time and resources, but without it, the risk of failure or prolonged delay close to planned implementation is exacerbated.

It is essential to win popular and political support. This can make the introduction of measures more widely understood and accepted. The role of the media is essential, and it is necessary to work with the media to ensure balanced coverage. Evidence by example, through well-publicised pilot projects, can also be a powerful tool. The sequencing of measures can be planned, in certain circumstances, to encourage acceptance where a package of measures includes some that will be seen as positive and others that will be seen as negative by certain interest groups. In a special case of this general principle, the allocation of revenues from new charging schemes to investment in improved transport services has also been used successfully in some cases.

Earmarking revenues to provide tax relief elsewhere (within or outside the sector) could probably be used to a similar effect.

In recent times, in many EU countries, specific policies have been adopted to support the development of STSs, relating to:

1 Good coordination of the use of space allocated to transport infrastructure.
2 Transit optimisation.

3 Supporting public cycling and walking.
4 Improving transportation management on public roads.
5 Increased application of new technologies and supporting actions.
6 Internalising external costs.

1 Good coordination of the use of space allocated to transport infrastructure
Land-use and transport coordination is a key element of European efforts to improve sustainability. Local and regional land-use strategies are seen as important ways of managing transport demand and transport impacts. In each country, land-use location policies are designed to reduce travel time and facilitate transit, cycling and walking – an approach referred to as 'short term' land-use strategies. Specific land-use strategies include revitalising existing centres, redeveloping warehouses and brownfields, placing high-density development near transit, developing in and contiguous to existing transit-served centres and planning for compact, mixed-use suburban development to support transit service. Although recent suburban development is lower density and more auto-oriented than development in the urban core and older suburbs, new development is typically provided with a mix of uses at a density that makes walking and cycling practical for many trips and that can be effectively served by transit.

2 Transit optimisation
Transit improvements are another key element of the European sustainability strategy, although in each of the countries we visited, transit mode share has declined overall. The relative decline in transit reflects the fact that an increasing share of the population has access to private automobiles, and ownership and use are growing at faster rates than in the United States. As in the United States, transit subsidies have been a topic of considerable debate, and subsidies have been reduced in recent years, but are still provided as a matter of social and environmental policy. Specific strategies to improve transit service include the development of expanded bus priority lane systems, high-quality architecture and landscaping at transit stations and stops, door-to-door service planning (including planning for walking and cycling access as part of transit planning), improved intermodal transfers and high-quality customer information services.

3 Support public bicycling and walking
More intensive land-use patterns in European cities and suburbs make short trips such that cycling and walking are a realistic option for many. Infrastructure investments that create safe and comfortable facilities for using these modes further support cycling and walking. Even though winters can be harsh in these countries, cycling is recognised as an important mode of transport, particularly for short trips. Extensive bicycle systems, bicycle parking and bicycle in-transit facilities have been established. Traffic controls, including signage and signals, are designed to accommodate slower bicycle speeds and acceleration and to improve bicycle visibility and safety. In urban centres, high-quality pedestrian spaces are plentiful, and many

more are created by widening sidewalks, traffic calming, creating vehicle-free or vehicle-restricted zones and expanding sidewalks at intersections to facilitate pedestrian crossings. Streets full of pedestrians enjoy a sense of vitality and safety (carried by much lower street-crime statistics than in comparably sized American cities). Suburban cities and rural villages are also improving bicycle and pedestrian infrastructure, with traffic calming a common practice.

4 Improving transportation management on public roads
Although there is considerable emphasis on alternative modes of transport, motorways, cars and freight vehicles are increasingly important for European transport, and most policy efforts focus on managing them rather than on efforts to stop their use. Fuel taxes often create a reaction to seek alternative modes of transport, but such taxes have not discouraged levels of fuel consumption or vehicle purchase. Heavy traffic is found in most urban centres, on main roads and in tourist areas. Traffic-related air pollution affects not only urban areas but also crops and forests. Beyond using fiscal policy to influence travel choices, government agencies have also been testing public information campaigns that try to persuade the public to travel as cheaply and as cleanly as possible. Europeans are pursuing highway safety as a major objective of their sustainable development programmes. Strict laws against driving under the influence of alcohol or drugs are a reality in many EU countries. Both speed management and traffic calming are part of newer programmes aimed at improving safety and, in particular, reducing the number of serious accidents on public roads. In addition, the management of freight vehicle operations and regulation with the aim of improving their safety is a topical issue. New designs for highways are seen as important ways to make transport more sustainable, and in a number of projects Europeans are experimenting with designs to improve urban liveability while providing quality services. Streets and highways are being built and rebuilt to reduce negative impacts, in some cases by undergrounding major facilities. Parking is also placed underground in many areas and is priced according to the resulting (high) cost. Joint development of air rights and partnerships with nearby developers and property owners are used to help finance these costly projects. Traffic calming is used extensively in residential neighbourhoods and on main streets in commercial areas; facilities are made of high-quality materials and are well designed and landscaped. Efforts are also being made to help protect biodiversity through good planning, siting, design and maintenance practices. For example, for both new facilities and reconstruction projects, designers are creating animal corridors. The construction and reconstruction of bridges pay attention to the aquatic ecosystem and ensure fish and amphibian movement. Other examples of project design aimed at supporting a healthy environment include using bioengineering techniques to create aesthetically pleasing and environmentally friendly structures; selecting materials to reduce noise and other environmental impacts; and incorporating recycled materials into structures and sidewalks.

5 Increased application of new technologies and supporting actions
New technologies also play important roles in developing a sustainable transportation system. Vehicles powered by alternative fuels are being tested for both transit and personal vehicle trips to reduce pollution and carbon emissions. In addition, Intelligent Transport Systems (ITS) technologies are being promoted to help drivers plan journeys more efficiently, avoid congestion and travel at speeds that reduce congestion and improve safety. Improvements in freight vehicle technology are being sought, and incentives to reduce truck emissions include both emissions pricing and restrictions on the use of 'dirty' trucks in sensitive areas. Road pricing is discussed as a way of properly reflecting the social and environmental costs of car use, but, as in the United States, it is approached with considerable caution because public support for it is mixed at best. Indeed, high fuel taxes have been the subject of several strikes and political debates in recent years. A rail network that is somewhat less conducive to goods movement than that of the United States makes Europeans relatively dependent on road freight, although rail and multimodal improvements are increasingly supported. At the same time, advanced improvements in logistics and complementary operations are being systematically implemented to move goods smoothly from ports to markets, maximise the capacity of existing facilities, reduce congestion, improve safety and reduce the need for facility expansion.

6 Internalisation of external costs
It is clear that, overall, the total external costs caused by transport exceed those resulting from user charges, the difference being borne by society.

The process of internalising external costs in order to develop a STS can in this context be compared to deregulation, with the caveat that it can only be successfully applied in a truly efficient market. A basic principle to be adopted in order to achieve this precondition is the requirement that economic instruments must be applied to the phenomenon and not to the activity that generates it.

The use of economic instruments to control pollution is mandatory, as traditional measures are practically ineffective from both static and dynamic perspectives (some traditional measures may lead to the introduction of costs that are too high for some users).

The economic instruments that can be used to internalise external costs have the following common elements:

1 They act through financial incentives.
2 The polluter is able to act voluntarily only on market signals, corrected by this process or by setting prices for common property resources (public goods).
3 Local or central authorities are actively involved, institutionally, legally, economically, administratively, etc.
4 There is a will to maintain or improve environmental quality through the application of economic instruments.

The essential aim of an internalisation policy is to increase the efficiency and equity of transport systems. The aim is, therefore, to ensure that prices reflect the true value of costs, so that individuals and businesses are able to make the right decisions about how to move goods and people across time and space on an economically efficient basis.

The introduction of economic instruments can have a number of advantages in preventing damage to the environment and health, including

- The promotion, among polluters, of appropriate behaviour to protect the environment by reducing/eliminating the level of pollution.
- Maintaining permanent incentives for the introduction of new anti-pollution technologies as they become available on the market.
- Creating the conditions for the development of new, efficient and cleaner technologies.

For the successful implementation of measures specific to a policy of internalisation of external costs, certain principles must be respected, including

- The level of taxes imposed should reflect, as closely as possible, the evolution of the costs to which they refer; this will lead to a decrease in the value of externalities and an increase in the benefits to society.
- In order to encourage users to adopt the most appropriate measures to reduce externalities, an efficient system should introduce differentiated levels of charges according to specific characteristics (level of pollution, type of vehicle, etc.).
- The tax structure should be clear to users.
- The tax system should not discriminate between different modes of transport.
- The economic instruments used should seek to ensure that, for all modes, the price of travel reflects as closely as possible the level of total social costs.

The introduction of an efficient and fair tax system will lead to a substantial increase in economic competitiveness, since by reducing externalities due to transport, unjustified financial losses will be reduced (eliminated). For example, the reduction of congestion reduces the time wasted by industrial enterprises whose activities are linked to the need to transport materials.

At the same time, the use of a fair tax system leads to the accumulation of revenue that can be used to reduce unemployment or create new jobs, for example, by putting it back into the economy. Leaving aside the obvious merits of these instruments, however, it is clear that they can only be used where polluting activity can be monitored. Unfortunately, many aspects of externalities due to transport are imperceptible and thus unrulable, being specific to travel time, route, type of vehicle, etc. In order to calculate the marginal level of external costs, the state must be able to monitor the activity and the resulting level of polluting effects for each individual vehicle.

In the long term, a policy to develop a STS should aim at improving monitoring systems. In the short term, however, it is necessary, first and foremost, to adopt those measures aimed at introducing economic instruments designed to reduce externalities.

## 1.4 Are Romania, the other Eastern and EU countries on the right track regarding sustainable transport?

Thanks to the growth in transport activity worldwide, sector-specific energy consumption has been steadily increasing. Although, at least in recent years, numerous policies have been adopted with a view to reducing energy consumption in the transport sector, in particular through the construction of more efficient vehicles and the use of alternative, less polluting fuels, the trends have continued to increase.

In the European Union, for the period 1990–2020, increases in final energy consumption in transport have predominantly been recorded, with a peak of more than 30% in 2019 (see Table 1.1).

At the sectoral level, compared to 1990, transport recorded the highest increase in final energy consumption, from 220.7 Mtoe in 1990 (24.3% of the total for the economy) to 252 Mtoe in 2020 (28% of the total for the economy) according to EC, (2022a). In the same period, final energy consumption of households increased from 26.5% in 1990 to 28% in 2020, while in the industrial sector, final energy consumption decreased from 34.2% in 1990 to only 26% in 2020.

The comparisons presented in Table 1.1 have been made for the period 1990–2019, because from 2020, due to the health pandemic, transportation activity has been restricted, thus the data are less relevant.

*Table 1.1* Evolution of Final Energy Consumption in the Transport Sector in the EU 1990–2020

| | | | | *Mtoe* | | | | | |
|---|---|---|---|---|---|---|---|---|---|
| *Sector* | *1990* | *1995* | *2000* | *2005* | *2010* | *2015* | *2019* | *2020* | *2019/ 1990 (%)* |
| Transportation | 220,7 | 235,9 | 262,9 | 281,6 | 280,0 | 272,8 | 289,0 | 252,0 | 130,9 |
| Rail | 7,5 | 7,5 | 7,4 | 6,8 | 6,3 | 5,4 | 5,3 | 4,7 | 70,7 |
| Road | 201,6 | 218,0 | 243,2 | 260,2 | 260,6 | 255,6 | 270,0 | 238,2 | 133,9 |
| Air | 5,0 | 4,3 | 5,7 | 5,8 | 5,6 | 5,4 | 6,5 | 3,1 | 130,0 |
| Naval | 5,2 | 5,1 | 5,2 | 5,6 | 5,0 | 3,9 | 4,2 | 3,6 | 80,8 |
| Road in total sector (%) | 91,3 | 92,4 | 92,5 | 92,4 | 93,1 | 93,7 | 93,4 | 94,5 | 102,3 |

*Source:* Data and processing after (EC, 2022a, p. 119).

Tonne of oil equivalent (toe or toe) is a conventional standardised unit for measuring energy consumption, defined on the basis of the consumption of one tonne of oil with a net calorific value of 41,868 kilojoules/kg (Mtoe = 1 million toe).

From the data presented in Table 1.1, it is clear that road transport is clearly the largest energy consumer in the transport sector, with a share, at all times, of more than 90% of the total for the economic branch analysed.

Moreover, the share of energy consumption from road transport in the total sector has continuously increased during the period under review, with a maximum increase of more than 2% (between 1990 and 2020). The increase, even if it seems statistically insignificant, is high, because the specific market has annual energy consumption of hundreds of millions of tons of oil equivalent (toe).

The concept of sustainable transport emerged in response to the harmful consequences for the environment and human health of this activity in the last half of the 20th century.

In fact, unsustainable transport can be characterised by excessive energy consumption, major environmental pollution, important contributions to the deterioration of the population's health and often unsatisfactory activities for the different economic and social groups it serves.

In this context, the European Commission's Green Paper on Security of Supply of November 2000 noted the important role of transport in increasing energy demand and $CO_2$ emissions.

Therefore, in September 2001, the European Commission adopted a new White Paper on Community Transport Policy. The document proposed measures to revise transport policy to make it more sustainable and avoid the huge economic losses caused by congestion, pollution and accidents.

A Constitution for Europe was adopted in October 2004 and a new White Charter was launched in 2011, entitled 'Towards competitiveness and resource efficiency of transport systems'.

The shift to low-carbon mobility initiated by the 2011 White Charter has subsequently been complemented by various legislative initiatives adopted at EU level.

For example, the 2016 initiative entitled 'A European Strategy for Low Carbon Mobility' highlights the requirements on which action should focus:

- Digital mobility solutions.
- Fair and efficient transport pricing (which should better reflect the negative externalities of transport).
- Promoting multimodality.
- Developing an effective framework for low-emission alternative energy.
- Upgrading infrastructure for the use of alternative fuels (electricity charging stations).
- Increasing interoperability and standardisation of EVs.
- Improve vehicle testing programmes.
- The adoption of a post-2020 strategy for all road transport, supported by research and investment efforts.

In December 2019 and in the above-mentioned context, at the EU level, the EGD was adopted, which set the guidelines for various European policies for the next five years. The document is linked to a number of legislative and non-legislative initiatives in multiple areas such as environment, climate change, energy, industry, transport, agriculture, digitisation and the financial sector.

The EGD is a new growth strategy to transform the European Union into a fair and prosperous society with a modern, resource-efficient and competitive economy. It is, therefore, necessary to promote and increase investment in modifying the tools and technologies underpinning such paradigm shifts. Digital technology can indeed contribute to reducing global emissions, improving quality of life and reducing society's environmental footprint, for example, by optimising energy consumption and monitoring emissions in transport. In order to achieve climate neutrality by 2050 and the ambitious goal of zero pollution for a toxic-free environment, all sectors of the economy, including transport, need to undergo intensive transformation processes.

In this context, the EGD announced the adoption of a proposal for stricter air pollutant emission standards for internal combustion vehicles (Euro 7).

To accelerate the 'decarbonization' of transport activities and road transport in particular, the European Commission proposed in July 2021 an amendment to the Regulation on $CO_2$ emission performance standards for cars and vans to ensure a clear trajectory towards zero-emission mobility.

In addition, in December 2020, the Commission adopted the *Sustainable and Intelligent Mobility Strategy* and, in May 2021, the *Zero-Emission Action Plan*. According to these plans, transport should become much less polluting, especially in urban areas, and Euro 7 is seen as an essential part of the transition to real sustainable mobility.

Euro 7 is the newest pollution standard and is part of Europe's complex and ambitious 'Fit For 55' project, based on the EGD.

From 2035, the decision was taken to stop producing cars powered by internal combustion engines, requiring only new zero-emission cars to be put on the market. The German amendment, however, stipulated that production of cars with internal combustion engines can continue after 2035, provided they are powered by synthetic fuels (e-fuels). Synthetic fuels could be considered neutral in terms of air pollution if the energy used to produce them and their components is generated from renewable sources (RES).

Unfortunately, however, manufacturing e-fuels is very costly and energy-intensive. Running such fuels in a vehicle powered by an internal combustion engine requires about five times more renewable electricity than running a battery electric vehicle.

In all these circumstances, the Euro 7 standard aims to support the transition by 2035 to electric cars or, according to the latest requirements, to cars with internal combustion engines that are powered by environmentally friendly fuels.

This Euro 7 standard also includes regulations for electric cars. Specifically, these new requirements relate to the production of the batteries needed to run these cars. Specifically, the battery must retain at least 80% of its charge capacity for the first five years. At the same time, at least 70% of the charge capacity should last up to eight years of operation or 160,000 km. So, as well as greater confidence in the long-lasting performance of electric batteries, the European Union also wants to limit consumption of the raw materials from which they are made. Even so, more than 20% of cars and vans and more than half of heavier vehicles are expected to continue to emit pollutants in 2050 (EC, 2022b).

To summarise, we can say that since 2001 the European Union has developed a series of legislative documents to support the development and modernisation of sustainable, energy-efficient and environmentally friendly transport. The most important documents adopted were as follows:

- 2001 – White Paper: *European transport policy for 2010: time to decide*, COM (2001) 370 final.
- 2004 – Communication from the Commission to the Council, the European Parliament, the European Economic and Social Committee and the Committee of the Regions: Towards a Thematic Strategy on the Urban Environment, COM (2004) 60 final.
- 2005 – Communication from the Commission to the Council and the European Parliament on a Thematic Strategy on the Urban Environment, COM (2005) 718 final.
- 2006 – Communication from the Commission to the Council and the European Parliament: Keep Europe moving – Sustainable mobility for our continent, COM (2006) 314 final.
- 2006 – Communication from the Commission: Action Plan for Energy Efficiency: Realising the Potential, COM (2006) 545 final.
- 2007 – Commission Green Paper: Towards a new culture for urban mobility, COM (2007) 551 final.
- 2009 – Communication from the Commission to the European Parliament, the Council, the European Economic and Social Committee and the Committee of the Regions: Action Plan on Urban Mobility, COM (2009) 490 final.
- 2011 – White Paper: Roadmap to a Single European Transport Area – Towards a competitive and resource efficient transport system, COM (2011) 144 final.
- 2013 – Urban Mobility legislative package: Towards competitive and resource efficient urban mobility together, COM (2013) 913.
- 2015 – The Paris Agreement (2015), evidenced by the Paris Declaration on Electro-Mobility and Climate Change – Requirements for Action, COP21-CMP11.
- 2019 – EGD, COM (2019) 640 final.
- 2020 – Sustainable and Intelligent Mobility Strategy – Putting European Transport on the Road Ahead, COM (2020) 789 final.
- 2023 – Fit for 55 (adopted April 2023).

The support for the development of sustainable transport and, implicitly, the efficiency of energy consumption in transport in Romania has been backed by documents, for the most part, produced after the country's accession to the European Union in 2007.

In Romania, according to the data presented by the NBR, the transportation sector has developed significantly in the last two decades, its rise being interrupted only by the COVID-19 pandemic. The analysed sector has gradually come to generate a consistent trade surplus, which was somewhat smaller in 2021, that is, 1.8% of GDP (NBR, 2022).

According to the same NBR report, 'a structural factor with an adverse effect on the external position of the automotive sector concerns the insufficient expansion of the national vehicle fleet.

In this context, automobile imports (for final consumer demand or investment) have induced a large negative contribution to the trade balance, growing strongly, especially before the onset of the financial crisis, and then recording more modest developments, but still on an upward trend in the period 2013–2019.

In spite of this, Romania continued to rank last in Europe in terms of the number of cars per capita; the aforementioned rise in imports has essentially only resulted in a narrowing of the gap between the second last and last place (Romania's) in the above-mentioned hierarchy.

This structural influence is expected to continue in the period to come, given the need for economic growth and mobility.

Of course, the perceptible pressure on the existing road infrastructure, which puts the idea of a shortage of means of transportation in a counterintuitive light in the first instance, once again emphasises the need to improve the road network' (NBR, 2022).

According to the analysis for Romania, over the last 15 years (2006–2020), the situation of transportation-specific energy consumption highlights two main trends.

The first refers to the halving of final energy consumption in the transport sector in the last 15 years, and the second to the fact that in Romania, more than 95% of transport-specific energy consumption is recorded in the road transport sector.

This is the main reason why, in the following analysis, the main focus will be on this mode of transport.

Clearly, with higher energy consumption, the environmental damage caused by transportation activities has also increased.

Even though, especially in recent years, there has been an increased focus on the manufacture and operation of low-emission or even zero-emission vehicles, the level of GHG emissions has increased, amid the continued growth in the ageing vehicle fleet and, more importantly, the increased demand for mobility.

Moreover, in Romania, the level of GHG emissions increased by more than final energy consumption between 2006 and 2019. This can be explained by the characteristics of inefficient vehicle fleets, in terms of pollution and energy consumption, which are specific to Romania's transport fleets. In other words, the age of some transport vehicles (especially rail and naval) and the predominant use of non-environmentally friendly vehicles have led to this situation.

In brief, the transport sector is an increasing source of GHG emissions, which are unfortunately growing faster than the sector's energy consumption. Therefore, *the future energy* strategy for the sector must focus on tackling the direct emissions generated by transport (especially road transport, the main pollutant in the economic sector) as well as the indirect emissions generated by alternative energy sources used in transport (electricity, hydrogen, etc.).

For these reasons, energy reductions in the Transport sector have unfortunately been much more modest than those desired under the implementation of specific policies. For example, from the latest available official reporting, it has been observed that the following savings in energy consumption have been realised through the measures adopted at the 2018 level (MEEMA, 2020):

- Urban public transportation: 940 toes.
- Bucharest Metro: 3915 toes.
- Rail: 973 toes.
- Naval transport: 276 toe.
- Air transport: 665 toes.
- Renewal of public and private vehicle fleet: estimated 100,000 toes (but not reported).

Taking also into account the estimated maximum for fleet renewal, at the 2018 level, Romania achieved a reduction in transport-specific consumption of 106,769 toe, which is just over 44% of the value proposed in 2014.

The greatest reductions in energy consumption were achieved in 2018 through the renewal of the vehicle fleet, the modernisation and extension of the Bucharest metro network, energy efficiency works on the railway infrastructure and the modernisation of urban public transport.

Despite this setback, Romania has continued to set itself increasingly ambitious targets for reducing energy consumption.

Thus, it has set as a priority objective for 2020 a 19% reduction in domestic primary energy consumption, as forecast in the reference scenario using the PRIMES 2007 model, with a primary energy saving of 10 million toes in 2020, compared to the domestic primary energy consumption forecast for 2020 of 52.99 million toe (EC, 2021).

In this context, the national domestic primary energy consumption target for 2020 required by Article 3(1) of the Energy Efficiency Directive became 42.99 Mtoe. Achieving this target leads to a final energy consumption of 30.32 Mtoe.

According to the same document, the savings resulting from the energy efficiency measures implemented in the transport sector were expected to reach 412.7 thousand toes by 2020, which is almost 15 times the actual consumption realised in 2014–2016 or 40% of the total consumption expected for the entire period 2014–2020.

Analysing the evolution of energy consumption in the transport sector and the savings achieved, we can state that, according to the data available at the moment, the energy savings achieved in the transport sector amounted to 0.12% in 2015 and increased in projection by almost 53 times by 2020, reaching a value of 6.35%.

For renewable energy, a target of 14% of final energy consumption in the transport sector has been adopted at the European level for 2030 for each Member State (EC, 2020a). The plan's projections indicate an accelerated electrification of the transport sector, based on the set of priorities identified and on the assumption that the costs of light EVs will converge with those of internal combustion cars by 2024.

Thus, by 2030, about 700,000 private electric cars (including hybrids) and about 650,000 charging points (of which about 40,000 in fast and semi-fast charging mode) are expected to be in circulation.

The 2030 target takes into account the current level of the share of renewable energy sources (RES) of 6.56% in 2017 and implicitly the efforts needed to reach a value of 14.2% in 2030.

In the case of Romania, of the total vehicle fleet of more than 8.6 million vehicles registered in 2019 (EC, 2022a), 1.3% were classified as green (electric and hybrid). In fact, at the end of 2022, there were about 9.9 million vehicles, of which more than 23,000 were electric and almost 120,000 hybrids (Predescu, 2023).

In view of the situation, from 2010 until 2050, when GHG emissions must be eliminated throughout the European Community, a programme for the Romanian transport sector to reduce this type of emissions could be as shown in Table 1.2.

In order to achieve the targets imposed by the European Commission on the gradual reduction of GHG emissions, the national authorities have drawn up several action plans in the field of transport, one of the most important being the sustainable urban mobility plan, which is on the parliamentary agenda and was not approved at the time of writing this study. Through this document, the Romanian authorities aim (MDPLA, 2023):

- Making public transport faster and more accessible than transport by personal vehicle. Increasing the use of public transport by optimising public transport means (buses, trolleybuses, trams) and the infrastructure necessary for their proper functioning.
- Transition towards multimodal travel and mobility as a service.
- Encouraging alternative forms of transportation (cycling, car-pooling, car-sharing, etc.) through urban planning and the development of adequate cycling infrastructure (bicycle lanes, storage racks, special bicycle carriages/compartments on the metro and trains, etc.) and the expansion of pedestrian zones especially in large urban agglomerations.
- Discouraging the development of new parking networks for private cars. Parking should be used and considered as a key lever for the transition towards sustainable mobility patterns.
- Renewing and developing public transport fleets with clean vehicles.

*Table 1.2* Trends and Outlook for Specific Transport Indicators in Romania for the Period 2010–2050

| *Indicator* | *U.M.* | *2010* | *2015* | *2019* | *2030*[a] | *2040*[a] | *2050*[a] |
|---|---|---|---|---|---|---|---|
| Level of GHG emissions[b] | Million tons $CO_2$ equivalent | 15,8 | 17,2 | 19,5 | 13,2 | 6,9 | 0,0 |
| Level of GHG emissions compared to 2010 | % | 100,0 | 108,9 | 123,4 | 83,5 | 43,7 | 0,0 |
| Final energy consumption from conventional sources[c] | Mtoe | 5,0 | 5,6 | 6,6 | 4,5 | 2,4 | 0,0 |
| Level of reductions of energy resources through specific national programmes | Mtoe | - | 0,0067 | 0,2937 | 2,1 | 2,1 | 2,1 |
| Vehicle fleet[d] | Million units | 5,49 | 6,60 | 8,68 | 9,00 | 9,50 | 10,00 |
| Polluting vehicle fleet | Million units | 5,49 | 6,60 | 8,67 | **5,90**[e] (*2,87*)[f] | **3,10**[e] (*1,50*)[f] | **0,00** |
| Clean vehicle fleet | Million units | 0,0 | 0,0003 | 0,005 | **3,1** (*6,13*) | **6,4** (*8,00*) | **10,00** |
| Annual rate of replacement of the polluting fleet by environmentally friendly vehicles | Thousand units | - | - | - | **281** (*557*) | **330** (*187*) | **360** (*200*) |
| Annual green vehicle purchase costs[g] | Billions of euros | - | - | - | **8,4** (*16,7*) | **8,3** (*4,7*) | **7,2** (*4,0*) |

*Sources:* Data and processing after (EC, 2022a).

[a] Author's estimates.
[b] For the period 2019–2050, an average GHG emission reduction rate of 0.63 million tons $CO_2$ equivalent per year has been assumed.
[c] Final energy consumption is shown for vehicles using fuels that produce direct pollutant emissions (by burning the fuels). For the period 2030–2050, the estimates have been made in line with the total GHG emission target values that transport has to meet.
[d] For the period 2030–2050, a scenario of a 'calming' of the growth of the vehicle fleet (by about 100,000 additional units per year) has been projected.
[e] The Green Scenario – the calculation was carried out according to linear regression taking into account the variables of the level of GHG emissions and the number of vehicles producing this level.
[f] Blue Scenario – the calculation took into account that a vehicle using conventional fuel (gasoline or diesel) produces about 4.6 tons of carbon dioxide per year (RFIDTIRES, 2023).
[g] For an electric vehicle, an average value of 30,000 €/year 2030, 25,000 €/year 2040 and 20,000 €/year 2050 was considered. The value of €30,000 for the year 2030 was estimated taking into account the average value of the ten cheapest electric/hybrid cars on the Romanian market at the beginning of 2023 (Automarket, 2023).

- Continuous updating of legislation to support sustainable mobility in line with new requirements and technical/procedural/conceptual changes.
- Urban regeneration of neighbourhoods with a focus on the '20-minute service' city.
- Counter urban sprawl through transit-oriented development and land-use planning.
- Intensify efforts towards inclusive transportation services and infrastructure.
- Support urban logistics policies as a priority.
- Optimise the flow of goods and people by focusing on the completion of TEN-T networks.
- Further develop, diversify and integrate metropolitan transport.
- Updating the PMUD plans taking into account the wide variety of new technologies and changing travel patterns.

Given the simulations in Table 1.2 it is evident that the 'zero' GHG emissions target from the perspective of the transport sector is extremely difficult, if not impossible, to achieve. If we consider only the replacement of the fleet of polluting vehicles with clean ones, it is clear that this is not achievable in less than three decades.

In support of this assertion, we emphasise the fact that in 2022 in Romania 11,954 electric cars were purchased, double, however, compared to the previous year (Socariciu, 2023). This value is, however, much too low to reach the targets proposed for the following years. Thus, the value in 2022 is about 23.5 times lower than the value estimated in the BOLD scenario for 2030 and 46.6 times lower than in the ITALIC scenario, also for 2030. Therefore, we consider that it is impossible that in eight years the growth in sales of green vehicles will increase by as many times to reach the levels of the scenarios presented.

Also, analysing the data in Table 1.2, for the replacement of the current car fleet with a fleet of green cars, the annual financial effort of about € 7.2–8.4 billion (in the BOLD scenario in Table 1.2) or between € 4.0 and 16.7 billion (in the ITALIC scenario) corresponds to an annual minimum of 2.5% of GDP (BOLD scenario) and a maximum of 19.5% (ITALIC scenario) taking into account only the GDP level in 2022, estimated at € 286.4 billion (Bancherul, 2023).

It is no less relevant the fact that, according to the data presented above, in the European Union as a whole, 700,000 environmentally friendly vehicles are expected to be in circulation by 2030, while the proposed plan for the renewal of the Romanian fleet alone should bring the total number of vehicles in circulation in the European Union to over 3 million. This is impossible to put into practice.

The low income of the population and the financial inability of the state to support, from budgetary resources, make the action to replace the current fleet unfeasible.

It is no less relevant the fact that, according to the data presented above, in the European Union as a whole, 700,000 environmentally friendly cars are expected to be in circulation by 2030, while the proposed plan for renewing the Romanian fleet alone should bring the total number of vehicles in circulation in the European Union to over 3 million. This is impossible to put into practice.

The low income of the population and the financial inability of the state to support, from budgetary resources, make the action to replace the current fleet unfeasible.

Moreover, the current solutions for replacing the fleet of vehicles equipped with internal combustion engines (oil fuel) with other types of vehicles, especially EVs, are not a viable solution. This can be explained by several arguments:

- Cars do not pollute in operation, but the energy consumed pollutes where it is produced (at source).
- Vehicle autonomy is low, and increasing it requires technical solutions, mainly to increase the service life of the batteries, which use increasingly polluting materials and manufacturing technologies.
- The cost of such vehicles is high and cannot easily be borne by individuals. As an example, on the Romanian market in 2023, the average purchase price of an electric/hybrid car was over 30,000 euros, taking into account the ten cheapest models available in the national offer (Automarket, 2023).

Regarding the share of renewable energy, the European Commission recommended that Romania increase the target for 2030 to a share of energy from RES of at least 34% (EC, 2020b).

In order to achieve these targets, Romania has set out to develop a series of policies and measures aimed at both reducing energy consumption and encouraging the use of RES in the relevant sectors – Heating and Cooling, Electricity and Transport – maximising the convergence between the different actions envisaged. As a result, Romania targets a primary energy consumption of 32.3 Mtoe and a final energy consumption of 25.7 Mtoe, thus achieving energy savings of 45.1% for primary consumption in 2030 and 40.4% for final energy consumption.

Furthermore, Romania shall achieve a cumulative new energy savings equivalent to 10.12 Mtoe between 2021 and 2030.

## Notes

1 According to World meters, by October 2024, the world's population had already reached 8.18 billion people (World meters, 2024).
2 Malcolm Rupert Cutler is a well-known American ecologist and authorised director of the Wildlife Society, (The Wildlife Society) in the United States of America (Prabook, 2024).
3 ETS (Emissions Trading System) stands for Emissions Trading Scheme.
4 According to Article 3 of Directive 2003/87/EC of the European Parliament and of the Council of 13 October 2003 establishing a scheme for greenhouse gas emission allowance trading within the Community and amending Council Directive 96/61/EC (EU ETS Directive) (EP, 2003).

## References


Auto Bild, *How many cars there are in the world and how many there will be in 2050*, 2024, https://www.auto-bild.ro/stiri/cate-masini-sunt-in-lume-si-cate-vor-fi-in-2050-72426.html

Automarket, 2023, *Cheapest electric/hybrid cars*, https://www.automarket.ro/masini-noi/topuri/cele-mai-ieftine-masini-din-romania/electrice-hibride/

Bancherul, 2023, https://www.bancherul.ro/stire.php?id_stire=21963&titlu=avansul-pib-pe-2022,-confirmat-la-4,8-la-suta.-rezultatul-nominal,-cu-3-miliarde-euro-sub-a%C8%99tept%C4%83ri

CBM (CECCAR Business Magazine), *IATA: Airlines expect the number of trips to double in the next 20 years*, 2024, https://www.ceccarbusinessmagazine.ro/iata-companiile-aeriene-se-asteapta-la-dublarea-numarului-de-calatorii-in-urmatorii-20-de-ani/a/NTQ2MTMyMTA0MzQ5NzE2M4yJxvQNQ5Za1CSZhjVinBY

Delucchi M. A., *External costs of transportation in the U.S*, Institute of Transportation Studies, UC Davis, 2010, http://escholarship.org/uc/item/13n8v8gq

Dexonline, *Resilience*, 2024, https://dexonline.ro/definitie/rezilien%C8%9B%C4%83

Dubey R., Gunasekaran A., Sustainable transportation: An overview, framework and further research directions, *International Journal of Shipping and Transport Logistics*, 2015 Vol. 7 No. 6, pp. 695–718.

EC, EU Energy and Transport in figures, Statistical Pocketbook 2010, ISBN 978-92-79-79-13815-7, 2010.

EC, *Mobility and transport: Transport in the European Union. current trends and issues*, 2018a, link: https://ec.europa.eu/transport/sites/transport/files/2018-transport-in-the-eu-current-trends-and-issues.pdf

EC, *Annual accident report 2018*, 2018b, https://road-safety.transport.ec.europa.eu/document/download/680f0863-aa0a-44f5-b3b8-39fe139924b0_en?filename=asr2018.pdf

EC, *Handbook on the external costs of transport version 2019*, Luxembourg, ISBN 978-92-79-79-96917-1, 2019.

EC, *Sustainable and smart mobility strategy-putting European transport on track for the future*, (COM (2020)789 final), 2020a, https://www.eumonitor.eu/9353000/1/j9vvik7m1c3gyxp/vlehreuol0wv

EC, *National integrated energy and climate change plan 2021–2030*, 2020b, https://energy.ec.europa.eu/system/files/2020-04/ro_final_necp_main_ro_0.pdf

EC, *Statistical pocketbook 2021-EU transport in figures, ISBN*: 978-92-76-40099-8, 2021

EC, *EU transport in figures*, ISBN 978-92-76-76-53698-7, 2022a.

EC, *Commission proposes new Euro 7 standards to reduce pollutant emissions from vehicles and improve air quality*, 2022b, https://ec.europa.eu/commission/presscorner/detail/en/ip_22_6495

EC, *Delivering the European green deal*, 2024a, https://commission.europa.eu/strategy-and-policy/priorities-2019-2024/european-green-deal/delivering-european-green-deal_en

EC, *EU transport in figures-2024*, ISBN 978-92-68-19812-4, 2024b, https://op.europa.eu/en/publication-detail/-/publication/c99d5d85-8c3a-11ef-a130-01aa75ed71a1

ECMT (European Conference of Ministers of Transport), *Sustainable transport policies*, 2000, https://www.itf-oecd.org/sites/default/files/docs/00sustain_0.pdf

EP (European Parliament), *Directive 2003/87/EC of the European Parliament and of the Council of October 13, 2003*, p. 32, 2003, https://eur-lex.europa.eu/legal-content/RO/TXT/PDF/?uri=CELEX:32003L0087&from=EN

EPA (United States Environmental Protection Agency), *EPA – Environmental justice implementation plan 1996*, 1996.

EUROSTAT, *Greenhouse gas emission statistics – Emission inventories*, 2019, https://www.sipotra.it/wp-content/uploads/2019/06/Greenhouse-gas-emission-statistics-emission-inventories.pdf

Fistung D., *Teorie economică, ecologie, legislatie* (Transportation. Economic theory, ecology, legislation), ALL BECK Publishing House, Bucharest, ISBN 973-94435-03-3, 1999.

Goldemberg J., The promise of clean energy, *Energy Policy*, Vol. 34, no. 15, pp. 2185–2190, 2006.

Gray P. G. and Frost J. C., The impact of catalysis on clean energy in road transport, *Energy & Fuels*, Vol. 12, no. 6, pp. 1121–1129, 1998.

Leuenberger D. Z., Bartle J. R. and Chen C., Sustainability and transportation, *Public Works Management & Policy*, Vol. 19, no. 4, 2014, https://www.researchgate.net/publication/319102281_Sustainability_and_Transportation

MDPLA (Ministry of Development, Public Works and Administration), *Proiect Lege privind mobilitatea urbană durabilă (Draft Law on Sustainable Urban Mobility)*, 2023, https://www.mdlpa.ro/pages/proiectlegemobilitateurbanadurabila

MEEMA (Ministry of Economy, Energy and Business Environment), *Report on progress towards national energy efficiency targets*, 2020, https://energy.ec.europa.eu/system/files/2020-07/ro_annual_report_eed_2020_0.pdf

MFA (Ministry of Foreign Affairs)-Government of Romania, *Green transition/European ecological pact*, 2022, http://www.mae.ro/node/55118

NBR (National Bank of Romania), *Annual report 2021*, ISSN 1584-0913, 2022.

Prabook, 2024, https://prabook.com/web/malcolm_rupert.cutler/293535

Predescu M., *National car fleet increased by 3% in 2022. How many electric and hybrid vehicles are registered in Romania*, 2023, https://www.gandul.ro/actualitate/parcul-auto-national-a-crescut-cu-3-in-2022-cate-vehicule-electrice-si-hibride-sunt-inmatriculate-in-romania-19929004

RFIDTIRES, How Many Cars Are in The World? 2023, https://www.rfidtires.com/how-many-cars-world.html

Schrank, D. and Lomax T., *The 2007 Urban Mobility Report, Texas Transportation Institute*, The Texas A&M University System, 2007, http://mobility.tamu.edu

Shafiee S., Topal E., When will fossil fuel reserves be diminished? *Energy Policy*, Vol. 37, No. 1, 2009, pp. 181–189. https://www.sciencedirect.com/science/article/abs/pii/S0301421508004126

SLOCAT (Partnership on Sustainable, *Low carbon transport, transport and demand*, 2022, https://tcc-gsr.com/transport-demand/transport-demand/

Socariciu C., *Electric cars in Romania in 2023*, 2023, https://www.autocritica.ro/green-zone/masinile-electrice-din-romania-in-2023/

Sorrentino, M., Rizzo, G., and Sorrentino, L., A study aimed at assessing the potential impact of vehicle electrification on grid infrastructure and road-traffic greenhouse emissions, *Applied Energy*, Vol. 120, pp. 31–40, 2014.

STATISTA, *Number of passenger cars and commercial vehicles in use worldwide from 2006 to 2015*, 2019, link: https://www.statista.com/statistics/281134/number-of-vehicles-in-use-worldwide/

TIMOCOM, *These are the trends in logistics for 2024*, 2024, https://www.timocom.ro/blog/trenduri-logistice-pe-piata-transporturilor-2024-582338

VIRTA, *The global electric vehicle market overview in 2022: Statistics & forecasts*, 2022, https://www.virta.global/en/global-electric-vehicle-market

Worldmeters, *World population*, 2024, https://www.worldometers.info/en/

WBCSD (The World Business Council for Sustainable Development), *Mobility 2030: Meeting the challenges to sustainability*, 2004, https://www.wbcsd.org/Programs/Cities-and-Mobility/Sustainable-Cities/City-Business-Collaboration/SiMPlify/Resources/Mobility-2030-Meeting-the-challenges-to-sustainability-Full-Report-2004

# 2 Sustainable transport infrastructure, a new concept

## 2.1 Prerequisites and opportunities for transport infrastructure development

In 2017, UN Secretary General Antonio Gutteres stated,

> Investment in infrastructure will be essential. The world should adopt a simple rule: if large infrastructure projects are not green (sustainable), they should not get the green light. Otherwise, we will be locked into bad choices for decades to come.
>
> (ISI, 2018)

At the global level of infrastructure (obtained by including alongside the actual infrastructures the measures in which the basic, technological, scientific and human resources meet the needs of firms), according to the World Competitiveness Centre (WCC, 2023), in 2023, out of a total of 64 countries analysed from all over the world, Romania ranks 50th with 35.84 points,[1] behind the Czech Republic, 24th place (66.69), Hungary 38th (49.57) or Poland 39th (48.46) but ahead of Bulgaria, 54th place (29.22). The CMC's ranking of infrastructure competitiveness in 2023 was led by Switzerland, with 92.99 points (2.5 times more than Romania), followed by Denmark (92.71), Finland (87.60), Sweden (86.71), the Netherlands (86.19) and the United States (84.52).

Moreover, investment in transport infrastructure was severely hit by the Great Recession, leading to significant funding gaps. Spending on transport infrastructure still accounts for a significant share of countries' budgets, but the current trajectory points to a shortfall of around $350 billion per year (excluding spending on maintenance backlogs), a gap that triples when the scale of investment needed in transport infrastructure to meet the UN Sustainable Development Goals (SDGs) is taken into account.

As is well known, increased investment in building and upgrading transport infrastructure could help to improve connectivity between rural and urban areas and boost local economies. However, there has been a long-term decline in the availability of infrastructure, as the stock of public capital (closely linked to infrastructure) as a share of output has declined over the last three decades worldwide.

DOI: 10.4324/9781003743064-3

The impact of the 2007–2008 global financial crisis has been profound, but infrastructure investment has remained below pre-crisis levels in many countries (OECD, 2024).

To support infrastructure investment, governments may need to prioritise sub-sectors where infrastructure is poorest (such as the rail sector in Romania). Clearly, returns to infrastructure investment are higher where current endowments are lower.

Despite some structural changes triggered by the pandemic, such as shifts in the passenger/freight mix, passenger volume growth is expected to be mainly driven by Asia, where it is projected to grow by about 3% per year between 2015 and 2050 (Milani et al., 2021).

By comparison, in Europe and the United States, growth is expected to be more moderate, at less than 1% per year. However, the increasing global population and continued economic growth are expected to increase the customer base, demanding constant and intense passenger and freight transportation, which will ultimately lead to significant growth in total transportation demand.

A major driver of transport infrastructure spending is the breakneck pace of continuing urbanisation. According to World Bank data, the global population is expected to reach 7.8 billion in 2020,[2] with urban areas hosting the bulk of this boom. According to World Population Prospects data, produced under the aegis of the United Nations, between 1950 and 2018, the world's population grew from about 2.5 billion people to 7.6 billion, an annual growth rate of 1.6%. The urban population grew faster at an annual rate of 2.6%. In 1950, 30% of people globally lived in urban areas. By 2018, the figure had reached 55%. Experts predict that the share will continue to rise to 70% by mid-century.

As urban areas grow and expand around the world, so will the transportation infrastructure needed to connect them. In developing and high-growth countries, new transport infrastructure is needed to support the movement of people from rural to urban areas. In developed countries, existing infrastructure needs to be upgraded to increase resilience, reduce carbon emissions and cut maintenance costs.

In addition, to mitigate the economic losses caused by the health pandemic from 2019 to 2022, governments around the world have introduced stimulus programmes that have significant infrastructure shares. In the United States, for example, a $2 trillion plan has been introduced to modernise national infrastructure over the next decade.

Gains from investment in transportation infrastructure in some regions could be offset by losses in others. The construction of China's national trunk highway system, for example, has helped connect urban prefectures to become more attractive, while rural prefectures have shrunk in size (OECD, 2024).

From a public policy perspective, this means that any assessment of the overall benefits of transportation infrastructure must take into account trade-offs and demand.

Technological change, in recent years, implies a redefinition of infrastructure investment needs. On the one hand, the digitisation of society requires investment in broadband infrastructure and the expansion of data centre capacity. Estimates from the European Commission (EC) and the European Investment Bank (EIB) suggest that additional investments of around €55 million per year are needed to meet the goals of the EU digital agenda (OECD, 2024).

On the other hand, the changing urban mobility landscape implies the need to convert existing infrastructure, for example, parking spaces.

There are, however, marked differences in the availability and quality of infrastructure between countries, but these differences are being compensated.

For example, in the EU, the new Member States have largely caught up with the older members on many social, economic and environmental fronts. However, perceptions of infrastructure quality remain very low in many of these countries.

Road and rail networks are two important sub-sectors of the transport network. Investment in their expansion and maintenance is essential. The large size of many transport networks has made them increasingly expensive to maintain.

According to the Global Competitiveness Report, in OECD member countries in general, the perceived quality of rail infrastructure is relatively lower than that of road infrastructure (OECD, 2024). This suggests that roads are more reliable than railways.

France and Japan are two of the top three OECD countries in the world in terms of perceived high-quality road and rail infrastructure. The perception of road infrastructure quality in countries such as Austria, Belgium, Mexico, New Zealand and Norway is lower than the OECD average, while Japan and Switzerland have the highest perceived quality of rail infrastructure.

Unfortunately, many countries have very large backlogs of maintenance work needed on their transport infrastructure networks, so maintenance backlogs will absorb a considerable number of investible resources in transport infrastructure.

When roads do not meet quality standards, average speeds drop, and safety is compromised. For example, in Ukraine, where the perception of road quality is quite low, highway speeds are one-third to one-half of those in Western Europe, and fatality rates, per 100,000 inhabitants, reached 11.3 in 2013, well above the OECD average of 6.8 (OECD, 2024).

In Romania, the road network is of major importance for local and regional development. However, although covering the entire network of municipalities and cities, the road infrastructure is of poor quality and requires extensive maintenance and modernisation. For example, in the central region, of the 4,696 km of county roads, only 42% have electricity, and 33% are gravel or dirt roads. In addition, there are over 4,000 km of communal roads linking the centre and villages. However, only 10% have light coverage, and half of this is outdated; in addition, 80% of the communal roads are gravel or dirt.

In many countries, the rail system is the backbone of long-distance freight transport. Under-investment in rail could pose significant obstacles to economic development.

In some Eastern European countries, rail infrastructure needs considerable modernisation. Freight transport also uses the same rail line, considerably lowering the average speed of passenger trains, and the average age of locomotives and passenger wagons is over 40 years.

In Romania, rail infrastructure is deteriorating, and the rail fleet is outdated and inadequate to meet current requirements. For example, at the national level, 4,000 km of railways (30% of the network) have been in need of rehabilitation for more than a decade. This has led to an increase in dangerous spots and a 44% increase in the length of the rail network affected by speed restrictions, to the detriment of the quality of service provided. The average travel speed achieved by passenger trains, taking into account station conditions, is between 28 and 58 km/h (OECD, 2024).

For express trains, the average speed is only 20 km/h higher. The low quality of rail infrastructure means that most freight is carried by road, which in many cases are congested secondary roads.

For upgrading and developing transport infrastructure, the European Recovery and Resilience Plan foresee around $850 billion, while China has launched a $500 billion fiscal stimulus plan to boost infrastructure investment.

This plethora of market and interventionist forces creates invaluable opportunities for the transport infrastructure sector.

However, the cost to the environment is growing. Not only is the transport sector the largest contributor of GHG emissions (about 28%), but it is also not moving convincingly with the general decarbonisation trend. While other sectors, such as industry and electricity, are reducing emissions by 1.5% and 1% per year, respectively, the transport sector reports an annual increase of 0.8% of $MtCO_2$, with passenger cars accounting for the largest share.

Therefore, 'Europe should move towards a 100% renewable transport system for environmental, energy and sustainability reasons by 2050 and reach at least 50% by 2023' (ICONS, 2023).

In reaching this goal, it is necessary to develop solutions that cover the whole life cycle of transportation infrastructure. Its design can both influence and limit overall emissions: from construction, maintenance, operation and decommissioning, measures to support sustainable transport infrastructure must be introduced.

In this context, there is huge potential in infrastructure development for resource generation: much of the waste currently produced in transport infrastructure works can be reused and recycled to create new construction materials and products.

Another important approach is that investment in transport should shift decisively towards rail as one of the most efficient and pollutant-reducing modes compared to road, maritime and air transport. Rail accounts for only 2% of energy consumption in the transport sector, despite being responsible for 8% of passenger transport and 7% of freight.

Unfortunately, at the EU level, just by looking at the evolution of the main transport infrastructures, it is possible to observe the trend over the last three decades to support the development of infrastructures for the most polluting mode of transport, road, to the detriment of the others. Thus, between 1990 and 2021, the length of high-speed roads (motorways and express roads) almost doubled (Table 2.1).

*Table 2.1* Evolution of Main Roads in the EU and Romania, 1990–2022 (in km)

| *Indicator* | *Country* | *1990* | *1995* | *2000* | *2005* | *2010* | *2015* | *2020* | *2022* | *2022/1990 (%)* |
|---|---|---|---|---|---|---|---|---|---|---|
| Length of motorways | EU-27 | 38.995 | 44.889 | 51.530 | 59.474 | 67.450 | 71.585 | 74.555 | 76.100 | 195,1 |
| | Romania | 113 | 113 | 113 | 228 | 332 | 747 | 920 | 949 | 839,8 |
| Railway length in use | EU-27 | 214.725 | - | 202.952 | 202.457 | 202.424 | 201.068 | 202.557 | 202.131 | 94,1 |
| | Romania | 11.348 | - | 11.015 | 10.948 | 10.785 | 10.770 | 10.769 | 10.615 | 93,5 |
| Length of waterways | EU-27 | - | 37.458 | 39.658 | 40.783 | 40.738 | 40.601 | 42.007 | 45.419 | 121,5 |
| | Romania | 1.782 | 1.779 | 1.779 | 1.779 | 1.779 | 1.779 | 2.763[1] | 2.763[1] | 155,1 |

*Source:* EC (2023) and, for 2022, EC (2024b).Legend: 1 = for Romania, from 2019 onwards, the length of waterways has also included the length of navigation along lakes, offering this possibility.

In fact, Europe's rail network has declined dramatically over the last three decades, while investment in roads has increased (Elton, 2023). The length of Europe's highways increased by more than 95% between 1990 and 2022, by about 37,000 km. During the same period, the total length of railways decreased by 5.9%, falling by about 12,600 km.

This was mainly due to the fact that, between 1995 and 2018, European countries spent 66% more on roads than on railways. Subsequently, between 2018 and 2022, European countries continued the trend, albeit by a smaller amount, spending 34% more on road than rail expansion. Specifically, the EU-27 countries, Norway, Switzerland and the United Kingdom spent about €1.5 trillion on road infrastructure and only €930 billion on railways between 1995 and 2020 (Elton, 2023).

Most rail closures occurred in Germany (reduced by 6,706 km), Poland (by 4,660 km) and France (by 4,125 km). Despite this, these three countries still have the longest rail networks, followed by the United Kingdom and Spain.

At the opposite pole, between 2018 and 2022, Austria, Belgium, Denmark, France, Italy, Luxembourg, and the United Kingdom have invested more in railways than in roads.

In Romania, the funding gap was particularly wide, with the government spending 12 times more on roads than on rail.

Highways grew most in Ireland, Romania and Poland and least in Lithuania, Latvia and Belgium.

In Romania, as of 2022, according to Table 2.1, the length of high-speed roads increased more than eight times compared to 1990, reaching 10,075.5 km by the end of 2023, almost nine times more than in 1990 (Cozmei, 2023).

In terms of railroad infrastructure, Romania has recorded a major contraction in the length of tracks in operation, with about 6.5% between 1990 and 2022, above the EU average.

As we have pointed out, it is clear that, during the period under review, the EU clearly supported road construction at the expense of other types of infrastructure, which, in turn, support less polluting transport. The length of inland waterways has been increasing, though much less than that of high-speed roads, and railways have been steadily decreasing, with almost 57% of them electrified by the end of 2022 (EC, 2024b).

This situation has led the European Union to propose increasing the share of investment in rail infrastructure in the next EU budget in the election year 2024 in order to improve rail efficiency by rebuilding, upgrading and ultimately expanding the rail network across the European Union. The future EU budget should completely exclude funding for new roads and airports (EC, 2024a).

In fact, according to the Maastricht Treaty of 1993, which stipulated that the Trans-European Transport Network (TEN-T) is the instrument of infrastructure policy in Europe and an important tool for achieving cohesion and growth of European regions, the core network, organised in nine main corridors, has relied on rail, air and water networks and less on roads.

However, transport remains the main contributor to the climate imbalance of the EU economy. It is the only sector to have increased its emissions since 1990. The Green Deal laid the foundations for halting the trend, with, for example, the agreement to drastically reduce sales of diesel cars by 2035, a measure revised in 2024 due to the 'market crash' of electric vehicles.[3] However, while other sectors are *decarbonising* faster, transport is not making the same strides and will account for 45% of total EU emissions by 2030. This calls for important and much-needed additional measures and actions towards the EU's goal of achieving net-zero emissions by 2050.

To this end, in March 2023, the EU Parliament and Council agreed on binding national targets for the deployment of alternative fuel infrastructure for cars and trucks. The action foresees the commissioning of several recharging stations on major EU roads and the provision of electric charging areas for cars with at least 400 kW per 60 km by 2026, increasing to 600 kW by 2028.

According to the same document, there will have to be charging stations every 120 km for trucks and buses on TEN-T corridors by 2028, with a power of 1400 kW to 2800 kW, depending on the road. By the end of 2030, each safe and secure parking area for heavy goods vehicles must include at least one recharging station with a capacity of 100 kW.

In addition, by 2031, hydrogen refuelling stations must be available at least every 200 km along main roads. By 2021, however, there were only 136 hydrogen refuelling stations in the European Union as a whole.

These measures were also adopted on the basis that between 2000 and 2014, $CO_2$ emissions increased less in countries with higher investment in domestic transport infrastructure (OECD, 2024). This might seem counterintuitive, as transportation accounts for 30% of $CO_2$ emissions in OECD countries.

Part of the explanation is that the situation presented has arisen in countries where more resources have been invested in national infrastructure, modernising it and creating transport links adapted to new environmental trends, and where the share of rail investment has increased. Railways are, in fact, among the most energy-efficient and low-emission modes of transport.

Even with these trends, there have been periods that have 'disrupted' policies and strategies. For example, the health crisis has unfortunately created many financial problems for carriers and especially those in the air and rail sectors.

As containment measures have restricted inter-regional and international travel, rail companies have seen demand fall dramatically. In Canada, for example, domestic and international rail passengers in April 2020 accounted for less than 2% of April 2019 passenger numbers. By July 2020, passenger numbers had increased again, but remained at only around 16% of July 2019 passenger volumes. Even in Sweden, one of the countries with one of the least stringent containment restrictions, rail passenger volumes in 2020 were about 20% below the previous year's level and from then until the removal of travel restrictions, about 10%.

To cushion the shock to their rail sectors, OECD countries took action at the start of the pandemic by supporting the continued operations of transport companies, spending billions of USD on bailouts or absorbing losses on their balance sheets (where operators are public).

In contrast to the evolution of conventional rail, metro and high-speed rail lengths have increased substantially. Conventional railways (as opposed to metros, trams or high-speed rail) account for 94% of all rail kilometres in Europe, North America, China and Japan, but their length has increased very little between 1995 and 2016 in most regions (OECD, 2024).

The fastest growth has been in the length of subway networks. In China, India, Japan, Russia, Europe and North America, the length of the metro rail network increased by 4,800 km between 2000 and 2017. China and India have played a leading role, with 32 of the 43 cities where a new metro system was commissioned between 2010 and 2017.

The length of the high-speed rail network, most of which is located in China, Japan, South Korea and Europe, expanded from around 30,000 km to around 70,000 km between 2010 and 2017, with China *responsible for* most of the expansion. The share of high-speed networks in China in 2010 was about 36%. By 2017, this had risen to 63%. But in other countries, unfortunately, there has been no expansion of this type of network. In Canada and the United States, for example, contraction – as measured by declining passenger-kilometres per kilometre – was largely constant between 1995 and 2016, while globally the trend was the reverse, almost doubling over the same period.

Urban transport also requires more investment. Despite the relatively high share of urban rail transport in Europe (15% compared to 5% in China), additional investment is needed in European cities to close existing accessibility gaps and adapt urban transport to the needs of a low-carbon transition. Indeed, urban transport accounts for a large share of passenger transport, which in turn contributes 50% of total GHG emissions from the transport sector (OECD, 2024). However, only 35% of the countries that signed the Paris Agreement in 2015 have included public transport in their environmental action plans.

Investing in local transport infrastructure helps to better connect cities with surrounding rural areas. Local authorities can maximise the returns from investments in local transport infrastructure by promoting policies that foster creativity and the exchange of ideas. Such policy approaches have been initiated, for example, in the western region of Sweden, which *revolves* around the main city of Gothenburg. The regional authorities have invested in innovation and research infrastructure, for example, by creating science parks, and at the same time, strengthened the local transportation network, which has increased the return on both investments

Of course, the sector's most extensive road network should not be abandoned or neglected. Although the speed of construction of new road networks, especially high-speed ones, has slowed down in recent decades, the share of investment in this area has remained the majority. This has led to an interesting situation.

As the expansion of the highway network has slowed, the cost of maintaining existing infrastructure has risen. In the United States, for example, the share of maintenance outstripped investment.

As the length of the road network expanded and as the network matured, the level of maintenance costs, relative to new investment, increased. Existing studies highlight that not only are the required maintenance costs high, but also that large deficits materialise in this category of expenditure, with negative consequences for traffic and road safety (OECD, 2024).

Under these conditions, highway networks in most developed countries require substantial maintenance expenditure. Most Western countries built highways in the first decades after World War II. As budgets shrank, the deteriorating condition of these highway networks became apparent, especially after the Great Recession.

Many European countries have been forced to implement tighter budget constraints, which have contributed to reduced maintenance of public investment, especially in transportation infrastructure.

Therefore, maximising lifetime must be a guiding principle for new investments in transport infrastructure. From a public finance point of view, it is important to minimise the maintenance cost of each new kilometre added to the existing transport network so that the total maintenance costs do not grow as fast or faster than the rate at which the transport network is expanded.

Expanded databases can be used to forecast transport demand growth more accurately and to make more efficient use of updates on the state of the transport infrastructure. Traffic management systems can also benefit from more in-depth modelling made possible through the use of statistical and traffic data. For example, local decision-makers could use these data to design a system that makes it possible to avoid non-essential deliveries during peak hours, thereby reducing peak traffic in certain areas and at certain times of the day.

Automated vehicles could also increase the efficient use of roads. The major benefits of automated vehicles on highways will be in the areas of traffic safety and driver comfort.

In Romania, the different stages of development and the political–administrative conjunctures have had a major influence on the evolution of land transport infrastructures. Since 1900 and until 2022, the rail and road networks have experienced ups and downs in terms of size, but one thing has remained constant: the share of roadways has always been much higher than that of railways in the total land transport network.

However, policy in this area has been different, so that in certain periods, rail construction has been more dynamic than road and vice versa.

Specifically, between 1920 and 1990, the share of railways in total land transport infrastructure increased at the expense of roads. Since 1990, this trend has been reversed. It is also interesting to note that, in terms of total land infrastructure, railways have always been below 15%, with a historic low of less than 5% in 1920 and a peak of 13.5% in 1990. Today, the railways' share of the total network is only 10.9%.

After 1990, with the change in the communist regime and the replacement of the general economic relations, the road infrastructure policy changed substantially, with all national governments giving priority to the construction of motorways, to the detriment of infrastructure for other modes of transport and even to the detriment of modernisation/maintenance of the existing road infrastructure at the level of 1990.

In fact, between 1990 and 2022, the length of roads increased by 13,520 km, which is 18.6% more in 2022 than in 1990. Over the same period, the length of highways in operation increased from 113 km in 1990 to 994 km in 2022, an increase of almost 8.8 times.[4] This development says almost everything about the direction of national policy in the field of transport infrastructure construction. In our opinion, the major financial efforts made by all Romanian governments since 1989, mainly directed towards the construction of motorways, have not achieved their goal and, what is more, have seriously hampered the policies for restoring and modernising the other infrastructure networks (railways, waterways, etc.). The average annual rate of motorway construction of around 27.5 km does not support the policy in this area, which means that either the governments have the wrong strategy or there have been permanent obstacles to the implementation of this policy (lack of the necessary financial resources, non-compliant and unprovable projects, lack of competent builders, delays in the legislative processes concerning the jurisdiction of land for motorway construction, etc.).

## 2.2 Sustainable transport infrastructure: the key component of future mobility

The benefits of investment in transport infrastructure can be extended through complementarities and synergies between a group of assets, rather than standalone projects. Such complementarities are easier to find and link at the regional and local levels. With many large infrastructures completed, future transport investments will involve extensions or links to other existing infrastructures. Intermodally should be further promoted, as it could facilitate the movement of goods and people between transport modes and develop so-called 'nodal' infrastructure. This requires better coordination in the planning phase and complex interactions in the implementation phase.

### *2.2.1 Prospects for sustainable transport infrastructure development*

In order to improve effective access to the opportunities that transport infrastructure can offer, more investment in public transport services is needed to increase mobility capacity. Public transport services (trains, metro, light rail, trams, trolleybuses, buses, etc.) require policies and support measures that differ from privately provided transport. Increasing travel capacity can, however, be achieved at significant fixed costs, which are more advantageous at sufficiently high population densities. In such cases, investment becomes *strategic*. For example, the high costs of metro systems, compared to surface transport, make metros attractive only for parts of the city with sufficiently high population density.

In addition, the experience of developed and developing cities shows that promoting accessibility means more than just increasing travel capacity and investing in public transport. It is, in fact, a combination of urban policies aimed at making it easier to reach destinations without the need to use private transport and at low travel costs. As London's experience suggests, each city, each part of the city, is unique and requires tailored transport solutions to support socio-economic growth and development (OECD, 2024).

Transportation infrastructure creates economic benefits in different ways. Importantly, these modes also differ between infrastructure that connects regions and infrastructure that connects areas (neighbourhoods) within a city. Inter-regional transport infrastructure mainly facilitates the movement of goods, while intra-city transport infrastructure facilitates the movement of people. Since investments take place at different values, the deciding authorities as well as the necessary policies will be equally different.

De facto status and evaluation are crucial to address the challenges that decision-makers face when targeting and approving transport investments.

This requires rethinking traditional ex ante evaluation methods, such as cost–benefit analysis (CBA), by adopting more comprehensive evaluation criteria for prioritising transport infrastructure projects according to their objectives and the size of the areas they will serve. For example, a quantitative assessment of the impact of transport infrastructure investments on employment and productivity at different geographical scales is often beyond the scope of the standard CBA, which is currently the most widely used assessment tool for transport infrastructure projects.

The list of costs considered in a CBA can also be extended depending on the policy objective of the project. For example, given that investments in transport infrastructure may disproportionately benefit high-income households, increasing inequality should be added to project costs if one of the project's implicit objectives is to ensure that the benefits are distributed equally across the population.

The location of transport infrastructure investment is not random but carefully chosen to optimise economic and social objectives. A comparison of the economic development as a result of new infrastructure between investment locations and other locations will therefore lead to misleading estimates of the value realised for the connected locations. If investments are, for example, directed to areas that are expected to grow, it is difficult to assess how much of the resulting growth can be attributed to the investment itself and how much is due to other factors that have promoted growth even before the investment took place. What is missing is data on a 'fictitious comparable' state of affairs, that is, an assessment of the development in a region or city where the investment was made, compared to the situation if that investment had not been made.

The gains for a given region depend on how many new customers can be engaged through the improved transportation network. For example, simply building a motorway is not enough for economic development to materialise in a given region. The crucial question is what accessibility improvements the new infrastructure offers. In this context, the average level of income – representing the size of spending capacity – in the new markets of regions in direct contact with or linked to the newly created transport infrastructure is one way of measuring accessibility.

Accessibility and overall gains also depend on the wider transportation network. If the additional infrastructure connects to an existing network with good accessibility or improves access to existing logistics centres, ports or airports, it is more likely to generate significant benefits in terms of freight traffic volumes. For example, the area of Western Scandinavia, covering the regions along the west coast from Oslo through Gothenburg to Malmö and Öresund, is an important freight transport hub for Norway and Sweden. It is home to Norway's largest airport, Sweden's two largest ports and the connecting road infrastructure, which has been improved in recent years. In contrast, there is a lack of fast rail infrastructure linking Norway to Sweden and between Sweden's major cities. Against this background, rail transport has gradually lost its competitiveness to road transport, particularly on the Oslo–Gothenburg route.

Additional highways tend to create better market access, but the gains vary by region. Roads are essentially a means to fulfil a need, such as the exchange of goods between two places, but the construction of a road itself does not necessarily generate such a need, so it may fail to create a market in an area where one does not exist.

An important feature of infrastructure is that, once a road is put in place, economic growth in one area along the road can create negative effects on other connected areas. Therefore, transport connections can only create a sustainable stimulus by ensuring that all connected places benefit.

In a study for 28 European countries, the construction of road infrastructure is the main determinant of improved accessibility when fewer variable factors are taken into account (Adler et al., 2020). For the entire affected population, 66% of the increase in accessibility between 1990 and 2012 comes from improved road infrastructure. In contrast, for the variable factors (such as employment and GDP ), the increase in connected areas becomes less evident. For GDP , only 18% of the accessibility improvements can be attributed to new or upgraded roads, with the remaining 82% coming from growth in already connected places, and for access to jobs, the two aspects are roughly equally important.

The potential economic gains from investment in transportation infrastructure for a given region are proportional to changes in market potential, where one way to measure market potential (MP) is given by the relationship (Adler et al., 2020):

$$PP_r = \sum_{(m \in M)} \left( GDP_m / CT^{(d)}_{(r,m)} \right) \quad (2.1)$$

where:

$PP_r$ = Market potential in region 'r'
r, m = regions (with M = total regions analysed)
d = degree of product substitutability
TC = transportation cost
GDP = gross domestic product

Since spending in a given region 'm' depends on its income, one way to assess a measure of market potential is to sum all the regions that make up a country's GDP (M) after reducing them by the cost of transporting goods from region 'r' to region 'm'. The parameter 'd' shows the degree to which goods produced in one region are substitutable for goods produced in another region. The more substitutable the goods are—a higher 'd'—the market potential offered by nearby regions is greater compared to more distant regions.

Output in a region may increase because, thanks to improved transportation infrastructure, local firms can now start selling to economically richer regions further away. However, changes in market potential depend not only on changes in the geography of transportation infrastructure, but also on changes in the geography of economic activity. Firms located in a small region can, in principle, benefit from the construction of a new highway connecting to a larger one. However, if firms located in the larger region are more competitive in a larger share of the economic sectors in which they operate, consumers located in the smaller region will want to buy their products from them. A labour shortage in the larger region will raise local wages and will eventually be filled either by commuting or by actual immigration from the nearby smaller region. In both cases, the market potential will diminish in the smaller region, thus encouraging firms to relocate to the larger region.

The economic development of Chinese cities following the construction of the highway network since 1988 exemplifies the potential trade-offs inherent in infrastructure development (Baum-Snow et al., 2016).

The network was conceived with the idea of promoting economic activity in local prefectures, but had the opposite result. The 'preferred' cities – the largest prefectures located within a day's drive of the local prefectures in the area – specialised in the production of goods and services for which they enjoyed a comparative advantage. Production created jobs that produced more value added, and therefore, higher wages could be paid, which attracted workers to the major cities. They grew in employment and population at the expense of the prefectures in nearby areas, which shrank in size and specialised mainly in agricultural work.

Therefore, it is clear that the benefits of new infrastructure depend on the existing network, supporting larger gains in certain places and reducing infrastructure gaps (Adler et al., 2020). Estimates for Eastern European regions, for example, for the period 1990–2012, show additional benefits in terms of GDP and GDP per capita from improved market access.

This is consistent with a stronger expansion of economic activity in high-value-added sectors, capturing the expansion of manufacturing activities and the integration of regions in the new EU Member States into European (or global) value chains.

In contrast, market access has increased employment in Southern European regions, but these have not translated into GDP gains; that is, employment improvements seem to have mainly stimulated growth in low- and middle-income jobs, as GDP gains and GDP per capita gains are statistically insignificant. The benefits of improved accessibility arise mainly through integration with the European road network. However, the distance to certain economic poles is a factor that increases or decreases the benefits felt from infrastructure development. When access to certain areas of interest is improved over a distance that can be covered in at least three hours from the place of departure (by road), the results remain positive, that is, there are still benefits, but these are reduced to a quarter of those achievable at shorter travel distances.

However, the achievable gains from improved infrastructure are limited. A counterfactual exercise for EU countries considers ensuring 'perfect' accessibility in all regions by upgrading the existing network of secondary roads from existing secondary roads to motorways (the highest speed roads).

The results of Adler suggest that the level of gains, estimated in terms of GDP, is on average between 17% and 26%, with the highest values for Eastern European regions (Adler et al., 2020). Projections for employment and population growth are slightly higher, however. In general, the already well-developed motorway network provides a high degree of accessibility in many parts of continental Europe, but there are still places where the benefits of new construction could become substantial, particularly in Eastern Europe. Additional interventions should, of course, follow best practice and be subject to careful CBA in line with the experience of OECD member countries.

In addition, synergies between adaptation and mitigation actions should certainly be encouraged for the transport sector.

Adaptation concerns both transport infrastructure (mainly road and rail) and the activity itself. In terms of infrastructure, adaptation includes actions to maintain, manage, strengthen and protect it from climate damage. This, in turn, includes the review and improvement of long-standing practices, but also new practices and technological innovations (e.g. new materials resistant to extreme temperatures and climate-proofing of underground tunnels exposed to flooding).

By mitigation, consideration should be given to infrastructure relocation, especially where infrastructure is beyond its lifespan or where it may be severely damaged.

*Green actions* are other examples of complementary measures that can play a role in reducing the vulnerability of transport infrastructure to climate change. These include the development, maintenance or restoration of natural floodplains (river rehabilitation and restoration), allowing for controlled flooding of rivers and hill consolidation, and the creation of natural buffers along roads and railways by planting and maintaining trees.

The costs of such policies are variable and depend on the specific adaptation actions that are designed and implemented. The highest costs relate to complex interventions to rehabilitate old infrastructure or relocate it to safer areas.

Benefits include reducing the overall vulnerability of transport systems, reducing malfunctions, inefficiencies and service disruptions, especially in cases of extreme climatic events. Greater benefits are delivered when transport plans consider both climate change adaptation and mitigation, with actions aimed at increasing the resilience of transport infrastructure and services in low-carbon transition initiatives. Long-term structural changes in the organisation of transport and mobility in general (e.g. changes in transport behaviour and demand) can work in this direction. Identifying adaptation actions with benefits, in terms of mitigation, can offset costs and favour investment.

But how can the transport sector reduce its emissions in line with global climate goals while keeping pace with demand for transport infrastructure?

The answer is not simple, but what is needed is a conceptual framework that provides a comprehensive way for stakeholders to approach the new concept of sustainable transport infrastructure.

As is well known, transport infrastructure is an essential part of modern economies.

For many decades, transport infrastructure has been designed in such a way that it can be operated for as long as possible without major renovations or costly maintenance. Unfortunately, however, sustainability works over the lifetime of the asset, but presents challenges at the end of the asset's life in recovering materials and components to keep them in use.

In addition to normal operating *conditions*, special, extreme and cyclical conditions on transport infrastructure have an impact on the lifetime and quality of service provided.

For example, in the last few years, the disruptions caused by the COVID-19 pandemic and the conditions caused by climate change have hampered economic development but have stimulated unprecedented investment in transport infrastructure.

Some examples in this regard are eloquent (Flinders, 2023):

- In the US, the federal government passed an infrastructure bill that aimed to build and rebuild critical infrastructure, including the nation's most vital transportation networks. This ambitious bill, managed largely by the US Department of Transportation, planned to direct $1 trillion to fund construction, manufacturing initiatives and public works.

- In the EU, 107 transport infrastructure projects have been selected to be funded with more than €6 billion in grant programmes over the next few years to help improve connectivity, public transport and alternative modes of transport in both urban and rural areas.
- In the Asia-Pacific region, the Asian Development Bank has announced that $1.7 trillion will be needed annually for infrastructure development over the next seven years to meet its goals.

Mitigating climate change will require significant investment in sustainable transport infrastructure, not just in Europe, the United States and China, but also in developing countries, where outdated equipment needs to be replaced or brought up to modern standards to meet the 2015 Paris targets.

The carbon footprint of the transport sector and its subsequent contribution to climate change are indisputable. According to the International Energy Agency, the transportation industry is responsible for 21% of global greenhouse gas emissions (Flinders, 2023). Also, the International Institute for Sustainable Development (IISD) attributes 64% of global oil consumption and 27% of total energy consumption to transportation services, as well as 23% of total global energy-related $CO_2$ emissions.

The average time needed to develop/modernise transport infrastructure is generally a few years (about 2–5 years). Such a long lead time, together with the economic costs of doing the work, can act as a hindrance to the development of transport routes. The lifetime of any infrastructure is, however, several decades, depending on the level of maintenance and operating conditions (e.g. traffic load, natural conditions, etc.).

Unfortunately, however, it is not sufficient to repair and modernise existing transport infrastructure using the same methods and technologies with which it was built. These 'transport infrastructure remodelling' projects must be truly transformative in supporting environmentally friendly modes of transport and eliminating the sector's dependence on the use of polluting materials such as fossil fuels, which means that the entire transport network must be rethought, 'adapted' to the concept of sustainable development.

This gives rise to a terminological dilemma: 'sustainable transport infrastructure' or 'infrastructure for sustainable transport'.

Right from the outset, it can be seen that the two concepts have points in common, but also differences in approach. Sustainable transport infrastructure refers to any road that supports clean transport. In addition to this concept, 'sustainable transport infrastructure' also requires the sustainable construction of communication routes. In other words, the processes of designing, building and maintaining transport infrastructure must take account of the general factors of sustainable development.

Thus, from the choice of where to build the infrastructure, to the materials used, to the technologies and machinery used for construction and maintenance, all must be framed within the concept of sustainable development.

In this context, there are four possible 'infrastructure-activity' combinations for the transport system that can be considered in line with sustainable development principles:

1 Conventional infrastructure (classical or modernised) and general transport (without environmental impositions).
2 Conventional infrastructure and sustainable transport.
3 Sustainable infrastructure and general transport.
4 Sustainable infrastructure and sustainable transport.

Clearly, from a sustainable development perspective, solution number '4' is the ideal solution but as is well known, the ideal is never reached.

*It is essential, however, to recognise the specific role and contribution of infrastructure in the transformation to a sustainable society and a sustainable world.*

To be sustainable, infrastructure should provide necessary services at affordable costs while conserving natural resources and energy. Moreover, these services should be maintained and continuously improved to remain competitive in the global market.

In view of the above, we consider that ***sustainable transport infrastructure*** *is the totality of networks supporting the mobility of passengers and goods, designed, built and maintained in accordance with the principles of the circular economy and which must ensure a high degree of adaptability to the mobility demand of future generations without damaging the environment and the health of living beings.*

According to this new concept, the transport infrastructure system can be adapted to the requirements of sustainable development through at least three actions, aiming at:

- Redirecting a significant proportion of the current infrastructure networks to support sustainable transport (such as the creation of bicycle lanes and separate lanes for public transport in urban areas).
- Investing, as a priority, in the construction of new infrastructure to support sustainable development as a whole and less polluting transport in particular (e.g. construction of electrified railways in areas that do not significantly *harm the* environment, such as nature reserves, key agricultural areas, etc.).
- The use of recycled materials (e.g. bitumen and plastic mixtures), which are recyclable and easy to maintain and adapt to future changes in transport demand.

In the context highlighted above, there are ways of greening transport infrastructure and, at the same time, improving health and quality of life in general.

To become sustainable, however, transport infrastructure must meet four common sustainability criteria (environmental, social, institutional and economic), which can be achieved by prioritising five strategic elements:

- Climate resilience.
- Contribution to better inclusion of emerging communities.
- Utilising technologies that can support sustainable infrastructure development.
- Making sustainable transport infrastructure investment more attractive.
- Increasing flexibility to variations in demand.

In addition, in order to perform better than what is currently common practice, transport infrastructure operators need to have clear performance indicators and targets for all four defining elements of the sustainability concept.

The growing international recognition of the importance of sustainable infrastructure and the establishment of knowledge-sharing platforms have led in recent decades to a plethora of tools designed to ensure the sustainability of infrastructure projects. Historically, the most commonly used tools at the project level have been CBAs, environmental impact assessments (EIA) and environmental and social impact assessments (ESIA).

Although the original theoretical framework of CBAs did not include sustainability, later iterations have incorporated it, mainly through the assessment of 'environmental assets' (OECD, 2006). On the other hand, sustainability is a fundamental premise in EIA and ESIA theories, which are designed to influence formal decision-making processes with clear objectives and a transparent process (Sheate, 2010). However, the main function of CBAs, EIAs and SIAs is applied at the level of a single project rather than networks.

In addition to these general tools, many sustainability assessment and rating tools exist at the project level. Unfortunately, it has been found that there is a lack of guidance on incorporating sustainability concerns into the upstream policy and planning stages of infrastructure development, which limits the effectiveness with which sustainability can be incorporated at later stages (IBRD, 2018).

Strategic environmental assessments (SEAs) aim to integrate environmental considerations into strategic planning at the programme level and take into account their interconnections with social and economic impacts (OECD, 2006). SEAs can be particularly effective when applied to sectoral policy and reforms, and their effectiveness depends on their application early in decision-making processes (Sheate, 2010). However, despite their potential to contribute to system-level (multi-sectoral) planning, SEAs are often applied only to specific infrastructure sectors, in isolation from other related sectors, which can lead to the loss of cross-sectoral synergies. This is caused by the way planning processes are designed and does not reflect the potential of SEA to support more integrated planning processes (World Bank, 2011).

In recognition of the need for a more systemic approach, a rating system for sustainable infrastructure, called *Envision*, has been developed by experts at the Institute for Sustainable Infrastructure (ISI), which helps to analyse the implementation of sustainable infrastructure and incorporates best practices in the sector (ISI, 2018). According to Hall et al., related to this system, there is a capacity assessment tool for infrastructure (CAT-I) and another specific to national infrastructure systems (NISMOD), which uses a 'network of systems' approach to ensure that cross-sectoral interdependencies are identified and synergies optimised (Hall et al., 2016).

These form a complementary 'toolkit' that planners and policy makers can use to develop sustainable infrastructure.

From an *environmental* perspective, transport infrastructure could be planned, designed, built and operated with the aim of increasing its climate resilience, which includes mitigating climate impacts, protecting biodiversity and minimising pollution. Sustainable transport infrastructure should catalyse a virtuous circle in which finite materials are gradually replaced by renewable materials. Where possible, materials should be reusable, repairable, recyclable and recoverable.

From a *social* point of view, sustainable infrastructures should take into account externalities on vulnerable populations, preserve cultural heritage, protect human rights, improve quality of life, increase inclusion and make transport services accessible.

The *institutional* aspect focuses on the alignment of transport infrastructure with the overall goals of its country, such as specific pathways towards depollution, especially decarbonisation. Finally, the economic area covers the long-term financial viability of infrastructure and its contribution to job creation and economic growth.

For a clear understanding of what constitutes sustainable transport infrastructure, we emphasise that every project in this area must contain supporting elements for:

- Climate resilience: transportation infrastructure builders must be prepared for the increasing frequency of extreme weather events due to climate change. Over the next three decades, the McKinsey Global Institute forecasts that the global average temperature is expected to increase between 1.5 °C and 5 °C, and there is a greater than 15% probability of an increase in extreme temperatures during the summer months (Hall et al., 2016). As long-lived assets, transportation infrastructure will be particularly affected by the effects of these climate variations. For example, many airports are located near 'wetlands' (lakes and rivers), and a quarter of the world's 100 busiest airports are located less than ten meters above sea level. Some are even less than five meters. Because they are more vulnerable to flooding caused by heavy rainfall during storms, measures must be taken to protect these terminals when nearby water levels rise.
- Inclusion: transportation infrastructure is a public good and should promote inclusion in surrounding communities. According to the Global Infrastructure Hub, inclusive infrastructure can be defined as 'any infrastructure development that enhances positive outcomes in social inclusion and ensures that no individual, community or social group is excluded or prevented from benefiting from improved infrastructure' (Hall et al., 2016).

- Use of state-of-the-art technologies: digital technologies are essential in the transition of transportation infrastructure, in its myriad forms, to greater sustainability. In the rail sector, the deployment of advanced train control and signalling systems, such as European Rail Traffic Management System (ERTMS) level 2 technologies, which use wireless communications to monitor train movements, will ultimately enable the rail sector to achieve its European goal of making freight transport more sustainable, reaching 30% modal share from current levels of 18%, while reducing capital and maintenance costs. Digital applications, including optimised movement sequencing, smart metering and energy solutions, will also reduce the carbon footprint of airports, while the full digitalisation of the supply chain, including maritime transport (notably cloud and IoT technologies, advanced analytics tools to optimise scheduling and routing of cargo and the use of biofuels), will reduce the sector's emissions globally. For roads, the installation of electric vehicle charging stations is currently among the most widely used solutions in increasing the sustainability of roadways. Other initiatives, such as smart maintenance (i.e. optimising road maintenance through data analytics), reduce traffic congestion, bringing additional environmental, social and economic benefits.
- Productivity and value creation: To be more attractive to private capital, transport infrastructure assets need to be productive and delivered on time. This is a particularly challenging issue for the sector because of the high level of fragmentation. Stakeholders thus need to move towards a more circular economy, where the whole supply chain can be incentivised to care about the full life cycle of an asset. Instead of being a financial liability, sustainable transport infrastructure must create value through sustainable innovation of business models. Global institutions play a key role in setting carbon neutrality targets. Together with penalties, carbon taxes or premiums for innovations that support carbon reduction can make unsustainable infrastructure too large a financial liability for owners. Identifying the total cost of ownership and initiatives that can increase revenues or reduce maintenance overheads is also a critical component. Actors across the value chain can use digital platforms to improve visibility, streamline processes and collaborate to increase overall efficiency. Similarly, disadvantaged sustainable interventions can focus on value creation by reducing redundancies and leveraging the capabilities of smart infrastructure, data analytics and connectivity along the value chain. For example, condition monitoring and predictive maintenance, coupled with road user data, can reduce lane closures and optimise traffic flow to reduce bottlenecks, congestion and emissions from idling vehicles.
- Flexibility: the COVID-19 pandemic has highlighted the importance of flexibility for transportation infrastructure. They need to be able to cope with sudden and significant changes in demand – both predictable and unpredictable – and therefore need to become more adaptable. For example, sustainable transport infrastructure will need to be able to absorb sudden changes in traffic patterns, such as the shift in air traffic from passenger flights to freight experienced during the pandemic.

The four dimensions of sustainability plus the five strategic elements provide a solid definition of what sustainable transport infrastructure can achieve, financially, environmentally and socially.

Tracking performance with clear indicators and values, including the number of safety incidents or the cost of public transport, has benefits for society.

Similarly, the institutional performance of a transport infrastructure asset can be measured by tracking the asset's contribution to a country's overall objectives, such as meeting a specific decarbonisation target or regional development plan. Sustainable, long-term performance in this case is a competitive advantage that can attract many investors.

The whole process of 'making transport infrastructure sustainable' has to be found in the five stages of the transport infrastructure life cycle: planning, design, tendering, construction and operation.

#### *2.2.1.1 Planning and design*

Whether it is a new project or an intervention on existing infrastructure, sustainable concepts need to be applied from the planning and design phase. For new projects, all externalities – from environmental to social to economic impacts – need to be carefully studied and planned, and attention needs to be paid to the resilience of assets to climate risks.

For specific one-off interventions or improvements to existing infrastructure, an integrated planning process optimises the allocation of resources (both human and financial capital) in critical areas. Planners can identify the major environmental regulations and objectives of the area and align interventions with national and sectoral plans. For example, authorities in a particular location can focus on increasing the use of more environmentally friendly modes of transportation.

Key stakeholders need to fully understand the needs and interests of all parties and agree on performance indicators and benchmarks to track performance during construction as well as over the lifetime of the asset. For example, stakeholders could agree to implement risk mitigation actions to increase resilience to climate change, which may cost more initially but will reduce operating costs in the long run. Plans could also be developed to tap into the local workforce, providing equal employment opportunities for all.

Stakeholders can agree on ways to optimise and simplify the speed of delivery while maximising positive socio-economic outcomes, such as ensuring accessibility for all vulnerable groups. A holistic approach to value creation reduces capital expenditure in categories such as superstructure, electrical or major equipment, with the potential to generate savings of 10%–15%. Simultaneously, targeted interventions can go beyond business objectives, with a focus on whole life-cycle project cost and benefit, unlocking positive socio-economic impacts.

#### 2.2.1.2 *Bidding phase*

The main objective during the bidding phase is to establish optimal sustainable processes and strategies to shorten the overall duration of this phase.

The technical sustainability criteria used to evaluate bids need to be clearly defined; providing a set of tools for tender management and using a standard template for evaluating bids can speed up this process. An appropriate scoring of technical and economic criteria provides an objective way of selecting suppliers on the basis of the optimal combination of quality of works and services, environmental performance, utility requirements and price.

Conditions may require bidders to have a proven track record of positive sustainability impacts in order to optimise operations in line with energy efficiency and environmental standards, prioritise natural and renewable capital and employ local workers. Setting economic parameters helps to limit the overvalorisation of price differences between competing bids. Avoiding excessive discounting means, among other things, foregoing some cost savings, but too much discounting generally supports compromised quality of work.

#### 2.2.1.3 *Construction itself*

The materials used in construction require careful attention. Where possible, sustainable construction materials such as concrete elements with net-zero $CO_2$, $CO_2$-free steel with improved corrosion resistance or landscaping materials with low emissions of volatile organic compounds are preferred.

For example, paving roads with asphalt containing bitumen containing ground rubber from recycled tyres, compared to conventional asphalt, could reduce $CO_2$ emissions by about 30% over the whole life-cycle of the road. Asphalt recycling also reduces a significant amount of carbon emissions associated with the transportation and processing of materials needed for infrastructure construction. In addition, all $CO_2$ emissions embodied in the raw material can potentially be saved when bitumen raw materials are recycled. This alone accounts for about 40% of the carbon found in the production process of new asphalt, according to the European Asphalt Pavement Association (Milani et al., 2021).

In these efforts, material suppliers can experiment with ways to reuse construction waste. Even if the technology does not yet exist, the benefits of such a business could be significant in the near future.

During implementation, construction and logistics processes must meet the highest environmental standards to minimise carbon emissions. Digital tools must be used to optimise construction processes and reduce unnecessary carbon emissions from recycling, inefficient use of equipment and waste of resources and waste.

In order to keep materials and components in use, plans for deconstruction and replacement should be in place from the very beginning of the project.

Sustainability thus takes the form of longevity of materials, components and systems.

In this context, the concept of 'shear layers – the 6Ss' was first proposed by architect Frank Duffy in the 1970s and further developed by Stuart Brand in the 1990s (Evan, 2023). The purpose of the 'layered' visualisation of infrastructure composition is that the interconnectedness between elements in need of maintenance and replacement can be more easily analysed.

Using the 'layers' approach to transport infrastructure design reduces replacement costs, conserves resources and makes it easier for operators to continuously adapt transport supply to changing demand without building new infrastructure.

A layer-based approach is an important element of the kind of systems thinking that needs to be applied in the complex process in which infrastructure assets are developed, financed, designed, built, operated, maintained and reused.

The importance of this approach is that it reveals the potential for more sustainable choices throughout the whole life cycle at the design stage, rather than achieving sustainability at all costs.

For example, some layers of transport infrastructure, such as lighting and signage systems, are more suitable for 'replacement'. Others are better suited to maintenance or repair. 'Longevity' is left to the base layers.

For modernisation works, including on high-speed rail, where newer and more modern infrastructure has to be placed in a very narrow cross-section, a layer-based approach can help to design future maintenance much more efficiently. Looking at the elements, in layers grouped by maintenance strategy, allows planning and design in a systematic way for the whole life cycle.

In the same context, highways are facing many changes/adaptations to future demands: electric and hydrogen-powered vehicles, congestion avoidance systems, connected and autonomous vehicles on the road, etc. At the moment, changes in the road network would result in excavation and asphalting of the upper layers of the entire network. Therefore, elements must be designed for efficient replacement. The new policy in transport infrastructure design will prioritize the efficiency of this process by replacing layers in the transport infrastructure structure and adapting them to both user requirements and the process of material recovery and recycling.

An absolute novelty in the process of building sustainable roads emerged in 2001, based on the idea of building roads using plastic waste. The originator of this new methodology, patented in India in 2006, is Professor Rajagopalan Vasudevan (The Guardian, 2018).

Plastic roads are made entirely of plastic or mixtures of plastic with other materials. They are different from standard roads in that standard roads are made of asphalt concrete composed of mineral aggregates and asphalt.

However, there are currently no records of standard roads made exclusively of plastic.

Plastic composite roads do exist, however, and they demonstrate superior characteristics to standard asphalt concrete roads, in particular, in terms of higher wear resistance (Table 2.2).

*Table 2.2* Comparison Between the Properties of Plastic–Bituminous and Regular Asphalt Roads

| *No.* | *Properties* | *Plastic Roads* | *Ordinary Road* |
|---|---|---|---|
| 1 | Marshall stability value[a] | Higher | Lower |
| 2 | Softening point | Smaller | Larger |
| 3 | Penetration value | Greater than | Lower |
| 4 | Tensile strength | High | Low |
| 5 | Rutting level[b] | Low | High |
| 6 | Water seepage | No | Yes |
| 7 | Durability | Higher | Average |
| 8 | Appearance of cracks or holes | No | Frequently |

*Source:* Singh (2019).

[a] The principle of the test consists of determining the breaking force for a cylindrical specimen subjected to a compressive load applied on the generators. The compression is exerted at 60°C· Marshall stability is the load, expressed in kN, reached when the specimen breaks (CREEAZA, 2024).

[b] Rutting or *rutting* is the process of small depressions appearing on the surface of the roadway. Pavement rutting (shearing) can occur along the sides of the trench. Rutting is particularly noticeable after rain when it is filled with water. There are three basic types of rutting: mixed rutting, subgrade rutting and densification (Pavement Interactive, 2024).

The use of plastics also opens a new option for post-consumer (downstream) plastics recycling.

*Plastic road* construction action objectives:

- Reduce plastic waste. Not all plastic waste is recycled, as even recyclable waste is too much to be reused 100%. But bituminous roads are continuously constructed and can incorporate extremely large amounts of plastic waste. Reducing road construction costs in this context is a bonus.
- The estimated lifespan of a plastic road is two to three times longer than that of roads produced by traditional road asphalt paving processes.
- The estimated construction time of a new road can be reduced by about 70%.
- Plastic road is four times lighter than traditional road structures.
- The plastic road is 100% adapted to the circular economic system and is made of recycled plastic wherever possible.

Stages of *plastic road* construction:

- Collecting plastic waste.
- Cleaning by washing.
- Shredding to a uniform size.
- Melting the plastic waste at 165°C and mixing it with hot bituminous aggregates.

The whole process of making the road compound is otherwise very simple. Plastic waste is first shredded into small pieces using a shredder (Singh, 2019).

The aggregate mixture is then heated to 165°C and transferred to the mixing chamber, and the bitumen is heated to 160°C to result in a good bond. It is important to monitor the temperature during heating. The shredded plastic waste is then added to the aggregate. The aggregate is uniformly coated for 30–60 seconds.

Subsequently, the aggregate coated with waste plastic is mixed with hot bitumen, and the resulting mixture will be ready for use. The casting temperature of the resulting *mat* should be between 110°C and 120°C.

In order to reduce costs and improve road performance, alternative binders such as polymers, bio-oils, waste materials, emulsions and shredded rubber have been substituted for road bitumen in three ways, which differ according to the percentage of bitumen substituted and the function of the substitute.

Bitumen modifiers are widely used to improve pavement performance, including pavement properties such as durability, rutting, strength, softening point and viscoelasticity.

A number of plastic additives are used to improve the performance of road pavements, such as rubber latex, shredded rubber, styrene, styrene–butadiene, styrene–butylene (SBS), recycled polypropylene (PP), low-density polyethylene (LDPE), polyethylene (PE), ethylene–vinyl acetate (EVA) and polyolefin.

SBS is by far the most common polymeric additive used in bitumen mixtures, with the percentage of its use in the final compound being around 2%.

Synthetic and natural polymers were not used to modify bitumen until the 1930s in Europe and North America, even though the recipe had been validated a century earlier.

Polymer modification of bitumen is widely practised for high-capacity and heavily trafficked roads to increase the quality of pavements by improving pavement elasticity, material cohesion, moisture, strength, fatigue life and increased performance at high temperatures by decreasing the resistance of pavements to deformation and environmental temperature variation.

PE is considered to be one of the most effective bitumen modifiers for improving rutting performance. All polymers reduce the thermal sensitivity of bitumen, such as rutting at high temperatures and fatigue or cracking at low temperatures.

Three categories of polymers are used to modify bitumen: plasterers, elastomers and reactive polymers.

For example, polyphosphoric acid (PPA) is a polymer that has been used most often to modify the structure and properties of bitumen. Other blends used are EVA copolymer, glycidyl methacrylate (GMA), styrene–isoprene–styrene (SIS), PE, SBS, polyammonium phosphate (APP) and acrylonitrile–butadiene–styrene (ABS).

Plasterers include the following types of plastics: *PE*, PP, EVA and ethylene–butyl acrylate or a combination of these. *PE* is considered to be the most effective modifier of plasterer and polymer bitumen. It is clearly established that PE–asphalt binder modification improves performance through fatigue and friction resistance compared to unmodified roads. High-density polyethylene

(HDPE) and LDPE were analysed to determine which better improves road properties. HDPE (optimal at 12% bitumen mass) has been shown to improve pavement properties more than LDPE by increasing roadway stability, reducing density and increasing aggregate voids of air and minerals.

*The advantages of* plastic pavements are as follows (Civil Dream, 2023):

1 *Financial savings*

The immediate savings associated with plastic roads are the cost of materials used to construct them. Plastic roads are constructed from far fewer materials than paved roads and require less energy and resources to build.

Plastic roads can be built at up to 50% lower cost than standard asphalt roads and require less maintenance. In addition, they last longer than standard asphalt roads, greatly reducing future maintenance costs.

2 *Sustainability*

According to the World Economic Forum, plastic roads have proven to be three times stronger and more durable than standard asphalt. This means that they can withstand harsher weather and resist potholes, making them extremely durable.

3 *Flexibility*

Plastic roads can be reshaped and adapted to ground movements due to their special construction.

The benefit of plastic is that it allows roads to maintain their course by expanding or compressing as the ground changes, which helps to reduce the risk of cracking and collapse of asphalt roads caused by ground movement.

This property also allows the road to be built in more compact shapes using fewer construction materials, resulting in a more durable road surface.

Plastic roads also create jobs in local communities and promote economic growth. Because plastic roads are extremely easy to build, local governments can use the money saved to hire more workers and reinvest funds in the local economy.

Plastic roads have become increasingly tempting to build because of their durability and qualities. Built using 100% recycled plastic and offering a longer lifespan than standard asphalt roads, they require less maintenance, as plastic is insensitive to the elements, unlike asphalt.

By carefully considering the potential environmental impacts, plastic roads can be an effective way to reduce plastic waste and extend the life of roadways.

4 *Reducing carbon emissions*

The use of waste plastics in road construction contributes to the conservation of fossil fuels as well as reducing $CO_2$ emissions.

5 *Pollution*

In addition to reducing air pollution, plastic roads can reduce road noise. Plastic roads are smoother and more comfortable to use than asphalt roads because they reduce friction between tyres and the road surface. This can reduce noise levels, making it quieter and more pleasant to travel on plastic roads.

Finally, plastic roads are also aesthetically pleasing, which is a great addition to any landscape.

*Disadvantages of* plastic roads.

1 Noxious gases are released during road construction.
2 After the construction of plastic roads, it has been found that the first rainfall will result in a wet, sticky coating (laitance) on the surface.
3 The mixture of toxic substances in the plastic, in combination with the bitumen, could melt and result in the release of hydrochloric acid (HCl) vapours.
4 An undesirable effect of plastic roads is also the possibility of plastic microparticles entering the food chain. Vehicle wear and tear produce plastic micro-particles on roads.

When these particles mix with the air and water, they can be taken up by wildlife and plants, and continue their journey up the food chain. This can have harmful effects on the environment and can actually affect the health of all living things.

Finally, plastic roads are not biodegradable. Plastic roads can take hundreds of years to break down, making them a potential source of long-term pollution. The lifespan of plastic roads can be prolonged with careful maintenance, but over time, they will eventually have to be disposed of or recycled.

Despite these potential drawbacks, plastic roads can still offer many environmental benefits. By combining recycled plastic with other sustainable materials, plastic roads can, however, and even paradoxically, provide an environmentally friendly solution for roads.

#### *2.2.1.4 Exploiting sustainable transport infrastructure*

Beyond the construction phase, operational excellence involves continuously tracking and increasing the sustainability impact of infrastructure. All four dimensions – environmental, social, economic and resilient – need to be considered.

In this context, the actual contribution of infrastructure to institutional commitments to reduce GHG emissions, increase national competitiveness and improve safety must be monitored.

To track ongoing performance, key metrics are essential. Digital tools can be used to monitor business dynamics. Environmental data (such as GHG emissions, energy consumption, the ratio of renewable to non-renewable energy and waste management) should be included in this context in infrastructure condition and impact assessments. Similarly, tools such as end-user satisfaction surveys should be implemented to track inclusion levels and identify potential areas for improvement.

Using advanced data analytics and digital tools, infrastructure operators may be able to maximise savings and increase efficiency by acting on indirect costs, maintenance, inventory, risk management and workforce management.

For example, rail operators can reduce carbon emissions directly by operating trains at automated train speeds. Automatic speed control systems such as ERTMS reduce energy consumption by optimising train speed, which also increases infrastructure capacity. As a result, the carbon footprint for passenger and freight transport decreases.

Another example can be found in the transport infrastructure supporting the logistics sector. Automating equipment for trucks and cranes to minimise container movements in terminal yards improves efficiency and reduces waiting times for transport lines and trucks. Digitising port infrastructure using RFID technologies can lead to an 80% decrease in air pollution. Optimising the movement of containers within storage locations can reduce energy consumption by 6.5% when the maximum speed is reduced from 28 km/h to 25 km/h (Milani et al., 2021).

In the middle of the last century, Europe started research on smart road infrastructure, which is mainly focused on sustainability, eco-transport and safety, including the famous 'Eureka' plan and the Easy Way project. A few examples of this approach are conclusive. In 2013, *Drive Me*, an experimental road in Gothenburg, provided real-time information to vehicles via roadside-mounted sensors for functions such as automatic parking and adaptive cruise. In 2016, Norway's E8 highway used the vehicle–road connection to create an app useful for the autonomous driving of freight vehicles.

Most countries have used roadside cameras, millimetre-wave radar[5], etc., to gain a large-scale perception of the road environment and have combined different communication methods for information interaction between vehicles and infrastructure. Objectives range from automated driving, crash alerting, responsible (green) driving, safety improvement, etc.

Next-Generation Intelligent Highway (NGAI) has become a global trend to improve transportation efficiency and safety. Exploring the features and framework for NGAI is a real guide for '*upgrading*' the current highway system.

The development of NGAI is also a process of mutually promoting intelligent vehicles and smart roads, which need to be developed together with emerging technologies. Effectiveness depends both on the level of automation of the vehicle and the level of 'smart equipment' of the highway.

Similar to the automated vehicle (AV) automation level classification, the NGAI can also be divided into different intelligent levels, depending on their digital abilities. These levels can be assigned to parts of the network to provide AVs with guidance for future highway evolution. The basic idea about this categorisation scheme is to differentiate infrastructure by considering its digital functions, such as digital information, system perception and vehicle–road cooperative driving.

NGAI as a whole is composed of several intelligent infrastructures. It serves not only VAs but also traffic operators. Therefore, the intelligent layers of

NGAI are determined by both their data characteristics and their supporting functions.

There are five defined NGAI levels, starting from the conventional highway (which is level 1 – in short A1) to the fully intelligent highway (level 5, i.e. A5). The architecture of a smart infrastructure system can be developed using two methods: the object-oriented method and the process-oriented method.

In comparison with the process-oriented method, the object-oriented method mainly focuses on the objects in the system, with little characterisation of the process between the objects, which is more measurable, reusable and adaptable.

The first step in the development of the NGAI technical framework is to determine its basic elements. Current studies prefer to determine characteristic elements based on application scenarios or the type of information. However, such definitions are not suitable for the design of the NGAI higher-level framework due to the complex elements and functions in the NGAI system. By summarising the data requirements of each application scenario in NGAI, the basic components of NGAI are determined as four types of elements, namely highly dynamic, transient dynamic, transient static and permanent static (Liu et al., 2021). Their specific features are as follows:

- The highly dynamic elements mainly include different types of traffic states, including vehicles, roadblocks, trajectories and control commands.
- Transient dynamic elements mainly include road network conditions, including work zones, traffic congestion, weather conditions, toll information and signal control information.
- Transient static elements mainly include road network infrastructure conditions, such as skid resistance, road roughness and unevenness, damage and marking conditions.
- Permanent static elements refer to features that are stable and almost unchanged, including road geometry, landmarks, traffic signs and barriers.

The physical architecture defines the basic elements and major components in the NGAI system. For different application or system scenarios, the data requirements are distinct. These need to be specified primarily because they determine the choice of the perception method and the technical requirements of the perception devices.

NGAI reshapes the existing highway system using digital and smart technology to improve its functionality, such as traffic capacity, speed and driving safety. NGAI also explores the potential of current physical facilities. However, there are still many critical issues in practice, among the most important being:

- Currently, some NGAI functions are too novel ('ahead of their time'), leading to poor compatibility with the current system. Therefore, given the limitations of laws and regulations, highways can only be used for autonomous driving testing. The deployment of too many roadside facilities for VAs highlights the difficulty of realising NGAI at this stage. It is recommended to set the correct construction content according to regional characteristics and predominant users. The architecture of the NGAI needs to consider its sustainability to ensure that the proposed smart highway is applicable for different stages of development.

- The design and construction of the NGAI, at the moment, focus on a single scenario and lack a more general vision. Many projects are designed for a specific objective (e.g. autonomous driving, life cycle maintenance and information transmission), which makes the usefulness of the complete system not perceived. There is a need to fully integrate existing resources to create a model that can be used for different scenarios and any type of situation on the ground.
- As the construction of the NGAI is still at an exploratory stage, there are large differences in data types, formats, protocols and communication methods. Therefore, relevant standards should be established to ensure that NGAI applications with different architectures can be interconnected.
- There are certain issues in communication stability, data security and privacy protection, in particular in multimodal interactive environments. Therefore, studies and research, in the next steps, need to improve the robustness of NGAI against external cyber-attacks on the system and to protect user privacy.

### 2.2.2 *Planning, designing and delivering a sustainable and resilient transport infrastructure*

To be sustainable, infrastructure must provide necessary services at affordable costs while conserving natural resources and energy. Moreover, these services should be maintained and continuously improved to remain competitive in the specific market. However, in order to be continuously sustainable, a system for planning, designing and delivering a sustainable and resilient infrastructure needs to be implemented with the following main objectives:

- Set the standard for what constitutes sustainable transport infrastructure
- Achieving higher than minimum performance levels
- Recognising projects that contribute significantly to the sustainability process

In order to build such an analysis, assessment and decision model, the premises need to be defined initially. These are in line with the data presented at length above:

- Decades of neglect of the importance and role of transport infrastructure in the evolution of society have meant that massive investments in this area are now needed worldwide.
- Ageing and obsolete infrastructure needs to be replaced and modernised, while at the same time developing completely new infrastructure systems adapted to the current and, why not, future demands of society.

- Infrastructure is at the heart of the key challenge of the 21st century, which is the need to integrate into a more sustainable and resilient society.
- The classic, past standards and methods of transport infrastructure development strategies are no longer adequate to meet the needs of the future.
- In addressing sustainability and resilience in the face of changing variables, it is difficult to assess the full range of benefits and impacts across the broad range of social, environmental and economic factors.
- In the process of developing a transport infrastructure project, the most sustainable solution is often project- and context-specific and difficult, if not impossible, to predict in its entirety.

In this context, a paradigm shift in the planning, design and delivery of sustainable and resilient infrastructure must be considered.

Related to this approach, several attempts have been made worldwide in recent years to define a methodology for analysing and assessing the degree of sustainability for infrastructure projects. Among the most important, we mention the one developed by specialists from the ISI in the United States, called ENVISION (ISI, 2018).

The created methodology aims to promote sustainable performance and resilience of infrastructure, supporting all infrastructure stakeholders (stakeholders) in implementing more efficient infrastructure investments in terms of cost, resource utilisation and long-term adaptability.

Importantly, the methodology aims at analysing all stages of the full infrastructure life cycle, targeting:

I **Design**: in-depth assessment of the design and identification of additional improvements needed to achieve increased sustainability.
II **Implementation (construction)**: quantifying and assessing the levels of sustainability achieved by projects in reality.
III **Maintenance**: assessing the sustainable impact of projects throughout their life cycles.
IV **Communication and training**: to develop an organised system to train stakeholders and obtain the necessary support to achieve sustainability throughout the life cycle of projects.
V **Support for achieving continuity in sustainability**: supporting the reassessment and adaptation of design standards, construction procedures and development strategies for transport infrastructure projects.

The analysis in the methodology is based on a number of indicators grouped into five broad categories, namely:

1 **Quality of life (QoL)**: addresses the impact of a project on host and affected communities, ranging from the health and well-being of individuals to the well-being of the entire social structure. Quality of life focuses on assessing the extent to which infrastructure projects align with community objectives, are embedded into existing community networks and

will benefit the community in the long term. Community members affected by the project are considered important stakeholders in the decision-making process. The category is divided into three sub-categories: *well-being, mobility and community*.

2 **Leadership (management) – MA**: supports the introduction of a new approach to the design and development of infrastructure projects. This section encourages collaborative action and ongoing communication between all stakeholders involved in infrastructure projects, always aiming to achieve a truly sustainable project. The category is divided into three sub-categories: *collaboration, planning and economics*.

3 **Resource allocation (RA)**: addresses the quantity, source and characteristics of the resources used in the infrastructure project and their impact on the overall sustainability of the project. *Materials, energy and water* are the three sub-categories of resource allocation.

4 **The natural environment (NE)**: addresses how to understand the place of infrastructure within ecosystems and how to minimise negative impacts, while considering how infrastructure can interact with natural systems in a synergistic and positive way. These types of interactions and impacts have been divided into three sub-categories: *siting, conservation and ecology*.

5 **Climate and resilience (CR)**: aims at minimising emissions that may contribute to climate change and other risks in the short and long term, and ensures the resilience of infrastructure projects. In order to be resilient, infrastructure must contain all the necessary information for identification, have access to resources and be robust, redundant, flexible, integrated and inclusive. The category has two sub-categories: *emissions and resilience*.

Based on the criteria within each influence category, a total of 1,000 points can be awarded to each project analysed (ISI, 2018), with the criteria being scored differentially.

At the end of the assessment, the infrastructures (infrastructure projects) are categorised, from a sustainable development perspective, into five broad categories, according to the score obtained, out of the total, as follows:

- Below 20% – UNSATISFACTORY.
- Between 20%–30% – SATISFACTORY.
- Between 30% and 40% – GOOD.
- Between 40% and 50% – VERY GOOD.
- More than 50% – EXCELLENT.

The scores, given for each of the criteria, lead to the analysed infrastructure being classified in one of the following situations:

- **Improved**: higher than conventional performance. Slightly exceeds regulatory requirements.
- **Improved**: sustainable performance that is on track. There are indications that superior performance is within reach.

- **Superior**: sustainable performance at a very high level.
- **Green**: performance that has achieved essentially zero negative impact.
- **Sustainable**: performance that restores natural or social systems. Such performance receives the highest score.

Obviously, after being placed in one of these categories, decision-makers will have to allocate resources (financial, material and human) to improve the classification of the project analysed, at least to a higher class than the one resulting from the analysis. Preferably, the necessary measures should be taken (if physically possible – location, surroundings, relief, etc.) to place the project infrastructure in the 'Excellent' category, which means to obtain a minimum of 500 points in the next diagnostic analysis.

For each criterion, increasing the sustainable performance of an infrastructure project can be achieved by adopting specific measures.

### *2.2.3 Examples of green transport infrastructure*

Transforming abandoned communication networks (railways, roads) into pedestrian or bicycle networks is a process being considered in many parts of the world.

In the United States, the Poughkeepsie–Highland railroad bridge in Poughkeepsie, NY, built in 1889, was left unused after it was almost destroyed by fire in 1974. About two kilometres long, it was redeveloped as a pedestrian walkway in 2009, becoming the longest pedestrian bridge in the world, which has since attracted over a million visitors.

Also in the United States, the East Coast *Greenway* is an attempt to create a continuous network of trails for bicyclists and pedestrians from Maine to Florida (Waren, 2023).

In Atlanta, a quasi-green approach is being used to promote what is called the Beltline, a 22-mile (35.41-km) corridor that incorporates parks, trails and public transportation, along with areas of commercial and residential spaces. It is part of an effort to better manage city planning to avoid more auto-oriented sprawl.

This type of approach could be adopted more often by authorities or private investors in many locations in Romania. Attempts in this direction have been made, but not with much success. One example is the Luncoiu Viaduct at the entrance to the municipality of Brad (Hunedoara County), built between 1939 and 1946 and used until 1997, when a landslide of several tens of meters led to the definitive closure of rail traffic on the Deva–Brad railway (Guță, 2014).

The Luncoiu Viaduct was built of eight reinforced concrete vaults in the centre, 20 meters span, 217 meters long and built in curved and counter-curved with a radius of 300 meters. It served as an underpass for the Deva–Brad highway, above which it rises about 25 meters.

The Luncoiu Viaduct is about five kilometres from the Brad railway station, one of the most appreciated period buildings in the Zarand Country, and travellers can reach it on foot, by bicycle or even by off-road car on the old embankment. They can also hike up the surrounding hillside paths.

Recently, the Brad authorities have proposed a bike path on the Luncoiu viaduct, but unfortunately, this has not been realised so far.

Another important way to promote sustainable infrastructure is the creation of dedicated bus lanes so that buses can circulate without entering the general, much more intense flow of traffic in certain areas. Part of a concept called *bus rapid transit*, dedicated lanes for public transit have been around for decades and are being used in South America, Europe and increasingly in North America and Asia.

A study of the 14-mile (22.5-km) bus rapid transit corridor in Guangzhou, China, the country's third-largest city, concluded that the system will reduce $CO_2$ emissions by 86,000 tonnes (78,018 metric tonnes) per year in its first decade of operation. In addition, the study by China's Institute for Transport and Development Policy found that Guangzhou's dedicated bus system reduced emissions of particulate matter that cause respiratory illness by 4 tonnes (3.62 metric tonnes) annually.

The use of green building materials is also an important solution for developing sustainable transportation infrastructure.

As an example, there are many elements of the Brooklyn Bridge Park (USA) that make it environmentally friendly. First and foremost, the 85-acre project replaces an industrial area full of wharves and warehouses with restored public spaces for recreation, recreation and other habitats, which are also built with recycled, local materials. For example, granite slabs from the nearby Roosevelt Avenue Bridge were used to create park steps; stone from the Willis Avenue Bridge in the Bronx was incorporated to create a variety of seating and walkway features; and steel from some warehouses was reused (Waren, 2023).

Even in cases where building materials cannot be recycled and reused, there are options that are more sustainable than those used in the past. For example, so-called permeable pavers are a great benefit when it comes to managing stormwater runoff so that it flows back underground or collects in basins where it can be reused, instead of running off and getting lost or polluting water bodies.

Related to this theme of construction materials for transportation infrastructure, the concept of using plastic waste as a bitumen modifier in roads has recently emerged, giving rise to 'plastic roads' (Patil et al., 2023).

As has been presented previously, waste plastics are extensively used for bitumen modification (with less than 10% in the final composition). Virgin polymers have also been used to modify bitumen, aiming to reduce costs and improve road performance.

As the use of waste plastics in road construction targets waste plastics generated after the use and maintenance stages, plastic roads are considered a downstream measure. However, upstream measures that promote waste avoidance by targeting natural resource extraction and plastic production remain highly necessary, in line with international treaties and the waste management hierarchy, which prioritises prevention over reuse, recycling, energy recovery and disposal.

There are many areas worldwide where plastic roads have been built or are planned to be built in the future.

Australia, Indonesia, India, the United Kingdom, the United States and many other countries have used a technology that can incorporate plastic waste into an asphalt mix.

In India, Chennai was among the first cities, globally, to adopt the technology in a sustainable way,[6] when the municipality commissioned 1,000 km of plastic roads in 2004 (Singh, 2019). In Chennai, nearly 1,600 tonnes of waste plastic have been used to construct 1,035.23 km of roads in recent years, which includes NSC Bose, Hals, Ethiraj Silai Road and Sardar Patel Road.

Since then, all major city municipalities in India, including Pune, Mumbai, Surat, Indore, Delhi, Lucknow, etc., have experimented with this technology.

*Pune*: using the technology of bitumen mixed with plastic waste, Pune Municipal Corporation constructed a 150-meter stretch of road from Bhagwat to Navi Peth near Vaikunth crematorium in 2016. The other sample patches in Pune include Dattawadi Kaka Halwai Lane, Katraj Dairy, Magarpatta City HCMTR Road, Kavde Mala Road, Koregaon Park Lane No. 3 and Yeravada Sadal Baba Darga Road from Chandrama Chowk.

*Jamshedpur: the* Jamshedpur Utility and Services Company (JUSCO), which is a subsidiary of Tata Steel, has constructed a 15-km-long rolling locale in the said district using a plastic road, including a stretch of nearly 1 km in Ranchi, 500 m in Dhurwa and Morabadi, 3 km each in Chas and Jamtara and 500 m in Giridih.

*Indore*: since 2014, the Madhya Pradesh Rural Roads Development Authority (MPRRDA) has constructed about 35 km of roads in 17 districts using asphalt recipes made from blending bitumen with waste plastic.

*Surat*: the idea of using plastic bitumen–plastic mix was implemented in January 2017. The problem of potholes was significantly reduced as no cracks developed in areas where the roads were layered with waste plastic.

The technology has penetrated deep and found application even in remote areas such as Meghalaya, a village, where, in 2018, 430 kg of plastic waste was turned into a kilometre-long road.

In the United Kingdom, in January 2019, the Department for Transport announced a £1.6 million trial of a plastic road technology developed by MacRebur, a Scotland-based asphalt improvement company (Civil Dream, 2023).

In Pakistan, in December 2021, a one-kilometre stretch of Islamabad's Atatürk Avenue was completely reconstructed using the bitumen–plastic mix recipe. The project was a cooperation between the Capital Development

Authority and the Coca-Cola Company Pakistan. Ten tonnes of plastic packaging waste combined with asphalt was used.

At the same time as the development work of 'plastic roads', the 'solar roads' project was launched.

Sustainability is a key issue when it comes to road activities in the future, which is why several solutions are being sought to reduce the carbon footprint of infrastructure use. From self-generation solutions to efficient energy transfer from infrastructure to vehicle flow. The first group includes solutions for 'solar roads'.

'Solar roads' are those whose pavement is literally made of solar panels, protected and resistant to road traffic and all types of vehicles.

In other words, photovoltaics is not just a matter of saving energy in buildings, where its use is most well-known. On roads, solar panels convert solar energy into electricity, and their use can be varied: from building recharging points for electric cars, which run on those roads, to lighting their own roads.

Solar road projects currently exist in countries such as the United States, France, the Netherlands and China. And while it is true that it is expensive to implement and expand, or that collecting solar energy is not permanent, it has a number of advantages:

- Clean energy generation by taking advantage of roads that consume that space anyway.
- Cheaper electricity generation in locations near transportation routes, for example, gas stations or parking lots.
- Low maintenance of these installations along roadsides, as solar panels last at least 20 years with very little maintenance.
- With inductive charging, it is even possible to power electric vehicles using conductive coils that generate magnetic fields that can be used by vehicles in the areas of influence.

Another approach to sustainable transportation infrastructures is related to the use of new technologies, thus creating *Intelligent Transportation Infrastructures (ITS)*.

The concept of an intelligent highway has been proposed internationally since the 1940s (Liu et al., 2021). However, the concept was not successful at the time, being considered too *futuristic*.

It was not until 1992 that the Automated Highway System (AHS) programme was launched in the United States, and vehicle-to-vehicle (V2V) and vehicle-to-infrastructure (V2I) communication modules were built to reduce traffic congestion and improve driving safety.

At the turn of the current century, the Smart Highway project in Virginia (USA) was an important milestone for the intelligent transportation system in America. It integrated several technologies, such as those for network performance monitoring, energy recovery and automated driving, to create a 2.2-mile (about 3.5 km) smart highway.

In California, too, the Smart Corridor is one of the most iconic smart highway programmes in the United States, being developed on the I-80 freeway. This system integrated a variety of sensing and active transportation technologies through electronic identification and a dedicated short-distance, short-range communications system (DSRC) to improve driving safety and transit time reliability.

But ITS has been developed in other parts of the world, not just the United States. Since the 1990s, Japan has carried out extensive research and design of intelligent transportation systems. The public SmartWay trial, launched in 2006, has become a global ITS benchmark. The system contains a vehicle information and communication system (VICS), an electronic toll collection system (ETC), an advanced vehicle safety system (ASV) and other functional modules to enhance the user's travel experience. Japan has formed unified standards across the country and used ETC and VICS systems.

China has built several smart highways for automated/semi-automated driving and *connective*[7] vehicle *infrastructure connective systems* (CVIS). For example, in Zhejiang Province, technologies such as free-flow charging, wireless charging and partially automated driving are integrated to create a 161-km '*Super Highway*'. Travel speeds, along it, are projected to reach up to 120–140 km per hour (Liu et al., 2021).

In the Hunan region, 113 km of '5G+' smart highway is being built to realise the accurate detection of each road section and each vehicle, with the idea of creating a high-definition dynamic digital map for automatic driving.

Based on the BeiDou satellite navigation system (BDS), Guangxi Province is building an intelligent highway system with functions of vehicle guidance, safety monitoring and the digitisation of driver facilities.

Shandong Province has launched the first fully autonomous domestic smart highway, which uses advanced laser radars, microwave radars and other technologies to carry out comprehensive vehicle control. This programme can significantly improve traffic efficiency and safety for fully connected automated vehicles, but this type of infrastructure is currently in the testing phase in an open environment (no obstacles, bumps, etc). Although China has made tremendous achievements in the construction of smart highways, it has not yet realised a unified *matrix* and framework to promote the concept nationwide.

A new engineering first recently took place in China. A 158-km stretch of highway was completely resurfaced using robots and drones (ProtvNews, 2024). The machines were guided using advanced algorithms based on the high-precision BDS and a high-performance communication network that allowed real-time route planning and precise operational control. Drones played a crucial role in the project, used for mapping, monitoring progress, and even transporting materials to hard-to-reach areas.

In Europe, the EC adopted on 30 November 2016, the 'European Strategy on Cooperative Intelligent Transport Systems (C-ITS)', a milestone initiative towards cooperative, connected and automated mobility. The objective of the

C-ITS Strategy is to facilitate convergence of investment and regulatory frameworks across the EU (EC, 2024a). Through this regulation, it aimed to integrate vehicles into cooperative, connected and automated systems, with a view to planning for sustainable mobility or developing the concept of 'mobility as a service', including public transport and alternative transport modes to motorised, polluting ones, such as walking and cycling (scooters, skates, etc). The introduction of C-ITS will achieve a cost–benefit ratio of up to 3:1, estimated on the basis of cumulative costs and benefits from 2018 to 2030.

In support of this action, the EC designated a specific frequency band for traffic safety applications in 2008. The initial introduction of short-range V2V and V2I communications is based on technologies already available using that band and, where possible, will operate in continuity with 5G technology on a principle of complementarity. In 2016, Member States and the EC launched the Cooperative Pathways Platform to correlate C-STI deployment activities, jointly develop and share technical specifications and verify interoperability through cross-testing. Initially created for EU co-funded C-STI introduction initiatives, 'Cooperative Roads' has been opened to all interoperability testing activities.

## 2.3 Romanian transport infrastructure – from today to the future

By the year 2024, in its 16 years of EU membership, Romania has benefited from more than €80 billion (Mihălțianu, 2023). It is a net beneficiary because it contributed only €26.4 billion to the EU budget.

Of the €56.3 billion, considered as the 'net profit' of EU funding, €7 billion was absorbed for large infrastructure projects. This money has been spent overwhelmingly on high-speed road construction, but even with this permanent priority, by the end of 2023, the total number of modern roads will be just over 1,000 km, and the rail infrastructure will have only a few hundred kilometre modernised.

Relative to Romania's needs and compared to what Romania could have done if it had absorbed all the EU funding, this is far too little.

Among the major causes that have led to this situation, we mention just a few:

- Romania's tendering system for projects funded by EU money is extremely bureaucratic and causes years of delays. Bidders who lose tenders continue to sue the state and block construction sites. The Romanian judicial system is far too slow to react, and so projects usually start at least two years late. As a result, projects end up delayed even longer or, in some cases, abandoned altogether.
- Romanian politicians have been unwilling to prioritise EU-funded projects. Governors have pushed for direct award of contracts to private companies, paid from the national budget, instead of applying for EU-funded projects.

By modes of transportation, the Romanian infrastructure has had different evolutions, with the advantage being, as we said, the road infrastructure. Here are some details in this regard for each mode of transport.

*Roads*

For the sustainable development of the road transportation system, in close correlation with the connectivity needs and available funding sources, two levels of complementary road networks have been defined, with the role of creating a branched system that will efficiently provide and serve the Romanian population and economy. Improved connectivity in the South-East European region was also taken into account in establishing the network levels, as well as the linkage with European climate change mitigation policies.

The two identified road networks, based on the detailed analysis of the Romanian General Master Plan for Transport (MPGTR), are represented by:

- The primary road network, which constitutes the basic road structure of Romania in the national (road connectivity corridors) and European (TEN-T corridors in Romania) context. The entire TEN-T Core network and most of the TEN-T Comprehensive network overlap with the primary road network.
- The secondary road network is both a complementary network to the primary network, serving as a feeder to it, and a connecting network between it and the national and county road networks.

**The primary road network** connects all development regions and the most important economic growth poles in Romania.

On the other hand, the primary road network also includes the top 18 cities in Romania, in relation to population level (over 125,000 inhabitants). Thus, the primary road network directly interconnects Bucharest, Cluj-Napoca, Timişoara, Iaşi, Constanţa, Craiova, Braşov, Galaţi, Ploieşti, Oradea, Brăila, Arad, Piteşti, Sibiu, Bacău, Baia Mare, Buzău and Dr. Tr. Severin – important socio-economic and demographic centres.

The primary road network overlaps with the entire European TEN-T Core network and most of the Romanian TEN-T Comprehensive network, thus contributing to the development of general road mobility, both at the national and European level.

**The secondary road network** is complementary to the primary road network, providing functional links between it and the national and county road networks.

The secondary road network serves the county municipalities (other than those located along the primary road network) and connects them to the primary road network. Out of the total of 42 county municipalities in Romania, 30 are on the primary road network, and the remaining 12 are indirectly connected to it via the secondary network. Thus, all county municipalities in Romania are directly or indirectly connected to the primary road network.

The Romanian authorities have drawn up a list of 40 highway and express road projects belonging to the primary network for the immediate future (up to 2030), totalling more than 3,257 km in length and with an estimated cost of around €37.4 billion (RG, 2022).

The Romanian authorities' plan also includes several elements to support the development of sustainable road transport, but they are largely not in line with the aim of achieving sustainable infrastructure. Some of the most important ones are as follows (RG, 2022):

- Changing road pricing policies from a 'vignette' to a distance-based approach.
- Implement green charging principles.
- Discourage the registration of cars older than 15 years.
- Increase the value of scrappage schemes for polluting vehicles.
- Extended tax incentives for zero-emission vehicles.
- Support for the expansion of electric car charging infrastructure.
- Using EU Green Public Procurement criteria for all aspects of transport.

Among the most important of these objectives, we consider to be those relating to electric vehicle charging infrastructure and the one requiring the use of EU green public procurement criteria for all aspects of transport, which includes materials and techniques for infrastructure construction.

Against this background, the national policy on electric charging points for zero-emission road vehicles includes:

- 1,836 already existing charging points.
- 2,896 high-power (at least 50kW) recharging points to be installed on the national road/highway network by 2030 (to be financed by the Recovery and Resilience Plan – PNRR).
- 13,283 recharging points financed by the 'Local Fund' component of the Ministry of Development, of which:
  - 4,000 recharge points in county municipalities.
  - 1,876 recharging points in other municipalities.
  - 1,600 recharging points in cities.
  - 5,724 recharging points in rural areas.
- 2,000 recharging points financed by the 'Renewal Wave' component of the Ministry of Development.
- Minimum 10,000 high and normal power recharging points financed from national sources, other EU funds (including Cohesion Policy) and/or private sources (including concessions).

Unfortunately, Romania's road infrastructure does not plan to build sustainable networks, which, in addition to facilitating green transport (e.g. electric charging facilities, dedicated lanes for public transport or by non-polluting means), also provide for the use of technologies and materials that correspond to the circular process of developing and operating such networks.

However, there is also a laudable target of shifting 30% of road freight transport over distances of more than 300 km to other modes of transport, such as rail or inland waterways, by 2030. This should exceed 50% by 2050, using efficient and environmentally friendly freight corridors.

Achieving this objective will require the development of an appropriate rail infrastructure that is economically efficient, leads to the sustainable use of energy resources and reduces negative environmental impacts.

Strictly related to the road corridors, at the end of 2023, Romania will have approximately 1,075 km of motorways and express roads in operation.

By comparison, Bulgaria – a country half the size of Romania – has a network of motorways open to traffic measuring some 861 km in October 2023.

To this should be added 7.5 km of express roads. Thus, in total for Bulgaria, we are talking about 869 km of high-speed roads.

Hungary has 26 high-speed roads – 13 motorways and 13 express roads. Hungary passed the symbolic milestone of 1,000 km in 2007. At the end of last year, Hungary's network of motorways and expressways amounted to some 1,855 km, with more sections to be opened this year. So, Hungary, with almost 1,900 km of high-speed roads and an area 2.5 times smaller than Romania, has an average annual growth rate of more than 56 km of high-speed roads, almost double the rate in Romania.

Poland has an incredibly large network of high-speed roads, which it has developed especially in the last 30 years. According to the latest data, Poland's high-speed road network will measure almost 5,000 km in December 2023.

Of these, about 1,850 km were highways, and the remaining more than 3,000 km were express roads.

Serbia, although not in the EU, has in recent decades embarked on a major campaign to modernise its road infrastructure and build high-speed roads. As a result, by the end of 2023, Serbia will have a motorway network measuring just under 1,000 km. However, Serbia is almost three times smaller than Romania.

In conclusion, Romania has the lowest density of high-speed roads in relation to the surface area of its national territory, in the riparian zone, not considering Ukraine and the Republic of Moldova (Table 2.3).

In the face of the massive targeting of financial resources by all national authorities in the period 1990–2024 towards the construction of high-speed road networks, the achievements have been extremely poor, considering that countries that have not benefited from European allocations, such as Serbia, are almost three times better off than Romania in this respect.

However, Romania's motorway network exceeded 1,000 km, reaching 1,005.32 km at the end of 2023 (130km.ro, 2024b).

Adding open express road routes, Romania's total network of motorways and express roads reached 1,074.02 km.

*Table 2.3* Comparisons Between the Size of the High-Speed Road Networks in Romania and Some of the Neighbouring Countries at the Beginning of 2024

| *Country* | *Area ($km^2$)* | *Length of High-Speed Road Network (km)* | *Density (km/100.000 $km^2$)* | *Position* |
|---|---|---|---|---|
| Romania | 238.400 | 1.075 | 451 | 5 |
| Bulgaria | 111.000 | 870 | 784 | 4 |
| Hungary | 93.030 | 1.900 | 2.042 | 1 |
| Poland | 312.680 | 5.000 | 1.599 | 2 |
| Serbia | 88.360 | 1.000 | 1.132 | 3 |

*Source:* author's calculations after Mihălţianu (2023).

From the point of view of the policy constantly proposed by the Romanian authorities – but not supported in practice – to build the Romanian high-speed road network quickly and consistently, this development has been a failure. From an ecological perspective, however, we can say that it has been useful in slowing down the deterioration of environmental factors by providing new segments of infrastructure for road transport, the most polluting mode of the sector.

In the near future, between 2024 and 2027, a network of about 691 km of new motorways and 86 km of express roads is planned to be opened, that is, a total of 777 km of high-speed roads (130km.ro, 2024c).

A further 553 km are planned and could be completed by 2030, reaching a total of more than 2,220 km of motorways (Mihălţianu A., 2023), increasing the density to about 931 km/100,000 $km^2$ of national territory. This means that by 2030, Romania will not be able to exceed Bulgaria's 2023 level in terms of high-speed road density (784 km/100,000 inhabitants).

In other words, even with the most optimistic forecasts, the density of modern, high-speed road networks will not, even in the next decade, undergo a major development that is satisfactory for the constantly rising demand for road transport. From a sustainable development perspective, this is a positive development, but at the same time, transport infrastructure funding should, in future, also focus on the construction of sustainable road networks.

With regard to *railways*, in the last 30 years, Romania has managed to modernise less than 700 km of the more than 10,000 km of national railways, mainly with non-reimbursable European funds. Thus, since 2007 to date, around €5 billion has been invested to modernise 500 km of railway networks out of a total of around 2,400 km of railway track belonging to the TEN-T Core network. Apart from the EU funding, renewal works carried out with national resources have been extremely low, with an average of about 12 km of single track/year (compared to a rate of about 800 km/year before 1990). This has led, over time, to a backlog of repairs/modernisations, currently estimated at 9,800 km (single track) on the whole rail network, with consequences in the large number of speed restrictions affecting the performance of the rail system.

In order to propose projects to remedy this situation for the period 2021–2030, the Romanian authorities have classified the components of the rail network into several categories:

- The primary rail network, which constitutes Romania's basic rail network in the national (rail connectivity corridors) and European (Romania's TEN-T corridors) context.
- The secondary rail network, which consists of both a network complementing the primary network and a connecting network between it and the tertiary network.
- The tertiary rail network, which is made up of the complementary railways that connect directly to the secondary rail network and provide the systemic link between small towns and the other networks.

The primary and secondary rail networks mostly overlap with the TEN-T rail network (Core and Comprehensive). The TEN-T Core network is entirely covered by the primary network in Romania, while most of the Comprehensive network is covered either by the primary network (Cluj-Napoca–Oradea–EP. Bihor, Videle–Giurgiu, Buzău–Făurei) or by the secondary network. In addition, two sections of the secondary network (Constanța–Mangalia and Pitesti–Craiova, both links to the TEN-T Core corridor) are proposed to be part of the TEN-T Comprehensive network, based on the 2022 network review.

The Romanian authorities considered it necessary to shift the paradigm towards a more practical approach whereby, instead of costly, higher-speed-oriented modernisations that in most cases mean rebuilding lines on new alignments, the new concept will be based on rehabilitating existing lines on the current alignment using high-productivity rail modernisation trains.

Thus, the expected productivity rate is 300 km of track (single track)/year/train (RG, 2022).

This method will also meet the interoperability requirements for the 22.5 t axle load applicable for TEN-T lines. However, a number of other complementary works (which these work trains cannot carry out) will be the subject of separate, smaller work packages (rehabilitation of bridges, culverts, tunnels, signalling/ERTMS and other works to increase the attractiveness of rail transport: CF stations and stopping points).

Using the new technology, it can be estimated that the modernisation of the TEN-T Core rail network in Romania and some segments of the TEN-T Comprehensive network could be achieved in about seven years at a cost of €2.1 billion (€180 million for the purchase of two work trains under the 2014–2020 IMPOP, and €1.9 billion for materials and manpower under the 2021–2027 ETP).

For the next period of a maximum of 5–7 years, railway projects should focus on ensuring the modernisation of all main links between Romania's provinces, allowing speeds of up to 160 km/h on each route.

The main guidelines for the future development of Romania's rail infrastructure should take into account:

- The renewal of the current infrastructure, necessary to allow the replacement of components with a high degree of wear and/or faults and to keep the probability of new failures low.

  In view of the high level of renewal backlogs, which has led to an advanced state of degradation of the infrastructure, for the next period, according to the Rail Infrastructure Development Strategy 2021–2025, a programme is needed to gradually increase the annual renewal (restoration) rate of current and direct lines to a level of at least 550 km/year (CNSF, 2023).
- Carry out, as a matter of urgency, routine repairs to the installations related to the lines in operation.

  The purpose of this type of action is to increase the running speed permitted by the infrastructure in order to close the gap between this speed and the projected speed of the rail network. The purpose of routine repairs to installations is to prevent subsystems from being taken out of service, which would have the effect of limiting train speeds.
- Routine maintenance of the whole railway system, necessary to keep the probability of failure of infrastructure components low.

  Routine maintenance is the performance of a set of procedures for the replacement of infrastructure components. The replacement of components is usually carried out before failure and concerns components whose wear is sufficiently advanced that the probability of failure is considered unacceptable. In terms of running speeds, the purpose of routine maintenance is to prevent the difference between the design speed and the maximum speed permitted by the railway infrastructure from increasing. This action must therefore be permanent and continuous. In order to increase the maximum speed permitted by the railway infrastructure, routine maintenance must be coupled with other actions such as repairs to defective infrastructure elements, renewals and/or upgrades that include elements to increase speeds.
- Infrastructure upgrades, which include projected speed increases, have the most significant impact on increasing the actual commercial speeds of trains.

  As stated above, this action aims to ensure an increase in the speed of movement, including an increase in the projected speeds of the line, within the limits of maintaining the 'conventional trains' category with a maximum speed of 160–200 km/h.

Public financing of railway infrastructure is complementary to the financing from the infrastructure manager's own revenue, which is the financial result of services provided to rail transport operators and other ancillary economic activities of the company.

The main component of the infrastructure manager's own revenue is the revenue from the infrastructure user charge (IUC). A policy aimed at making rail transport more competitive in the transport market, while ensuring a level playing field in this market, can lead to full cost recovery on a purely commercial basis.

TUI is the way to implement the 'user pays' principle with regard to the use of rail infrastructure. TUI is also a form of internalisation of external costs oriented towards rail infrastructure, but there is a discrepancy between the similar non-application of the user-pays principle for road transport.

Small steps in the modernisation of Romanian railways have only been made in the last few years.

Thus, in December 2023, 75.5 km of modernised railways were put into service in order to ensure the movement of trains at a speed of 160 km/h on the Coşlariu–Simeria section and part of the Braşov–Simeria route. For the same route, in May 2023 and March 2024, a further 91.6 km on the Coşlariu–Sighişoara section were put into service.

In conclusion, we can state that, once again, the Romanian authorities, despite the appearances of supporting transportation infrastructure development projects in accordance with the requirements of the concept of sustainable development, are doing exactly the opposite.

Thus, if only for the construction of new roads, until 2030, about €45.4 billion is allocated (€37.4 billion for motorways and express roads and €8 billion for the secondary network), and for the development/modernisation of railways, €29.55 billion is allocated. This corresponds to about 65.1% of the funds foreseen for road infrastructure.

Moreover, as hitherto, motorway construction still accounts for the lion's share, with 82.4% of all road infrastructure projects or 50% of all road and rail infrastructure projects combined.

This does not seem like a major change of vision by the Romanian authorities.

However, there is a project that fits in with the policy of supporting rail transport by promoting the 'metropolitan train'.

The metropolitan train is, in fact, a local or regional railway system of general economic interest in the field of public passenger transport, which provides a transport service in and around large urban agglomerations, using conventional railway lines, dedicated railway lines, a combination of these or a combination of railway lines and those intended for tram or metro vehicles.

The metropolitan train, as established in Central and Western Europe, runs within the city as a short-distance regime, with frequent stops (especially at the intersection with major boulevards or at the convergence with other urban public transport – metro stations, bus trolleybuses and trams), and outside the city on a medium-distance rail system, with less frequent stops (at the junction with major national or county roads – where park and ride parking can be provided – or near residential, commercial, leisure or logistics developments). In the European examples of S-Bahn or RER (Vienna, Zurich, Paris, Munich and Berlin), they often link the central areas or main stations of major cities (sometimes running on the urban metro network) with the localities in the metropolitan area, even up to 40 km away.

The main features of the metro rail infrastructure and service are as follows:

- Timetable adapted to passenger flows, with multimodal correlation.
- Uses existing rail infrastructure or dedicated suburban lines.
- Has single platforms, on the current line, linked to each other or to complementary transport networks by under- or over-crossings.
- The stations or stops are unstaffed (ticket purchase or passenger information is done through integrated computer systems).
- The journey (commute) generally takes less than one hour, with a commercial speed of 40–60 km/h.

For Romania, according to the analyses carried out by the competent authorities, the main cities in which the metropolitan train can be implemented, using the existing railway infrastructure or newly created connections, are as follows: Bucharest, Cluj-Napoca, Brasov, Iasi, Timisoara, Sibiu and Constanta.

In this context, the necessary steps to implement the metropolitan train concept are as follows (RG, 2022):

1 Introducing the metropolitan train concept in Romania's transportation strategy in order to provide the necessary strategic framework for financing from non-reimbursable European funds.
2 The creation of a complex database to provide the framework for the application of the criteria, carried out in the context of the realisation of this Programme.
3 Prioritisation of the metropolitan train projects for all the county municipalities using a complex multicriteria analysis, within the framework of this Programme.
4 Providing support to the beneficiaries for the implementation of the metropolitan train concept in Romania.
5 Implementation of metropolitan train projects, realisable in two stages:

    a Implementation of projects on existing railway routes that foresee works of low complexity with short implementation times, not significant modernisations of the railway line. This may include rail renewals, Quick Wins projects, modernisation of existing platforms and stations, ensuring accessibility for persons with reduced mobility in station areas and vehicles, non-discriminatory access to points of economic, social and cultural interest, environmental protection, increased safety and security for passengers and residents in the railway area, etc. Also, in this phase, the purchase of rolling stock will be eligible.
    b Phase II: implementation of projects with significant interventions (modernisation, electrification, capital repair, reconstruction of disused lines and construction of new lines to provide a balanced service for the whole territory concerned) + purchase of rolling stock.

In relation to *naval transport*, the national infrastructure is dominated by the Danube River, with a length of 1,075 km on Romanian territory and bordering the Black Sea, along a 244 km coastline.

The length of the inland waterways in Romania totals 2,763 km, of which, as already mentioned, the Danube River is 1,075 km, the secondary branches of the Danube (524 km), the Danube-Black Sea Canal (64 km), the Poarta Albă–Midia Năvodari Canal (28 km) and the Bega Canal (40 km). Romania's inland waterway system is concentrated on the Danube River, the secondary navigable branches of the Danube and the two canals.

The main problem of transport on the Danube is the lack of predictability caused by shallow depths (less than 2.5 m) at certain times of the year. For this reason, especially in international logistics, transportation on the lower Danube is avoided.

With a time horizon of 2030, the Romanian authorities have set the following objectives to improve shipping (RG, 2022):

- Ensuring minimum navigation depths on the Danube. It is envisaged to ensure the reliability of waterways, in particular the Danube and its navigable channels, by ensuring navigation depths of 2.5 m on the Danube all year round, as well as strengthening/defending the banks and modernising the lock equipment.
- Development and modernisation of navigable channels. The main targets are the completion of works on the Danube–Bucharest Canal and (pre)feasibility studies for new sections of waterways, in particular on the Prut and Siret rivers.
- Development and modernisation of the Port of Constanta in order to reach new cargo flows and increase its competitiveness
- Creating a primary network of Romania's river ports and focusing investments for its development. The river ports will be systematised in such a way as to create a primary network of ports whose hinterland corresponds to an economic potential, as well as a secondary network of ports that ensure the accessibility of specialised goods. The ports on the primary network are as follows: Constanța, Sulina, Tulcea, Galați, Brăila, Cernavodă, Călărași, Oltenița, Giurgiu, Corabia, Calafat, Dr. Tr. Severin, Orşova, and Moldova Nouă. The ports in the secondary network are as follows: Bechet, Turnu Măgurele, Zimnicea, Fetești, Medgidia, Basarabi, Ovidiu, Luminița, Măcin, Hârşova, Isaccea, Mahmudia, and Chilia Veche.

It should be pointed out that the Romanian authorities do not currently have a strategic plan for the development/modernisation of Romania's naval infrastructure. The realisation of such a document is foreseen for the future, and until then, for the ports of the primary network where no individual investment preparation projects have been launched, the volume of investments for the next 10 years will be limited to €100 million, and for the ports of the secondary network in the same situation, the volume of investments will be limited to €50 million. This adds up to €150 million for the next 10 years (unless the sectoral strategy with other proposals intervenes in the meantime), which corresponds to an annual budget of €15 million, compared to half the annual budget of a municipality on the Danube bank, Turnu-Severin, which in 2023 had a budget of €30 million (Screciu, 2023).

Romania has no modern waterways. In the last 33 years, it has failed to dredge and maintain a navigable channel along the Danube for sea vessels, while the port of Constanţa, which could have been the most important in the region and which could act as a main connection for the traffic of goods coming from China, as well as for the north–south transport routes, is still in urgent need of modernisation.

For this reason, the crisis in Ukraine has, as usual, caught us unprepared to cope with the enormous flow of goods – in reality, a relatively modest flow compared to the volumes that could be transported if the eastern route desired by the Chinese were to pass through.

In this endeavour, in 2022, China opened a rail freight connection – the China Railway Express – linking Wuhan with the port of Constanta and the rest of Europe (Mihălţianu, 2023).

As the war in Ukraine continues, China may be even more interested in having an alternative route to the existing ones through Russia and Turkey, activating both the Ukrainian and Romanian segments once the military conflagration ceases.

In this global context, major international projects for cooperation on the development of sustainable and economically and militarily efficient transport infrastructures should not be neglected.

This is the case of the 'Three Seas' initiative. After the launch of the war against Ukraine by Russia, the geopolitical importance of intensified cooperation and better connectivity in the North–South axis corridor has increased dramatically. The Three Seas Initiative (TSI3M) serves as a political platform to facilitate this cooperation to create closer ties in the fields of transport, energy, military mobility and digitalisation within and beyond the three seas regions.

The major objectives of the initiative at the moment are as follows:

- Further development of critical infrastructure projects on the North–South axis in the areas of transportation, energy and digital networks.

  Against this objective, it is envisaged to facilitate strategic infrastructure interconnection and investment projects in the region on the North–South axis, thus realising the 3SI objectives of developing the economies of the participating countries, promoting convergence with the rest of the EU and the transatlantic partnership.
- Attracting private investment and developing public–private partnerships.

  The current reality of the challenges and disruptions in the Three Seas region calls for continued efforts to attract much-needed private investment and support from international financial institutions at the appropriate scale required by the differences in infrastructure development between the countries of the Three Seas region and the rest of the continent.

For the development of the naval infrastructure, at the end of 2021, the Romanian authorities approved the 'Investment Program for the Development of Transport Infrastructure for the period 2021–2030', in which €4.46 billion is allocated to the naval infrastructure.

Regarding *air infrastructure*, Romania has 27 airfields and 6 operational heliports. Of these, 9 are major international aerodromes (airports), 7 are secondary international aerodromes (airports), 11 are national aerodromes (airports), and 6 are national heliports.

The investment programme in the field of air infrastructure, as in the naval sector, is built at a minimal level for the period 2021–2030, with a view to carrying out specialised studies for the development/modernisation of each air terminal.

Until then, airports have been divided into three categories: primary, secondary and tertiary.

The primary network comprises airports of Constanţa, Iaşi, Bucharest, Craiova, Sibiu, Cluj-Napoca and Timişoara.

The primary network airports are required to develop cargo and passenger terminals for domestic and international traffic. Airports in the primary network will need to have intermodal rail and road infrastructure links.

The secondary network consists of regional and small regional airports: Tuzla, Tulcea, Bacău, Suceava, Brasov, Tg. Mureş, Caransebeş, Baia Mare, Satu Mare, Oradea and Arad. Secondary network airports are required to develop passenger terminals for domestic and international traffic and may develop cargo terminals if socio-economic analyses support their economic viability. Secondary network airports must have an intermodal link.

The tertiary network shall consist of domestic aerodromes dedicated to domestic traffic, in particular general aviation, school, sport and leisure activities, which are not located in the localities served by an operational airport. It is composed of the aerodromes of Ploieşti–Strejnic, Piteşti, Bistriţa and Deva.

Until the development plans for each aerodrome are finalised, at the airports of the primary network where individual investment preparation projects have not been launched, the volume of investments for the next 10 years will be limited to €200 million, except for the Bucharest airports. For airports on the secondary network in the same situation, the investment volume will be limited to €100 million. For aerodromes on the tertiary network, the volume of investment is limited to €25 million.

Consequently, except for Bucharest airports, for the time being, the Romanian authorities have foreseen a total investment for the modernisation/development of Romanian aerodromes of up to €325 million, that is, €32.5 million per year.

Summing up, all investment projects in the field of transport infrastructure in Romania, according to the projections of the authorities in the field, have been allocated, for the period 2021–2030, the amount of €75.43 billion.

Some brief conclusions:

- For the coming years, the Romanian authorities aim, as a priority, to develop/modernise the road infrastructure at the detriment of other modes.
- The differences in investment values between road infrastructure and naval or aerial infrastructure (except for the Bucharest airports and subject to the further elaboration of strategies on these two modes, which may modify these values) are so large that they do not require any further comment.
- It is worth noting that the total investment in rail infrastructure has been quite significant, about 40% of the total for the entire sector.
- In relation to Romania's GDP, the annual investment allocation for the modernisation/development of transport infrastructure represents about 2.6%.[8]

## Notes

1 In calculating the infrastructure competitiveness score, 106 specific indicators were used, of which 20 were for basic infrastructure (such as surface area, population, length of roads, railways, shipping, energy infrastructure, etc), 17 specific to technological infrastructure (such as investment in telecommunications, number of internet users, cyber security, etc), 20 for scientific infrastructure (such as research spending, number of innovations, patents, etc), 28 for health and environment and 19 for education (author's elaborations after WCC (2023)).

2 As of September 2024, the world population has reached 8.178 billion (Worldmeters, 2024).

3 According to the British publication 'INews', more and more drivers are switching from electric cars. It is predicted that more than 56% of electric vehicle owners will opt for an alternative fuel type, while gasoline will dominate the market with 30%. There is also great 'turbulence' in the electric vehicle market, making it unattractive for the future. A study by Auto Trader shows the value of used electric vehicles falling 23% in 2023. Furthermore, in February 2024, it was reported that Apple had dropped its plans to enter the electric vehicle market. Moreover, Tesla's sales and shares were declining. The value of used electric vehicles was also down. By comparison, that of vehicles with internal combustion engines remained stable.

4 As of September 2024, 1,112.1 km of highways were in operation in Romania (130km.ro, 2024a).

5 In terms of vehicle safety, millimetre-wave radar plates are widely used in automotive automatic braking systems, which can avoid traffic accidents by detecting the surrounding environment in real time.

6 As we present in the current study, *plastic roads* are a paradox. Plastic is not an environmentally friendly element but recycling it and reusing it in the mixtures used for road surfacing or even building roads from recycled and recyclable plastic, is, in our opinion, a sustainable approach, taking into account the fourth element of unsustainability, namely resilience.

7 By Infrastructure Connective System we mean a dual infrastructure–vehicle system that simultaneously receives and transmits data and information in real time between the two receivers (infrastructure and vehicle).

8 Romania's GDP for the year 2022 was €286.4 billion (Gheorghe, 2023).

## References


130km.ro, *Lista proiectelor de autostrazi si drumuri expres din Romania* (List of motorway and expressway projects in Romania), 2024a, https://www.130km.ro/autostrazi.html

130km.ro, *Review highways and highway projects*, 2024b, https://www.130km.ro/calendar.html

130km.ro, *Opening calendar for highway projects*, 2024c, https://www.130km.ro/calendar.html

Adler M. et al., '*Roads, market access and regional economic development*', OECD Regional Development Working Papers, OECD Publishing, Paris, 2020, https://www.oecd.org/en/publications/roads-market-access-and-regional-economic-development_8b9eca17-en.html

Baum-Snow N. et al., *Highways, market access, and urban growth in China*, 2016, https://www.theigc.org/sites/default/files/2016/08/Baum-Snow-et-al-2016-Working-paper.pdf

Civil Dream, *Plastic roads*, 2023, https://dreamcivil.com/plastic-roads/

CNSF (National Railways Supervisory Council), *Study on the development of rail transport in line with the implementation of the Rail Infrastructure Development Strategy 2021–2025*, 2023, https://www.consiliulconcurentei.ro/wp-content/uploads/2023/11/Studiu-evolutie-transport-feroviar.pdf

Cozmei V., *Interactive road map of Romanian highways*, 2023, https://monitorizari.hotnews.ro/stiri-infrastructura_articole-26771562-harta-interactiva-autostrazilor-din-romania-deschis-inaugurat-2024-scenariul-optimist-realist-blocaje-majore-cum-stam-comparativ-bulgaria-ungaria-sau-polonia.htm

CREEAZA, *Tests on asphalt mixtures and bituminous mixes*, 2024, https://www.creeaza.com/tehnologie/constructii/IcircNCERCARI-ASUPRA-MIXTURILO182.php

EC, *EU transport in figures – Statistical pocketbook 2023*, ISBN 978-92-68-68-06324-8, 2023

EC, *Delivering the European green deal*, 2024a, https://commission.europa.eu/strategy-and-policy/priorities-2019-2024/european-green-deal/delivering-european-green-deal_en

EC, *EU transport in figures-2024*, ISBN 978-92-68-19812-4, 2024b, https://transport.ec.europa.eu/document/download/ee264fc5-ec49-4751-9d92-08c038856ce1_en?filename=MI-AA-24-001-EN-N.pdf

Elton Ch., *Europe spent more on roads than rail in the last 25 years*, 2023, https://www.euronews.com/green/2023/09/19/europe-spent-more-on-roads-than-rail-in-the-last-25-years-these-10-countries-bucked-the-tr

Evan M., Designing sustainable transportation infrastructure, 2023, https://www.arup.com/perspectives/how-a-layers-based-approach-can-help-design-more-sustainable-transport-infrastructure

Flinders M., *How to build more sustainable transportation infrastructure*, 2023, https://www.ibm.com/blog/transportation-infrastructure/

Gheorghe L., *Romania in the EU – Key data: Top growth based on imbalances*, 2023, https://cursdeguvernare.ro/romania-in-ue-date-esentiale-crestere-de-top-bazata-pe-dezechilibre.html

Guţă D., *The curse of the most spectacular railway viaduct*, 2014, https://adevarul.ro/stiri-locale/hunedoara/blestemul-celui-mai-spectaculos-viaduct-feroviar-2329962.html

Hall J. W., Tran M., Hickford A. J. and Nicholls R. J., *The future of national infrastructure*. Edited by Jim W. Hall, Martino Tran, Adrian J. Hickford, and Robert J. Nicholls. Cambridge University Press: Cambridge, 2016, https://www.cambridge.org/core/books/future-of-national-infrastructure/7D4DF0295A9D8A7304E6C87204BAA0EA

IBRD (International Bank for Reconstruction and Development), *IDBG framework for planning, preparing, and financing sustainable infrastructure projects: IDB sustainable infrastructure platform*, 2018, https://publications.iadb.org/en/idbg-framework-planning-preparing-and-financing-sustainable-infrastructure-projects-idb-sustainable

ICONS, *Moving towards a more sustainable transport infrastructure with the LIAISON project*, 2023, https://www.icons.it/news/moving-towards-a-more-sustainable-transport-infrastructure-with-the-liaison-project/?cn-reloaded=1

ISI (Institute for Sustainable Infrastructure), *ENVISION – Sustainable Infrastructure Network. Version 3*, Washington DC, USA, ISBN 978-1-7322147147-0-5, 2018, https://sustainableinfrastructure.org/envision/envision-v3-release

Liu C., Du Y., Ge Y., Wu D., Zhao C. and Li Y., *New generation of smart highway: Framework and insights*, 2021, https://www.hindawi.com/journals/jat/2021/9445070/?utm_source=google&utm_medium=cpc&utm_campaign=HDW_MRKT_GBL_SUB_ADWO_PAI_DYNA_SPEC_X_X0000_Nov2023Commissioned&gad_source=1&gclid=Cj0KCQCQiA5rGuBhCnARIsAN11vgT9OZGL2IsiNd36T4VqCPuLRd28uydjjr6uNGWWf02aNRpDXlVdjyCmoaAjO7EALw_wcB

Mihălțianu A., *Romania's infrastructure: why we don't have modern highways and railways, but also why Romania in 2030 will be very different*, 2023, https://pressone.ro/infrastructura-romaniei-de-ce-nu-avem-autostrazi-si-cai-ferate-moderne-dar-si-de-ce-romania-din-2030-va-fi-mult-diferita

Milani L., Mohr D. and Sandri N., *Built to last: Making sustainability a priority in transport infrastructure*, 2021, https://www.mckinsey.com/industries/travel-logistics-and-infrastructure/our-insights/built-to-last-making-sustainability-a-priority-in-transport-infrastructure

OECD, *Cost-benefit analysis and the environment: Recent developments – Executive summary*. 15-27, 2006, http://www.oecd.org/env/tools-evaluation/cost-benefitanalysisandtheenvironmentrecentdevelopments2006.htm

OECD, *Transport infrastructure trends and regional development*, 2024, https://www.oecd-ilibrary.org/sites/512e786f-en/index.html?itemId=/content/component/512e786f-en

Patil P., Stankevich N., Tsydenova N. and Diana Z., *Plastic waste in road construction: A path worth paving?*, 2023, https://documents1.worldbank.org/curated/en/099003105242322144/pdf/IDU0743eda4a0091c049000acd506d2f0b9ae123.pdf

Pavement Interactive, *Rutting*, 2024, https://pavementinteractive.org/reference-desk/pavement-management/pavement-distresses/rutting/

ProtvNews, *Şoseaua de 158 de km asfaltată doar de roboți și drone. Niciun om nu a muncit pe şantier* (The 158 km road paved only by robots and drones. No humans worked on the construction site), 2024, https://stirileprotv.ro/divers/soseaua-de-158-de-km-asfaltata-doar-de-roboti-si-drone-niciun-om-nu-a-muncit-pe-santier-video.html?utm_source=referrer&utm_medium=cross_extern&utm_campaign=Hotnews

RG (Romanian Government), *Investment program for the development of transport infrastructure for 2021-2030*, 2022, https://drive.google.com/file/d/1aNUgLyKwC5xeyQrJVgeo9usSCyXKaQ08/view

Screciu M., *Bugetul Severinului a fost aprobat în şedința de Consiliu Local* (Severin's budget was approved in the Local Council meeting), 2023, https://restartnews.ro/2023/01/31/bugetul-severinului-a-fost-aprobat-in-sedinta-de-consiliu-local/

Sheate, W. R., The evolving nature of environmental assessment and management: Linking tools to help deliver sustainability. In *Tools, techniques and approaches to sustainability*, 2010, https://www.worldscientific.com/worldscibooks/10.1142/7519

Singh S., *What is plastic road*, 2019, https://www.civilengineeringweb.com/2019/10/what-is-plastic-road-concept-objective-advantages-disadvantages.html

The Guardian, *The man who paves India's roads with old plastic*, 2018, https://www.theguardian.com/world/2018/jul/09/the-man-who-paves-indias-roads-with-old-plastic

Waren C., *10 Ideas for a green transportation Infrastructure*, 2023, https://science.howstuffworks.com/environmental/green-science/10-ideas-green-transportation.htm
WCC (World Competitiveness Centre), *IMD world competitiveness booklet 2023*, 2023, https://imd.cld.bz/IMD-World-Competitiveness-Booklet-2023/50
World Bank, *Strategic environmental assessment in policy and sector reform*, 2011, http://siteresources.worldbank.org/ENVIRONMENT/Resources/244380-1236266590146/Policy_SEA_WB.pdf
Worldmeters, World population, 2024, https://www.worldometers.info/en/

# 3 Transport activity from a sustainable development perspective

The concept of sustainable mobility (Rueda, 2010) has been developed in response to the economic, social and environmental issues associated with the extensive use of cars. In increasingly congested urban spaces, which continue to attract more and more people, the distribution of space between roads, parking lots, pedestrians and cyclists is disproportionate, as is the modal distribution of transport, with road transport becoming the most attractive to the detriment of other travel options.

Sustainable mobility objectives include (Mattauch and Leonard, 2012):

- Achieving a high level of service quality.
- Ensuring accessibility to/from centres of interest.
- Achieving high levels of transportation safety and security.
- Reducing chemical (including greenhouse) and noise pollution.
- Reducing energy consumption.

## 3.1 External costs as a fundamental element of transport unsustainability

### *3.1.1 General*

It is well known that all economic activity comes at a certain cost. In a broad sense, this cost can be called the total social cost.

In the case of transport, part of the total social cost is fully compensated for by individual expenditure (vehicle depreciation, fuel, etc.) or in the form of collective expenditure (infrastructure maintenance, insurance, etc.). However, a significant proportion is only partially compensated or not compensated at all (air pollution, noise, etc.). This difference is due to external costs, which have particular aspects in the case of transport (chemical and noise pollution of the environment, accidents, etc.). The existence of such costs can be highlighted by distinguishing marginal individual costs from social costs.

The difference between social and individual costs defines the marginal external cost or *externality of the* activity.

DOI: 10.4324/9781003743064-4

Most of the problems arising in the transportation system are the effects of market failures.

In this context, a major issue is the failure to take into account or even underestimate the costs of externalities and not to measure them correctly in transportation prices. This leads to inefficiencies, loss of natural and economic resources and even reduced social welfare.

An economic agent's activity and its efficiency depend on a multitude of internal and external factors. Externalities arise when an economic agent's activity does not depend exclusively on the variables under its control, but also on other variables outside the scope of its decision-making activity.

Thus, externalities arise because the producer (consumer) is unable to control all the variables in the production (consumption) function. The variables that generate externalities may be the subject of a private (individual) or public (collective) activity.

However, the effect of the externality is dependent on the specific behaviour adopted by the receiving agent. At the same time, the manifestation of the externality presupposes a transmission medium, that is, a support (physical or otherwise) for it.

Externalities generate interdependencies between those who produce them and those who bear their effects. The identification of interdependencies between producers, between consumers and between the two must take into account the conditions of expression – of manifestation of the behaviour at the origin of the externality.

External costs, also known as negative externalities, arise when the social or economic activities of one person (or group of persons) have an impact on another person (or group of persons) and when this impact is not fully paid for or compensated for by those who have caused it.

For transport, these types of costs are generally not borne by the transport user and are therefore not taken into account when a transport decision is made. Vehicles with high noxious emissions, for example, cause damage to human health, imposing an external cost. This can be explained in a simplified way by the fact that the impact on those who suffer damage to their health is not taken into account by the driver when deciding to use the vehicle (EC Delft, 2019).

Transport-specific negative externalities are costs imposed on society which, without policy intervention, are not taken into account by transport users. Thus, transport operators benefit from indirect subsidies that are economically and morally incorrect and that greatly unbalance the relationship between transport supply and demand, leading to welfare losses for society as a whole. It is therefore important to distinguish between (EC Delft, 2008):

- Social costs, which reflect the totality of costs generated by transport, among which we mention those specific to the provision and use of specific infrastructure, capital costs, costs of sustaining the activity (wages, insurance, etc.), accident costs and environmental costs.

- Private (or internal) costs, which are borne directly by the transport user, such as the cost of depreciation of vehicles, the cost of fuel consumed by the transport activity, overheads (salaries, per diems, etc.), transport fares and taxes, etc.

In this context, external costs result from the difference between social and private costs. Based on the economic welfare theory, transport users should pay all the marginal social costs that arise as a result of a transport activity. Taking into account marginal private costs (such as the costs of damage and destruction of the vehicle and the personal costs of the driver), optimal infrastructure charges should reflect the marginal external costs of using an infrastructure. These costs include infrastructure wear and tear costs, congestion costs, accident costs and environmental costs. However, only some of these costs are monetarily relevant.

In the short term, these costs are related to a constant infrastructure capacity. Thus, fixed infrastructure costs are not relevant for efficient pricing. In the long term, however, the change in infrastructure capacity due to the construction of additional infrastructure for the same traffic segment becomes relevant.

From an economic point of view, an infrastructure project is economically viable if the additional social benefits of the project exceed the additional social costs (EC Delft, 2008).

As the market does not provide an incentive for hauliers to take external costs into account, they only measure part of the total social costs when making a transport decision, which leads to sub-optimal (optimal, for the specific market equilibrium) outcomes.

By internalising these costs, negative externalities can be included in the decision-making of transport users. This can be done through regulation (e.g. command and control measures) or by providing appropriate incentives to transport users, that is, fiscal instruments (e.g. exemption from certain taxes and charges) or economic instruments, for example, trading of emission certificates. A combination of these two basic types of instruments is possible, for example, the introduction of mandatory annual taxes differentiated by vehicle emission classes.

The use of economic instruments to internalise external costs is generally considered to be an effective way to limit the negative side-effects of transport and/or to generate revenues for the government. Effective application of these instruments, however, requires detailed and accurate estimates of the levels of external costs.

This will, however, be presented in detail in future studies.

### *3.1.2 Types of external costs specific to transport*

As stated above, transport produces internal and external social costs – costs that are felt throughout society.

Externalities caused by transport arise when a user (transport operator) does not bear the full costs of the activity or does not benefit from the full benefits.

The distinction between internal and external costs depends on the nature of the specific actions. Thus, if a user pays for the use of a resource (infrastructure, fuel, etc.), the costs associated with this action are considered to be internal. Conversely, if the user influences the environment of other individuals without paying for this action (e.g. air pollution), the costs are considered to be external.

Table 3.1 presents a classification of internal and external social costs according to the nature of the specific transport expenditure. Of the many external costs specific to transportation, the most important are those associated with:

- Chemical air pollution (including effects on the ozone layer).
- Noise pollution.
- Congestion.
- Accidents.

In a market economy, economic decisions depend to a large extent on changes in price levels. If they do not also reflect existing shortcomings (lack of clean air, low natural absorption capacity of the environment in the case of air pollutants, etc.), individual decisions by producers and consumers will not be aimed at maximising benefits for society as a whole.

This underlines the need to pay the full social costs, both to make the transport system more efficient and to spread the effects of the activity fairly in the environment in which it takes place.

*Table 3.1* Classification of Social Costs of Transport

| *Nature of Expenditure on Transport* | *Social Costs* | |
|---|---|---|
| | *Internal (Individual)* | *External* |
| Own | Costs of fuel, vehicles, etc. | Costs borne by the company (e.g. free parking). |
| Infrastructure specific | Taxes on vehicles, fuel, etc. | Costs not borne by the company |
| Specific to accidents | Insurance, medical costs, etc. | Costs not covered (suffering, pain imposed on other individuals, etc.) |
| Specific to environmental pollution | Cost of personal injury to users | Costs not covered (air pollution, noise imposed on other individuals, etc.) |
| Specific to agglomerations | Costs related to the value of time | Costs of time losses imposed on other individuals |

The importance of transport-specific externalities varies considerably with the mode of transport under investigation, as well as with the time and place of investigation. Therefore, some caution should be exercised in generalising from case to case.

Table 3.2 highlights some of the main links between the causes and effects of transport-specific external costs.

In brief, the definition of external costs, specific to transportation, highlights some structural elements and particularities, among which we mention (UIC, 2012):

- *Accident costs*, which include values for material damage, administrative and medical costs, economic losses (mainly production losses) and immaterial costs, which are difficult to measure from a financial point of view, such as reduced life expectancy, suffering, pain, etc. Summed together, material and immaterial costs constitute the total costs of accidents caused by transport.
- *Traffic congestion*, which affects transport activity by increasing travel time, reducing reliability and increasing operating costs.
- *Air pollution* generated by transportation, which naturally increases the incidence of cardiovascular and respiratory diseases. It also has an effect on buildings and causes material damage by damaging crops or livestock. The most important air pollutants are particulate matter (PM10 and PM2.5), nitrogen dioxide ($NO_2$), sulphur dioxide ($SO_2$), volatile organic compounds (VOCs) and ozone ($O_3$). The total volume of greenhouse gas emissions (such as carbon dioxide – $CO_2$, or nitrous oxide – NO) from the transport sector is second only to the energy sector, with transport being the only sector to have seen an increase in this indicator since 1990.
- *Noise* caused by transport activities affects people both physically and mentally and produces adverse health effects. For example, at levels above 85 dB(A), it can cause hearing damage and significantly increase the risk of cardiovascular diseases, while at levels above 60 dB(A), it can cause increased blood pressure, hormonal changes and impaired sleep quality.
- Transportation causes *pollution of soil and water* through emissions of heavy metals and polycyclic aromatic hydrocarbons, and can reduce soil fertility, pollute drinking water and damage wildlife habitat.
- *Impacts on nature and landscapes*, including ecosystem damage due to interference by transport infrastructure, drastic changes in landscape appearance and diminution/loss of soil ecological functions such as rainfall absorption, biomass production and $CO_2$ storage.
- *The reduction in biodiversity* as a result of transport activities is also increasingly significant.
- *Additional impacts in urban areas are* felt mainly by the resident population and non-motorised traffic participants such as pedestrians and cyclists. Some examples in this respect are the increased travel time for pedestrians

*Table 3.2* Cause–effect Relationships for Transport-specific External Costs

| *Cause Cost* | *Cost Elements* | *Common Cost Elements* | *Main Cost Causers* | *External – Not Paid by the User* |
|---|---|---|---|---|
| Traffic congestion | Wasted time and missed opportunities. Transportation safety and environmental surcharges | Travel speed variations. Value of economically relevant lost time | Type of infrastructure. Traffic and throughput of a flow depending on: time, location, congestion and inter-infrastructure maintenance | Additional waiting times or lost opportunities caused by some users |
| Accident | Medical intervention or assistance, loss of life and economic losses | Statistical assessment of human life | Type of infrastructure, traffic volume, speed of vehicle and skills of vehicle driver | Accident costs not covered by insurance |
| Noise | Devaluation of nearby properties and reduced health | Estimating the costs of pollution by analysing property values and measuring willingness to pay | Density of urban agglomerations, traffic, day/night traffic intensity, emission levels, type and quality of infrastructure and vehicles | Almost entirely uncovered |
| Air pollution (particulate matter and noxious) | Health deterioration, years of life lost, real estate degradation, environmental degradation and biosphere degradation | Estimated years of life lost, property degradation, and long-term risks to biosphere | Density of urban agglomerations, characteristics of the area, level of vehicle emissions, type of infrastructure and travel speed | Almost entirely uncovered |
| Climate change | Climate change prevention. The cost of the rising global average temperature | Risk of long-term climate change. Acceleration and exacerbation of effects caused by climate change | Emission level depends on: vehicle type, speed, driving style and fuel type | Partially covered by fuel excise duties |
| 'Other' (soil and water contamination, degradation of natural environment) | Ensuring soil and water quality. Restoring and protecting biodiversity | Assessing the cost of countermeasures | Level of emissions. Area characteristics. Type of infrastructure | Not covered |

due to discontinuity of pedestrian routes caused by rail/road infrastructure, and the reduced size/non-existence of special space for cycling (electric scooters, more recently).

- Transport activities also produce *indirect negative effects* through downstream and upstream processes, such as emissions due to the production and distribution of fuel and electricity, as well as the production, maintenance and disposal/recycling of vehicles/components and infrastructure.

On the basis of the studies carried out so far, the most important externalities are congestion, accidents, chemical (especially air) and noise pollution. However, the importance of the specific costs of infrastructure damage should not be minimised.

It should also be emphasised, however, that transport is also a 'source' of positive externalities, the most important of which are stimulating/supporting other types of economic activity (production, trade, etc.), ensuring the mobility of goods and people between different locations, and making a major contribution to the general process of economic growth.

However, the values of external costs and benefits do not cancel each other out. On the basis of this principle, trends towards more efficient activities can be highlighted. Thus, in transport, and especially in road transport, an increase in economic efficiency can also be achieved by reducing the values of activity-specific characteristics (e.g. travel distance).

However, in the presence of negative externalities, there may be a 'market failure', in the sense of a reduction in the intensity of those activities, which result in an increase in the benefits that can be felt by society as a whole. This is why hauliers tend to continue their activity up to the point where the net marginal benefit over the last mile travelled is zero (the phenomenon of last-mile indifference). For this reason, it is possible that those affected by the negative externalities caused by transport may be willing to pay more to reduce these externalities than the hauliers themselves.

There is a view that transport activity can lead to external benefits that are greater than or at least equal to the negative externalities. To verify the correctness of this claim, however, a more detailed analysis is needed.

As stated above, transport has always made a major contribution to the development of society. At the same time, this activity has helped to diversify the range of goods on offer to communities, in many cases lowering their prices. It is no less true that transport activity has generated the development of regions and has been the cause of many jobs.

But do all these small examples represent external benefits compared to transport activity? In other words, if there are no external costs of concern, is it beneficial to 'indirectly subsidize' all transport activities in order to compensate for the alleged external benefits?

There is a belief that benefits due to infrastructure, for example, are generally confused with those of the transport activity itself.

The benefits of infrastructure can be measured in terms of reduced social costs as opposed to individual costs. In other words, these benefits are externalities because they arise 'outside the activity'. Even taking these effects into account, however, excessive use of infrastructure is still not justified.

At the same time, there are certain benefits felt by society (a fall in the value of certain goods) that can be obtained from transport activity, combined with the fact that its costs are low (not influenced by the value of externalities).

These lower delivery costs, for example, are financial benefits leading to an economic surplus. They are not, however, external benefits as long as they relate exclusively to those who pay the cost of transportation, irrespective of the stage of the economic processes. External benefits arise when individuals travel not for economic reasons but simply for pleasure.

The key issue that remains is to compare the level of positive externalities (benefits) with negative externalities (costs). At the same time, it is necessary to highlight those economic measures that are necessary to eliminate marginal external costs. For example, to this end, road hauliers pay numerous taxes, including fuel, motor vehicles, value-added tax (VAT), etc. However, specific effective taxes to offset the negative external effects of transport activity are not among the listed fiscal instruments. Therefore, environmental taxes aim at increasing economic efficiency and filling the gap between individual and social costs, while most increases in income taxes result in negative effects on society that are much larger than these increases (Maddison et al., 1996).

The effect, on the economy in general, of income tax increases is similar to that of a sack left floating after a millstone has been tied to it. For this reason, the state always seeks to minimise these trends. To maintain economic equilibrium, there is the possibility of increasing revenue by raising taxes on some goods while lowering taxes on others. For example, raising the level of taxes on transportation activity can only be done as long as it is environmentally justified. Moreover, economists do not agree with imposing taxes in order to increase overall revenue, as this is completely at odds with the economic efficiency to be gained in the transport activity.

From an economic perspective, the most important types of external costs specific to transport are as follows:

- *Total external costs* (expressed in monetary units – billions or millions of euros in the case of European studies), which refer to the total external costs in a given geographical area caused by the plenitude of transport modes.
- *Average external costs*, which are closely linked to total costs, are expressed as the ratio between total costs and a specific performance indicator (freight journey, passenger journey, etc.). At the European level, the values of these costs are generally expressed in euro cents/passengers × km (pkm), euro cents/tonnes × km (tkm) and/or euro cents/vehicles × km (vkm).
- *Marginal external costs* are defined as additional external costs caused by an additional transport activity. In the short run, these costs are related to a constant infrastructure capacity, while long-run marginal costs have to take

into account the construction of new infrastructure capacity. This means that, for example, short-run marginal costs caused by congestion are generally higher than long-run marginal costs caused by the same congestion phenomenon. In general, marginal external costs are expressed in the same units as average external costs. Since short-run marginal cost values are more relevant to the process of internalisation of external costs caused by transport, they appear most frequently in the latest global studies.

After a significant period of analysis, it has been accepted at the international level that the following components should be included in the category of external costs caused by transport (EAMA (European Automobile Manufacturers' Association), 2008):

- Congestion costs (loss of time and vehicle operating costs).
- Accident costs.
- Chemical air pollution costs.
- Costs of noise pollution.
- Costs associated with climate change (stratospheric chemical pollution).
- Costs of damage to nature and landscapes.
- Costs of soil and water pollution.
- Costs of downstream and upstream processes (related or collateral, such as the cost of pollution at the source producing the energy needed to run electric vehicles, etc).
- Additional costs in urban areas.
- Energy dependency costs.

At the same time, as already pointed out, there is a controversy as to the size and inclusion in the category of external costs of all costs caused by transport accidents. Often, in the case of accident costs, only those costs that are not covered by third parties (e.g. through insurance) are considered as external costs

Accident costs mainly include the following categories:

- Direct costs.
- Indirect costs (police, insurance and administration).
- Technical costs (loss of products, vehicles, etc).
- Humanitarian costs (pain, suffering, etc).

The total external costs obviously result from the sum of the individual cost components specified above.

Marginal social costs (MSC) are expressed as a change in the total external costs for all transport users when an additional user enters the system. They can be calculated mathematically by summing the total costs of all users in the system, or experimentally by direct estimates in the field or by using simulation models at the macro-level.

The marginal social cost components can be summarised by the following equation (Sansom et al., 2000):

$$MSC = MINFR + MOPC + MECC + MEACC + MEEC \tag{3.1}$$

where

MINFR = marginal cost of infrastructure use
MOPC = marginal operating cost
MECC = marginal external cost due to congestion
MEACC = marginal external accident cost
MEEC = marginal external environmental cost

In summary, the major types and particularities of external costs caused by the different modes of transport are presented in Table 3.3.

Some conclusions on the perception and connection between the different modes of transportation and the negative externalities caused by them emerge from Table 3.3. Thus, congestion costs are more significant for road transport, especially in urban areas, while for scheduled transport (rail, air), additional waiting times for the use of a travel time segment should also be considered in the case of non-scheduled operators. Accident costs are treated similarly for all modes of transport. The costs related to chemical, noise and climate change pollution depend largely on the distance to the polluting source, the people exposed and the density of the areas concerned, together with the type of fuel used in transportation. Impacts on nature and landscape are only relevant for the construction of new transport networks.

For all transport modes, infrastructure costs are defined by considering capital (e.g. new investments or funds for asset replacement) and operating costs (maintenance and management).

Operating costs are defined as the sum of all costs paid by the operator for the provision of transport services and include vehicle-specific costs, personnel costs and administrative costs.

The costs caused by infrastructure undersupply arise because users cohabit in the same system and use, under competitive conditions, a limited infrastructure supply (Miola et al., 2008). In short, demand exceeds infrastructure supply. These costs are identified with those specific to situations caused by agglomeration (welfare loss) and scarcity costs (infrastructure cannot be utilised due to overcrowding).

Excess infrastructure demand, depending on the mode of transport, the type of users, the characteristics of the infrastructure, the duration of journeys and related activities (e.g. maintenance, restoration of infrastructure), can be the cause of several negative situations, such as increase in travel times (the most important component for estimating congestion-specific costs); increase in vehicle operating costs (especially in the case of commercial transport); reduction in vehicle reliability due to excessive vehicle use; and occurrence of a

*Table 3.3* Particular Features of Different Types of Mode-specific External Costs

| *Type External Cost* | *Transport Mode* | | | |
|---|---|---|---|---|
| | *Road* | *Rail* | *Air* | *Naval* |
| Infrastructure deterioration | Individual transport is the main cause of congestion, especially congestion and bottlenecks | Scheduled transport causes reduced access to infrastructure, increased transport times and associated financial losses for unplanned operators | Scheduled transport causes reduced access to infrastructure, increased transport times and associated financial losses for unplanned operators | If there is no specific timetable for the allocation of slots in ports/canals, congestion can be caused by any of the ship operators |
| Accidents | The level of externality is assessed for each individual accident (there are individual and/or collective risks), the insurance covers only a part of the compensation of the damage caused to the victims (less the statistical value of life – VSL) | There is a difference between operators and accident victims. The insurance compensates only certain components of the effects of accidents on victims (excluding VSV) | There is a difference between operators and victims of accidents. Insurance only compensates for certain components of the effects of accidents on victims (without taking into account VSV) | Not relevant |
| Chemical pollution | Roads and inhabited areas are connected or in the vicinity | Electric traction and hydrocarbon (diesel) combustion have to be analysed differently | For higher areas, air pollutants have to be considered | It is difficult to assess pollutants specific to shipping activity within the radius of ports (there are usually several modes of transport as sources of pollution in those areas) |

(*Continued*)

*Table 3.3* (Continued)

| *Type External Cost* | *Transport Mode* | | | |
|---|---|---|---|---|
| | *Road* | *Rail* | *Air* | *Naval* |
| Noise pollution | Roads and inhabited areas are connected or in the vicinity | Rail noise is usually perceived as less disturbing than other modes (rail bonus). But this depends on the time of day and the frequency of trains. | Airport noise is more complex to analyse compared to other modes. | Not relevant |
| Climate change | All relevant CFC pollutants | All relevant CFC-type pollutants, taking into account the use of petroleum fuels (diesel) or electricity | All relevant CFC-type pollutants (in particular, for higher areas) | All relevant CFC-type pollutants |
| Habitat and landscape degradation | Differentiation between existing and future infrastructure (under construction) | Differentiation between existing and future infrastructure (construction of a high-speed network) | Not relevant | New navigation channels to be opened are considered |

*Source:* Miola et al. (2008).

shortage of time segments in the case of infrastructure with regulated access (reduction in opportunity costs for service providers due to unavailability of desired departure or arrival times). Conversely, excess demand can also lead to beneficial situations, such as increasing the value of investment in the construction of new infrastructure capacity. The so-called Mohring effect can also be mentioned in this context. This effect is associated with the emergence of positive externalities due to the supply of improved or additional transport infrastructure, enjoyed by users who have already used the old infrastructure before its modernisation/development.

Depending on the type of infrastructure, the effects of congestion can be separated into:

- Agglomeration at road junctions, railway stations, ports and airports through the emergence of so-called 'vehicle queues'. In these cases, the level of additional costs for the use of infrastructure depends on the capacity and reaction/processing time of the facilities in that area (existence of traffic lights and their management, for example).
- Flow (transport vector) agglomeration, characterised by the overflow of transport infrastructure capacities between different locations, which can be easily evidenced using 'SPEEDFLOW' diagrams or simulation models for short routes, analysing the interdependencies/influences of the speeds of moving vehicles on the transport flow.

Accident costs are related to the possibility of transport users being injured (or even killed) in an accident. However, external accident costs are only those social costs related to transport accidents that are not covered by well-defined insurance premiums. Therefore, the level of external costs depends not only on the number and severity of accidents, but also on institutional factors (such as the existence of a compulsory insurance system), which affect the extent to which the party responsible for the accident is obliged and able to compensate the victims.

The most important components of accident costs are property damage, administrative (institutional) costs, medical costs (real health sector expenditures oriented towards the effects of accidents), production losses (current and prospective) and risk values (monetary values associated with the effects of accidents on those involved in these events – pain, suffering, panic, etc.).

In the case of fatalities, the economic (financial) dimension of this event is established by using the Statistical Value of Life (VSV) as a reference. This is defined as the value for which people agree to negotiate income to avoid the risk of premature death (Miola et al., 2008). The marginal value of the reduction in the risk of death is determined based on the level of the Payment Agreement (PA) accepted by individuals to avoid such a situation. VSV scaling is considered appropriate in the event of an accidental death, whereas its use has been criticised for other causes of mortality, such as air pollution, because the total number of deaths attributable to transport-related air pollution cannot be truly determined (Rabl, 2005), and pollution is a contributory, not a leading, cause of death. For these reasons, it is considered more appropriate to refer to loss of life expectancy (LOS) in premature death due to exposure to air pollution and to monetise this impact using the value of a life year (VLY). Both VSV and VLY are estimated by specific calculation methods and models that are not the subject of this study and will be detailed in future studies.

Air pollution costs include the impact of different air pollutants produced by vehicles through fuel combustion processes, or at the source of the electricity needed to run electric vehicles. Consequently, the costs of air pollution can be divided into health costs (resulting from pollution and noise), climate change, nature and landscape costs and land and water pollution costs (such as loss of biodiversity).

Air pollution costs are caused by emissions of certain air pollutants such as particulate matter (PM), NOx, $SO_2$ and VOCs, and include costs specific to reduced health, damage to buildings (constructions in general), crop losses and costs specific to additional damage to ecosystems (biosphere, soil, water). Health damage costs are the most important category of air pollution costs. The state of research on these costs is much more advanced than for the other components, mainly based on estimates from the ExternE model, funded by several EU research projects.

The specific (marginal) costs of noise pollution (noise) are defined as the additional noise costs caused by adding a vehicle to the existing traffic flow. Two types of negative effects of transportation noise can be distinguished:

- 'annoyance' costs, defined as the noise levels produced by transport that cause undesirable social disturbances and cause additional social and economic costs to arise through any kind of restriction in the performance of recreation, relaxation, recovery, rest or discomfort (e.g. stress).
- Health costs: transportation noise can cause major health damage, such as migraines, stress, depression, etc.

The specific costs of climate change have a high level of complexity due to the fact that they are long-term and global, and the risk patterns are very difficult to predict. They need to be taken into account for reasons of intergenerational equity, as ignoring them would be tantamount to shifting the burden of coping with climate change to the next generation. As the effects of climate change materialise in the long term and the adjustment path is still unclear, climate change is analysed using two scenarios, one that coincides with the status quo and one in which the effects of climate change occur. There are, however, difficulties in assessing in a unified way the damages to be allocated to the transport modes. Therefore, a differentiated approach (both damage and avoidance strategies) is needed.

Climate change or the impact of transport on global warming is mainly caused by greenhouse gas (GHG) emissions, such as carbon dioxide ($CO_2$), nitrous oxide ($N_2O$) and methane ($CH_4$). To a lesser extent, emissions of HFCs (hydrofluorocarbons) from cooling installations (air conditioning, refrigeration, etc.) also contribute to global warming. In the case of aviation, other aircraft emissions (such as water vapours, sulphates, soot aerosols and nitrogen oxides) at high altitudes also have an impact on global warming.

Finally, other types of impacts caused by transport-specific external costs, which are less addressed in expert analysis and research, are as follows: habitat fragmentation, environmental quality damage caused by the construction of transport infrastructure (modal and multimodal terminals, etc.) and soil quality degradation caused by road traffic, in particular through deposition of heavy metals and polycyclic aromatic hydrocarbons (PAHs). These pollutants can cause plant damage and soil fertility decline along transport infrastructure and even pose a threat to human and animal health.

## 3.2 Methodologies and tools needed to assess the external costs caused by transport

### *3.2.1 General concepts and particularities*

As emphasised above, in a free economy, the value of goods and services is determined in the specific market.

Any activity generally influences the characteristics of environmental factors to a greater or lesser extent. Unfortunately, for environmental factors and their characteristics, there is no specific market, which makes it rather difficult to determine the monetary value of some environmental externalities.

The allocation of resources for investment in environmental improvement should, as in other areas, be based on the assessment and comparison of costs and benefits.

For these analyses, methods are generally used which seek to collect all the information available on the market needed to characterise the externalities or to determine certain preferences of the individuals of the groups involved in the situations under investigation.

The cost of improving the environment is, of course, expressed in monetary terms, but it must be situated, in terms of level, around the value placed by the community (public value) on the material and human resources used.

Under such conditions, the comparison of costs and benefits presupposes that the latter must also be expressed in monetary terms; this is the only way of applying the principle of marginalist economic theory, of equivalence between marginal cost and marginal benefit, in order to determine the level of resource allocation and maximise the total net benefit (taking into account the restrictive nature of the availability of resources).

Reality shows, however, that not all benefits can be measured in a practical way, although even if this possibility existed, allocating resources in a way that maximises net benefits would not be possible. This is because a large number of variables have to be introduced into the model, including 'special' variables such as political choices, equity, national security, etc., which is why it is justifiable to say that society does not seek to maximise net benefits at any cost.

Therefore, the measurement of benefits can only be a means to an end whose objective (expressed in a slightly elliptical form) would be to validate the improvement in economic efficiency through the environmental projects to be implemented, compared with the previous situation. The concept of 'benefit' in this context, however, has a special meaning, conferred by the fundamental principle that the basis for measuring benefits must be what people want, that is, the preferences of individuals.

The ambition to monetise the effects of economic activities is a long-standing one, but the establishment of an efficient and correct methodology for achieving this desire remains, in some areas, still in the wish stage.

In this context, estimating the social costs of transport is a fairly recent concern. The difficulties for such an approach have not been entirely overcome, even in this economic sector, and there are opinions that argue that it is impossible to assess all the influences produced by negative externalities specific to transport with the utmost precision.

Moreover, the difficulty in quantifying externalities is mainly due not to a lack of calculation methods but to the fact that there are many variable aspects depending on factors that cannot be measured in specific markets.

In order to highlight the negative effects caused by transport, it is necessary to go through a number of steps, namely:

1 Identify the effects: at this stage, so-called 'target objectives' or areas of analysis are defined, which may be specific natural resources, human communities, animal species, etc.
2 The adoption of quantification criteria, or specific indicators, highlighting the influence of adverse effects on the target objectives.
3 Identifying cause–effect relationships. In other words, the aim is to discover the causes of the effects identified in the initial stage and the function linking these two variables.
4 The adoption of 'quality standards' corresponding to various levels of more or less acceptable harmfulness. In transport and elsewhere, the most widely used standards are the maximum permissible concentrations of chemical pollutants or the maximum permissible levels of noise pollution, vibration, axle weight, etc.
5 The adaptation of aggregated indicators showing the cumulative effects on the areas concerned of the various types of nuisances (chemical pollution, noise pollution, depletion of non-renewable resources, etc.).
6 Establishing methods to highlight the initial conditions of the target objectives. In other words, the aim is to realise the initial status of these objectives before they change some of their characteristics due to the influence of harmful effects.
7 The creation of models for forecasting the evolution of the target objectives over different time periods under the influence of adverse effects of different intensity or type.

Unfortunately, in the case of transport, it has not yet been possible to go through all the steps for the many different types of effects. Even if the initial step is carried out for the full variety of harmful effects, it seems that the analysis can still be improved. Some effects, such as noise pollution, have reached level 5 of the method of analysis, but for others, such as air pollution or vibration, the forecasting models or even the quantification criteria are still unsatisfactory.

The reversible or irreversible nature of some phenomena must also be taken into account in the analyses carried out to assess the negative effects caused by transport. However, these concepts are random, depending on both the temporal and spatial characteristics of the effects.

Even if a procedure can be used to replace damaged features, for example, by replanting trees, the residual effect is irreversible for an elderly person who will not benefit from the advantages of a new plantation (the initial effect will be felt after a period of time), or for a person who will no longer be in the vicinity of the green area (replanting will be in a different area from the original one).

In order to create an obligation for transporters to pay for damage to environmental factors, it is first of all necessary to evaluate the characteristics of the environmental factors in monetary terms.

In this approach, according to the theory of Pearce and Markandya, environmental economics uses numerous methods to evaluate environmental factors (Pearce and Markandya, 1989).

The concept of total economic value is of paramount importance for the valuation of environmental factors, as it encompasses different types of benefits from protecting and enhancing them. By definition, the total economic value of a characteristic is the sum of the actual use value (or utilisation), the indirect value, the option value and the acquired (intrinsic) value.

Use value is the individual benefit that is obtained due to a given environmental characteristic.

Use value comprises the actual use value, that is, the benefits that actually accrue to the users of an environmental resource, and the potential (indirect) use value. Thus, the actual use value of wilderness areas corresponds to the benefits obtained by those who wish to provide pleasing sights for their viewers, effectively satisfying such a preference.

Independent of this, however, the natural heritage also has a value for those who do not, for the time being, wish to do something like the above, but wish to preserve this option for later. Indirect value is therefore the value of those features of the environment that result from the functioning of certain ecosystems.

The advantage of knowing that the natural heritage will always be there when this option is desired is, in fact, a monetary value: individuals normally agree to pay a certain amount to protect the natural heritage if they are convinced that their preferences will not change. This type of benefit is called option value and has the effect of correcting the level of user benefit, which is represented by consumer surplus and depends on the uncertainty of demand and/or supply.

Earned (intrinsic) value is related neither to actual nor potential use. It originates in the value conferred by the existence of an asset, a resource, over and above any possibility of benefiting from it, directly or indirectly, now or in the future. It is therefore a recognition that certain things have a value 'in themselves': even if it is not possible to identify a particular use for a plant or animal species, it is given an 'intrinsic' value. Clearly, this is on the borderline between economics and economics, which deals only with exchange value or use value.

Pearce and Turner explain the issue of intrinsic value by recognising a 'right to existence' of natural resources and a 'feeling of sympathy' for fauna and flora, and by this, 'existence values represent a frontier linking economists and ecologists, for they cannot be explained by any conventional motivation' (Pearce and Turner, 1991).

Sizing (evaluating) these values is done using the following methods:

1 Direct, observational.
2 Analytical.

a The direct observation method must be applied over different time periods and on specific target objectives. The instruments used must be adopted according to scientific criteria so that it is possible to construct strategies for eliminating/minimising harmful effects and protecting the environment. At the same time, it should be taken into account that this procedure does not allow a precise prediction of the evolution of the characteristics of the target objectives in relation to different adverse influences.
b Analytical method.

Environmental and health impacts are most commonly assessed analytically.

By sizing the levels of environmental and health characteristics for any type of living being, the negative externalities (external costs) caused by transport can be determined by sizing the damage caused by this sector to the environment in general.

The damage produced can, in turn, create cumulative effects that require a more detailed approach to determine the level of damage produced for each environmental characteristic

Cumulative effects generally occur because of:

- Point (baseline) effects repeat frequently in time and space and cannot be absorbed by the environment.
- The impact of one action (activity) combines with that of another. In general, this combination leads to the emergence of situations totally different from those of the effects analysed individually, and their assessment requires the collaboration or establishment of different institutional or legislative bodies (the best example is the need for collaboration between inter-state institutions in the case of transboundary pollution).
- Aspects analysed separately actually have a synergistic effect.

There are two broad categories of analytical approaches for assessing external costs (UIC, 2012):

I A 'technical', qualitative approach that examines the relationships between transport activities and environmental impacts based on scientific assessment/research. For example, emission levels of air pollutants can be measured by vehicle types and categories.
II An 'economic' approach, which is based on quantitative analysis of the impacts caused (e.g. highlighting the number of fatalities from transport accidents, quantities of crops destroyed and species lost) with the aim of monetising these aspects.

For non-commercial goods, external costs are determined by methods that essentially aim to determine the level of acceptability of compensation for people who are affected by air pollution, for example, caused by transport (i.e. the minimum level of compensation for which victims would 'voluntarily' agree to accept such damage).

Two methods are used in particular to determine the external costs of non-commercial goods:

1 Demand assessment, which aims to determine the level of the individual payment agreement (PA). The latter can be sized by means of analyses aimed at identifying individual preferences. In this context, consideration is given, for example, to the use of PA sizing methods in situations where different levels of pollution (higher or lower) could be achieved or to reduce the costs associated with accidents (through the use of Statistical Value of Life).
2 Supply-side assessment, the main method being the 'damage' cost assessment. This aims at estimating the monetary cost of remedying the 'damage' to the environment or human health and avoiding such damage in the future.

In this context, PA level valuation methods are oriented towards avoiding the demand for the occurrence of such phenomena, while damage restoration cost valuation methods are oriented towards sizing the supply of welfare restoration.

### *3.2.2 Main methods for assessing external costs caused by transport*

In recent decades, attempts have been made to assess as accurately as possible the effects of transport externalities on environmental and health factors, using various analytical models.

The main methods used have been based on:

- Analysing cause/effect functions: these models seek to establish a relationship between environmental influences and observable (measurable) effects.

- Assessing the level of protection costs: the method is based on how to adopt specific legislation necessary to prevent and halt negative effects due to externalities, assessing the costs of these actions.
- Assessing the level of hedonic costs: the method is based on the analysis of substitution markets, in which goods are subject to an exchange process influenced by the variation in environmental characteristics observed in the price of these goods. In other words, a substitute market is the market for a good that is influenced by environmental characteristics that are non-tradable (the most illustrative example is the real estate market, where the price/value of a building is influenced by air pollution, noise levels, etc.).
- Community opinion polling: in this model, survey technology is used, using questionnaires to analyse individuals' options to pay or accept a change in environmental characteristics.

The calculation methodologies currently used are very different, and comparisons of the results do not always allow common conclusions to be drawn. However, in many countries, particularly in the European Union, attempts have been made in the first place to measure the external costs of road transport in monetary terms, estimating that they are of a major dimension, even greater than the contribution of the activity itself to GDP (Rothengatter, 1994).

The basic method, which is considered to be the simplest, is to analyse the properties of environmental factors, considering them as part of the so-called 'hypothetical markets', and at the same time, the behaviour of individuals who are faced with the choice between certain goods and services. From such a position, we can consider the actual preference for a given good as the agreement to pay (AP) in order to acquire that good. The concept of PA thus 'gives' us a monetary indicator of preferences.

This approach is important because it emphasises two fundamental concepts related to the notion of advantage: the first based on the agreement to pay (AP) and the second based on the consent to accept (CA).

In economics, the concepts of harm or benefit are defined in terms of individuals' preferences to avoid (a harm) or to accept (a benefit). These preferences manifest themselves in the market, as we have seen in the form of AP or CA.

Although economic theory suggests that AP and CA do not differ at all, empirical studies often show significantly different results. In the environmental domain, situations arise very frequently where it is necessary to assess losses and then, theoretically, the interested groups may be asked how much they are willing to pay to stop certain losses (income, welfare), or how much they demand to receive as compensation for the risks of accepting a worse environment.

In these circumstances, there are two ways of measuring the benefits of environmental improvement and two ways of measuring the losses (damage) caused by environmental degradation:

- Agreement to pay to obtain a benefit.
- Consent to give up a benefit.
- Agreement to pay to avoid a loss.
- Consent to incur a loss.

The fact that there can be significant differences between APs and CAs underlines the difficulties in measuring benefits, as in many cases, it is more about preventing harm than obtaining benefits.

Therefore, it is possible that preventing the occurrence or recording of damage may not be justified if it is assessed according to AP to avoid it, but may be justified if it is assessed according to CA, compensation for the willingness to bear damage.

There are several models for estimating PA (CA), which can be broadly grouped into two categories (EC Delft, 2019):

1 Stated value (SV) assessment, or the Method of Random Appraisals (MEA), based on the use of questionnaires or surveys in which respondents are asked to provide their personal level of AP for the non-occurrence of environmental and health damage.
2 Indirect valuation (IE) using different methods such as the Hedonic Pricing Technique (HPT), the Transaction Cost Method (TCM) or the Household Expenditure Method (HEM). This type of analysis aims at deducing the monetary value of externalities caused by transport by analysing economic activities (transactions, for example) in other economic markets, such as the real estate market.

1 The MEA can be oriented towards the evaluation of two types of activities: unit assessment of a given target group (where respondents are asked, in concrete terms, the level of AP to achieve a given outcome) and testing alternative opportunities (where survey respondents have to choose the most convenient alternative in different situations and PA is inferred indirectly).

The MEA measures AP directly and also allows for the accurate identification of the discomfort-producing negative externality. However, the accuracy of the results obtained by using the SEM depends very much on the design/realisation of the questionnaires/surveys and the level of information collected and is somewhat unrealistic as it involves only hypothetical costs. The avoidance of strategic behaviour of respondents (taking partisan positions in support of certain solutions that are profitable only for them) is also a major obstacle in the use of these types of surveys.

The performance of the Randomised Appraisal Method (MEA) derives not only from the fact that it can be applied in any context, but also from the possibility of using the responses of some individuals, related to the level of AP, to assess, by extrapolation, the behaviour of other individuals not directly involved in certain situations. The MEA utilises a number of techniques such as the limiting approach (e.g. to highlight the maximum amount that respondents are willing to pay to maintain a certain situation, in particular a certain level of pollution), the enforced choice approach (e.g. to target the willingness to pay an amount to improve an existing situation, in particular to reduce certain levels of pollution) and the bargaining approach (where the respondent is offered increasingly higher amounts of payment up to a level where he/she refuses to pay). Each of these techniques has certain shortcomings.

The limitative approach presents a difficulty in obtaining answers (the weight of estimating maximum values for the AP), whereas the imposed choice requires the elaboration of several examples to make it useful in a real way. The bargaining technique may suffer from hidden effects if certain suggestions are offered regarding the AP level. Also, the questionnaire itself may affect the results of the analysis by its content (type of questions).

Another important problem related to the use of MEA is the lack of selectivity. It usually arises when respondents are faced with the problem of differentiating the value of certain components of goods (clean air, quietness of the area where they live, etc.).

The existence of non-selectivity is evident in this case. The socio-economic conditions of the individuals interviewed also influence the level of AP. All the techniques used by the MEA aim at identifying certain behavioural characteristics (preferences) rather than analysing the results of hypothetical experiments. This is the key advantage of this method over that based on hypothetical markets.

However, the MEA can only approximate the value of some characteristics (features) of the environment, and it is not possible to assess their exact value.

2 In the case of EI, the most widely used analysis aims at estimating hedonic prices based on price differentials arising in specific markets due to negative externalities specific to transportation. These analyses are carried out in order to determine the level of AP needed to reduce or eliminate the effects of these negative externalities.

The strength of the IE method is that it is based on actual market behaviour, where the individual AP level to avoid/diminish a given externality can be observed.

However, the results of TPH-based studies are sensitive to observed market conditions. In addition, a lack of knowledge among market actors about the damage caused by a particular type of externality can significantly affect the reliability of study results.

TPH, when analysing the construction market, involves the use of a statistical regression of a function of the form:

$$P_h = f\left(L_h, V_h, A_h, E_h\right) \tag{3.2}$$

where

$P_h$ = the value of real estate.
$L_h$ = the actual characteristics of buildings.
$V_h$ = variables characterising the neighbourhood (e.g. the building is located in the vicinity of a park).
$A_h$ = parameters describing accessibility conditions.
$E_h$ = environmental attributes (ecological factors).

This function will therefore be characterised by the number of dwellings, their location in space, and also by the environmental conditions of the area where they are located (defined as explicit variables).

The shape of the function has a significant influence on the results. If it has a linear evolution, the change in the variables will influence, proportionally, the change in the price of that real estate.

If the function has a concave allure, then the price of real estate will fall faster and faster as certain values of the environmental characteristics increase.

The regression considered is a partial with respect to the different characteristics of the environment considered. The coefficients attached to the variables can thus be considered as the marginal costs resulting from the 'acquisition' of an additional unit in relation to an environmental feature. These values are defined as 'implicit prices' of these characteristics.

It has often been observed that the relationship between the independent and dependent variables has a significant influence on the level of the resulting implicit prices for environmental characteristics. Moreover, many variables can influence the price of a property, but this aspect, unfortunately, cannot be fully understood. Therefore, considering that many specific variables of some environmental characteristics can be omitted, the resulting implicit prices will not be the true ones. This can only be avoided by including as many variables as possible in the model from the outset, so that in the final analysis, the absence of some otherwise unmeasurable influences can be compensated for.

All that can be achieved through the link between environmental quality and the price of goods is to emphasise the individual AP with respect to a minor change in a particular characteristic of environmental factors. For a major change, the only way to assess AP is by comparing the actions of individuals who allocate similar amounts of money in an effort to obtain additional quantities of favourable environmental factor endowments, and then identifying these quantities. This, however, requires the availability of data on separate markets (spatially and temporally), which is not considered to be highly relevant based on the data of most studies that stop after marginal AP calculations (Maddison et al., 1996).

There are, however, many theoretical, methodological and statistical difficulties in the application of the TPH, among which we mention the following points:

- **At the theoretical level**: considering the real estate market as a fluid market, where prices are known to valuers, and residents are perfectly aware of the effects of pollution and correctly evaluate the utility function, the value of consumer surplus is ignored, although it can reach very high levels (about 40% of the total value).
- **Of a methodological nature**: the choice of the function leads only to emphasising a link between the price of goods and their characteristics, considering as included in these characteristics also indicators relating to environmental characteristics.
- **Of a statistical nature**: the problem of separating factors, the dispersion of results and the possibility of incorrect data.

Another indirect valuation method is the Travel Cost Method (TCM), which estimates the use value of a given environmental characteristic, based on the estimated costs of travel to certain 'targets' and the number of trips/persons, specific to each area under analysis.

Even if the use of some environmental features is free of charge (e.g. air quality in a park), their beneficiaries have to pay a certain price for the cost of getting to the place.

The biggest difficulties encountered in the application of TCM are in the (monetary) valuation of travel time and total travel costs. In general, the calculation technique used by the TCM assumes that the cost of travel time to a given target is proportional to a certain percentage of average income. However, the values currently used unfortunately only refer to the time gained on certain routes (better road, higher speed, etc.) and not to the travel time to a particular objective.

The general conclusion drawn from the application of this method is that the behaviour of the various subjects analysed is identical when travel or other specific costs (parking, park entry, etc.) are increased. The difference between this method and the hedonic pricing method is the way in which the objective as a whole is evaluated.

If an environmental good is complementary to or substitutable for a number of market goods, the valuation method that can be used is that of household expenditure (GCM). This method mainly concerns the existence of certain preference priorities revealed by the level of individual consumption. In other words, it aims to evaluate the expenditure incurred in different households to reduce the negative effects due to transportation (air pollution, existing noise, etc.). Starkie and Johnson have calculated, for example, that 5% of the income of some households should be allocated annually to reduce the overall level of noise pollution by 14db(A). These results are in line with those reported by Walters, who indicated that between 2% and 7% of total household expenditure is needed to insulate dwellings to reduce road traffic noise (Walters, 1975).

The method requires knowledge of a multitude of data on individual consumption and environmental characteristics, but unfortunately, these data are rarely available.

The preferred option of specialists in activities aiming at the assessment of external costs specific to transport remains the damage cost assessment method. This approach considers the valuation of all the damage suffered by individuals in a given area over a given time period as a result of the existence of an externality (e.g. health effects caused by traffic noise). Since market prices are often unavailable, for the damage suffered, the AP of the individuals concerned, to avoid the damage or to accept the damage, is used as a real indicator of individual preferences.

Some individuals, however, are willing to pay more than the market price, which means that they benefit more than the market (price) expresses; the surplus is called 'consumer surplus'.

These considerations lead to the existence of the relationship:

$$\text{AP} = \text{Market Price} + \text{Consumer Surplus} \tag{3.3}$$

Theoretically, the monetary valuation of the advantage, which can be obtained by certain individuals, can be based on the following situations:

- Individuals manifest their preference for certain goods through AP.
- The market price, which the individuals agree to pay, is the initial indicator and, consequently, the total expenditure incurred is a first approximation of the benefit obtained.
- The AP is higher than the total expenditure, as individuals are willing to pay a higher price than the market price and thus receive an additional benefit.

In conclusion, it can be said that the use of all these techniques raises numerous problems in application. These can be both methodological and practical. From the analysis carried out, the most important are as follows:

- The problem of the distribution of effects among the various social groups involved (monetary techniques tend to favour the wealthier social groups).
- The problem of the distribution of effects between generations (an equity issue, calling for the introduction of the concept of sustainable development – sustainable transportation).
- The need to collect an extremely large amount of data for assessments.
- Differences in accuracy in the use of certain methods.
- The degree of public acceptability of the results of applying the methods.

In order to reduce the disadvantages presented, recent research has highlighted other alternative methods to assess the external costs of transport.

One of these is the application of the avoidance cost approach *(Avoidance Cost Method)*. The avoidance cost approach determines the external cost drivers (e.g. shadow/collateral prices) by determining the costs of achieving a given policy objective (e.g. EU $CO_2$ reduction targets). This is done by estimating an avoidance cost function (in the form of a curve), which provides a proxy or guidance for achieving a given level of environmental quality. It is thus envisaged to estimate the marginal cost of increasing the environmental quality (e.g. by reducing an additional tonne of $CO_2$). Based on this cost curve, the minimum cost of achieving the objective is estimated, within a policy type aiming at reducing the effects of negative externalities caused by transport. It is assumed that this policy objective reflects collective preferences with respect to the externality in question, and hence that the level of the minimum cost to achieve this objective is a good indicator of the (collective) PA to avoid the externality damage.

A viable solution for the correct sizing of PA levels, in any situation in which a given region, geographical area or country may find itself, and being able to cover possible unavailability of data from primary sources, is the one offered by the *Transfer Value Method*. It can also be used to transfer input and output values to other areas (countries, regions) or other time periods.

Often used in the dimensioning of negative externalities, where reliable figures on the costs of damaging or avoiding environmental damage are not available, is the *Replacement Cost Method (restoration of damaged areas)*. This technique is aimed at estimating the value of a negative externality through the costs of replacing/remediating the negative effects caused by the externality. Unfortunately, this method may lead to an overestimation, as it is not always economically efficient to restore all the effects caused by negative externalities. On the other hand, the replacement cost approach may underestimate the true value of external costs, as it is not always possible to remedy all damages.

There are different approaches in using the replacement value method, namely (EC Delft, 2019):

- Unit value transfer with subcategories:
  - Simple transfer of unit values.
  - Transfer of unit values concurrent with income adjustment.
- Transfer of functions, with the following categories:
  - Benefit function transfer.
  - Multi-criteria analysis (multi-analysis).

The unit value transfer procedure is the transfer of primary data from the original location directly to the new location. This can be either a direct transfer (simple unit transfer) or through minor adjustments such as exchange rate, inflation and income (unit transfer with income adjustments).

The benefit function transfer approach is to estimate a function that relates the unit value to the characteristics of the original location in order to be able to transfer the information to another location. Although theoretically superior due to the fact that it takes into account more information, the benefit function transfer approach tends to be more complicated to implement in practice, as it requires information about the characteristics of the new location. For example, the willingness to pay for noise abatement can be estimated based on data on income, population density and age. When approximating the willingness to pay for noise abatement in a different region, it will be necessary to collect statistical data, specific to these indicators, from the new analysis area.

For these reasons, the transfer of unit values together with income adjustment is preferred for its simple, transparent and reliable results.

The method of transferring unit values while adjusting income is an option recommended by a large amount of research carried out in this area in recent times.

Transferring unit value from the home country to other countries considered in the analysis requires some adjustments, controlling for differences between locations, such as:

- Price differences: controlling price differences is essential to minimise errors when transferring values between locations. The recommended approach is to use exchange rates in current prices to account for the 'real cost of living'. Where appropriate, adjustments can also be made for differences in living standards between regions within the same country.
- Income differentials: an important issue in converting values between countries is to take income differences into account. The common approach is to multiply unit values by the ratio of income in the country of origin to income in the country of analysis.

On the basis of all these specifications, two major approaches are generally considered for the assessment of the external costs caused by transport, namely the bottom-up (bottom-up) and the top-down (top-down) approaches, both with their advantages and disadvantages.

For example, the estimation of marginal costs is usually based on bottom-up approaches, which take into account specific aspects of traffic conditions (with reference to specific case studies). Under this approach, the effects of pollutant emissions on health, water and soil are monetised using the targeted (customised) impact method. Thus, in the case of road transport, the impact of a single vehicle could be determined by taking into account its technical characteristics together with local environmental data (so that its impact on pollutant concentrations can be estimated). Then, using special cause-and-effect functions, it will be possible to determine the increase in the concentration of a particular pollutant caused by the use of that vehicle. In a similar way, all the values of the impacts caused by the analysed vehicle on the different types of pollutants will be determined, with the total impact value being the sum of all the unit values of the pollutants in the analysed area. The uncertainties in the accuracy of the impact estimates (and consequently in the assessment of external costs) derive from the non-linear evolution of the cause-and-effect function and the difficulty in determining exactly the contribution of each pollutant type to the pollutant mix in the area under consideration.

In order to derive national averages of external costs, the estimation is based on top-down approaches using existing data at this level.

It is important to emphasise that each approach has advantages and weaknesses, so that neither is superior to the other. In particular, the top-down approach is more suitable for assessing average external costs. This type of cost is more representative at a general level and allows comparisons between modes of transport. On the other hand, the cost function needs to be simplified, and it is difficult to allocate different types of costs in certain traffic situations to different categories of vehicles participating in the traffic, which makes the determination of marginal costs equally difficult.

In this context, the bottom-up approach is more appropriate for assessing marginal costs. The method is more precise, and the necessary differentiations can be made more easily. However, the method is more costly (expensive) and more difficult in aggregation processes in the context of averaging at the national level or for particular transport routes.

Therefore, in practice, a mix of bottom-up and top-down approaches (based on one's specific data) should be used. Thus, for each component element of the value of external costs, aggregate monetary valuation techniques have been used over the last decades (Miola et al., 2008).

In the case of *infrastructure-specific external costs* (caused both by infrastructure deterioration and by reduced usable capacity), there is a consensus on the basic approach to evaluate time losses caused by travel vector characteristics (interurban road transport), bottlenecks and waiting times (urban road and air transport) and to use the opportunity cost method for limited routes and time periods (rail and air). For this cost component, marginal cost estimates are considered more appropriate than average cost estimates.

Indeed, marginal costs are clearly more appropriate for transport policy making. They are determined according to actual traffic conditions. Theoretically, the marginal external costs caused by congestion/traffic congestion, for a given traffic volume Q, can be calculated according to the formula (EC Delft, 2008):

$$MEC_{cong}(Q) = \frac{VOT^{*}}{V(Q)} \varepsilon \tag{3.4}$$

where

VOT = time value (EUR/vehicle × h).
Q = traffic level at the time under consideration (vehicles/h).
$V(Q)$ = traffic–speed function (km/h).
$MEC_{cong}$ = Value of marginal external costs specific to congestion.
$\varepsilon$ = elasticity of demand coefficient (vehicles × h).

In order to calculate this type of external cost, the following inputs are needed: the value of time savings, the flow as a function of speed and the elasticity of demand with respect to traffic conditions.

The value of time savings is typically 50–150% higher than the normal travel time. Traffic–speed functions describe how average speed is influenced by traffic intensity, defined as the number of vehicles per hour. They are normally calculated, through simulations, against a 'model situation'. The elasticity of demand, in line with the characteristics of the supply of capacity, can be calculated using sophisticated mathematical models, which make it possible to take into account the reactions of users to a given situation.

With regard to the monetary evaluation of *external costs caused by accidents*, most of the studies carried out so far refer to the total value of these costs. Fewer, however, are the studies that have attempted to assess the marginal external costs specific to accidents for different traffic values.

From this perspective, the following calculation formulae are theoretically used (EC Delft, 2008):

$$\mathrm{TC} = \mathrm{A}(\mathrm{a}+\mathrm{b}+\mathrm{c}) = \mathrm{r} \times \mathrm{Q}(\mathrm{a}+\mathrm{b}+\mathrm{c}), \tag{3.5}$$

$$MC_j = \frac{\partial A}{\partial Q}(a+b+c) = r(E+1)(a+b+c) \tag{3.6}$$

$$\mathrm{MC_J} = \mathrm{MC} - \mathrm{PMC_j} \tag{3.7}$$

$$E = {}^{\partial r}\!/_{\partial Q}\,\frac{Q}{R} \tag{3.8}$$

where

A = number of accidents.
a, b, c are the statistical values of the lives (VSV) of those involved in the events.
TC = total external costs specific to accidents.
j = type of pollutant.
r = A / Q represents the risk of being involved in an accident (Q = traffic level at the time under consideration (vehicles/h)).
E = risk elasticity.
MC = marginal external costs associated with accidents.
PMC = marginal external costs or private marginal external costs.

A more complex formula has been proposed by Jansson as follows (EC Delft, 2008):

$$\sum_{i=1}^{6}\sum_{n=1}^{4}\left(a^n + b^n + c^n\right)\sum_{j=1}^{7} r_{ij}^n X_i \tag{3.9}$$

where

$i$ = mode of transportation (vehicle type).
$n$ = severity of the accident (fatalities, seriously injured, slightly injured, property damage only).
$X_i$ = distance travelled by vehicle 'i'.
$r_{(ij)}^{(n)}$= the probability of an accident of severity 'n' between vehicles 'i' and 'j', where 'i' is the victim.
a, b, c = model constants.

The use of 'top-down' models for the assessment of average and total crash costs requires national crash statistics plus insurance data. In this context, the material and administrative (insurance) costs of injuries, medical costs, production losses and the value of risks at the societal level are considered.

This approach compares the size of total social costs with the amounts covered (and not covered) by insurance. From this perspective, external costs are considered to be mainly those related to production losses or the statistical value of life (VSL).

As far as *air pollution* is concerned, most studies, methods and methodologies have been carried out to assess specific external costs (marginal, total or average). It is generally agreed that a 'bottom-up' analysis is more useful when using characteristics specific to the area of analysis, cost variations related to the time periods analysed or technological dimensions.

If we refer to the impact on human health, by using the Pathway Approach, the 'footprint' of the impact on health can be estimated, and the costs of the damage caused can be estimated in monetary units per tonne of the various types of air pollutants (Maddison et al., 1996).

Synthetically, the formula is as follows:

$$\Delta H_{ij} = b_{ij} \cdot POP_j \cdot \Delta A_{jt} \qquad (3.10)$$

where

$\Delta H_{ij}$ = health changes in year 'i' due to the pollutant 'j'.
$t$ = fuel type.
$b_{ij}$ = the slope of the specific health function.
$POP_j$ = population exposed to pollutant 'j'.
$A_{jt}$ = concentration of pollutant 'j' emitted by fuel type 't'.

Using the resulting values, so-called 'cause-effect functions' can be constructed, which show the influence of a given amount of a pollutant on health status. The evolution of the function curve will therefore show the change in health status (effect) due to the change by one unit in the quantity of a given type of pollutant emitted (cause).

In the second phase, the aim is to assign an economic (monetary) value to the impact of the pollutant on health. This is accomplished using the relationship:

$$P_i \cdot \Delta H_{ij} = P_i \cdot b \cdot POP_j \cdot \Delta A_{jt}, \tag{3.11}$$

where $P_i$ = the economic value (price) attributed to changes in health level 'i' due to the influence of pollutant 'j'.

The value $P_i$ is commensurate on the basis of the size of the PA in relation to the change in health status due to pollutant 'j'.

Once the calculations have been made, for each pollutant emitted and due to the use of a given fuel type, the values are summed up, resulting in the total external cost. In turn, this cost will be divided by the amount of fuel consumed (expressed in litres), resulting in the average external cost per litre of fuel consumed, which is as follows:

$$CE_{pam} = CE_{paT} / F(\text{monetary units} / 1), \tag{3.12}$$

where

$EC_{pam}$ = average external cost due to air pollution, in monetary units per litre (monetary units/l).
$EC_{paT}$ = total external cost due to air pollution.
F = total amount of fuel consumed (l).

It is assumed that there is a linear trend relationship between the change in the amount emitted of a given pollutant and the level of health status. This means that for each additional litre of fuel consumed (defined as a 'marginal amount'), the same effects will occur on average. In other words, the average external cost is the marginal cost per litre of fuel consumed.

It is very difficult to assess the level of the specific external costs of transport *noise pollution*. Over the last decades, many experts in the field have tried to arrive at a universal calculation formula, but so far it has not been achieved.

However, in order to highlight the link between traffic flow and noise level, Lamure suggests the following expression (Lamure and Lambert, 1993):

$$L_{ec}\left[dB(A)\right] = a + 20\log s - 10\log d + 10\log q \tag{3.13}$$

where

a = constant.
s = velocity.
d = distance.
q = traffic intensity (average traffic per hour).

The decreasing function of the traffic flow represents the marginal increase in the noise pollution level. If the level of negative effects (disturbance) is linear with the level of decibels emitted, then the marginal external cost is below the average. However, the specific function is generally non-linear. Because of these two opposite trends, in the analyses to be made, the marginal and average external cost values are taken to be equal.

Obviously, the noise level varies depending on the type and condition of the vehicle, the period of travel, the state of the infrastructure, etc.

According to Van Gent and Rietveld, the noise produced by a heavy vehicle is equivalent to that produced by about 10–25 cars (van Gent and Rietveld, 1993), whereas according to Mayers, this ratio (identical to that between motorcycles and cars) is 1 to 10.

There are also differences depending on the degree of motorisation and the level of the exposed population.

To calculate the specific external costs of noise pollution generated by transport and, in particular, by road traffic, studies carried out in recent years have used several methods, the main ones being:

1 Analysis of changes in house prices.
2 Evaluation of PA to reduce high pollution levels.
3 Analysis of the expenditure needed to protect against noise pollution.

1 Using the techniques for analysing the change in house prices, the conclusions of the studies developed over time have been different, as the traffic intensity, the type of vehicles (the technology used for noise damping differs from type to type) and other characteristics of the environment in which the investigations were carried out have been continuously changing.

   The evidence of increasing noise protection costs and perceived annoyance led Lamure to conclude that the depreciation of house prices increases as a percentage with increasing noise levels (Lamure and Lambert, 1993).

   Considering, however, only a linear depreciation ratio will underestimate the level of damage in highly exposed areas.

   Another important issue is to determine the maximum level of pollution at which these impairments occur. Recently, various researchers who have attempted to estimate the external costs of noise pollution have adopted a threshold value below which no pollution is considered to exist. Unfortunately, this level is not identical, with many values, ranging from 55 to 65 dB(A) during the day, being adopted by authors. This has led to situations that do not correspond to reality. Thus, for example, adopting a limit level of 65 dB(A) would neglect a range that 'annoys' the population. This was the case in the studies carried out by German researchers who, by changing the reference level from 55 dB(A) to 65 dB(A), showed a variation in the depreciation of real estate values of between 1% and 3%. It can therefore be concluded that this method will show lower levels of deterioration than the actual ones, as it does not take into account the disappearance of consumer surplus.

2 In the case of using PA valuation techniques to reduce high levels of pollution, the work of F.J. Langdon can be mentioned, which emphasises the cost of reducing noise within reasonable limits, according to a certain degree of dissatisfaction created by the exposed persons (Langdon, 1978). He concludes that the average value of the cost depends not only on the level of pollution but also on certain socio-economic (e.g. income of those exposed) or psychological (sensitivity, irritability, etc.) characteristics.

German researchers have also shown that, in addition to noise levels, PA is also influenced by:

a Size of personal income (e.g. for values above 75 dB(A), the PA ratio between individuals with an income above 2000 euros and those below 800 euros/month is 1/3)
b The degree of information of the individuals interviewed (influences the PA level by almost 20%).

3 Numerous authors have used the method of analysing the expenditures necessary to protect against noise pollution in order to estimate the external costs of noise pollution.

For Scandinavian countries, for example, the average external cost per person has been calculated for different noise pollution levels (Table 3.4).

There are notable differences: values about six times higher in Denmark than in Sweden; the maximum ceiling is set at 70 dB(A) in Sweden and Norway and 75 dB(A) in Denmark.

In general, the studies for the three countries followed the rule of doubling the cost value for a 5–10 dB(A) increase in sound level (exponential function).

As for the estimation of external costs specific to *climate change*, as specified above, the approach is very difficult. This is due to the complexity and total lack of knowledge of the interactions between humans and natural systems, of the effects that must (can) be taken into account in the specific analyses.

*Table 3.4* Values of Noise-specific External Costs in Selected Scandinavian Countries (in EUR/person)

| *Noise Level (dB(A))* | *Sweden* | *Denmark* | *Norway* |
|---|---|---|---|
| 55–60 | 30,6 | 215,5 | 60,5 |
| 60–65 | 122,0 | 432,5 | 120,9 |
| 65–70 | 306,2 | 884,8 | 241,9 |
| 70–75 | 612,5 | 1.828,4 | 483,9 |
| >75 | 612,5 | 3.774,9 | 483,9 |

*Source:* Miola et al. (2008).

As Stern says:

> Climate change is a unique challenge for economics: it is the largest and most extensive market failure ever observed. Therefore, economic analysis must be comprehensive, long-term oriented, and make particular use of risk and uncertainty theories taking into account the possibility of major, non-marginal changes.
>
> (Stern, 2006)

A more user-friendly method is to assess the costs of avoiding $CO_2$ emissions. This method is based on cost–benefit analyses aimed at obtaining a minimum cost of reducing greenhouse emissions by a certain amount. The approach can be useful, for example, in the realisation of a specific type of environmental policy, developed for different levels of applicability (local, regional, national, European, global, etc.). Such an approach can also be used for the transport sector as well as for the whole economy. For these reasons, this methodology has been used in recent decades in many specialised studies such as UNITE and ExternE (Miola et al., 2008).

At a general level, the issues that should be considered in the climate change impact assessment process are as follows:

- The variety of impacts that climate change implies.
- Accounting for the differences in the changes (damages) caused by climate change from a socio-economic-ecological perspective, at different levels and in relation to certain baselines (considered as reference points).

In line with these aspects, the analysis of the economic costs of the impacts of climate change, as well as the costs and benefits of actions to reduce greenhouse gas emissions that cause them, can be carried out by:

- The use of disaggregation techniques, which take into account the physical impacts of climate change produced on the economy, human life and the environment by assessing the costs of different technologies and strategies to reduce greenhouse gas emissions.
- Using economic models, including integrated assessment models that estimate the economic impacts of climate change. Macroeconomic models are also used to assess the costs and effects of the transition to low-carbon energy systems.
- By comparing the current level and specific future developments of the 'social cost of carbon' (the cost of the impact associated with one additional unit of greenhouse gas emissions) with the marginal abatement cost (the costs associated with one unit of reduction in real emissions).

These impacts were quantified (monetarily) by taking into account previous studies using different estimation techniques, ranging from general equilibrium models to those estimating the costs of deterioration.

Utilising many of these techniques, several analyses have been developed in recent years that have highlighted that (Miola et al., 2008):

- Extreme climatic conditions, analysed in isolation, can produce costs, estimated at about 0.5–1% of annual world GDP.
- Water availability and crop yields in southern Europe will decrease by 20% if a 2 °C increase in global mean annual air temperature is achieved.
- The costs of 'stabilizing' $CO_2$ emissions at 500–550 ppm are high. In fact, they could reach almost 1% of global GDP annually by 2050.
- The 'social cost of carbon' is $85/tonne of $CO_2$.

In fully understanding the importance of these values, it should not be forgotten that transport is one of the most significant economic sectors responsible for the production of greenhouse gases. Thus, as of 2018, in the EU, transport was responsible for the production of some 1,115.9 million tonnes of $CO_2$ equivalent, representing 25.4% of total EU-wide emissions (processing after EC, 2020).

Estimating the specific costs of *habitat damage* can be done by two methods. The first is to assess the value of the damage produced, in particular the costs felt through the deterioration in human health caused by air, water and soil pollution. The second method is based on estimating the costs necessary to remedy the situations created in polluted areas (restoration and/or replacement of contaminated soils, water purification and decontamination, air pollution abatement, etc.).

## 3.3 A new methodology for assessing the statistical value of life is needed to correctly dimension the external costs caused by transport

One of the most interesting areas of empirical research in economics is related to the issue of the heterogeneity of individuals and the economic and political relevance that goes with it. Considerable progress has been made in exploring aspects of heterogeneity, but the field is still open to new approaches.

Among the most important aspects of the heterogeneity of the Statistical Value of Life (VSV) are the age and income elasticities of the VSV. The income elasticity is particularly important because it is essential to understand international differences in VSV and associated differences in safety-net policy, as well as to indicate the adjustments that policy makers should make to limit VSV values as a result of rising incomes.

To this end, the proposed methodology assumes that, for each individual, it is necessary to attach a utility function related to living conditions at the national level, very well determined by the value of GDP/capita and life expectancy (E), as follows:

$$\mathrm{VSV} = \mathrm{f}\left(\mathrm{E}, \mathrm{GDP}\right) \tag{3.14}$$

These two variables include, in our opinion, the most important characteristics that could influence the size of the statistical value of life.

GDP per capita most faithfully reflects the annual value of the country's economic performance, and in relation to each individual, it is the only useful monetary indicator in this approach. In addition, life expectancy reflects the time interval between birth and death, during which the life of each individual in the area analysed is influenced by the value of the country's annual economic performance. Therefore, VSVs differ for individuals who have equal life spans but live in different countries.

Under this assumption, our formula will be:

$$\mathrm{VSV_i} = \mathrm{E_i} \times \mathrm{GDP_{(i)}}\left(\text{monetary units}\right), \tag{3.15}$$

As an average value over a given period of time, the formula will be:

$$\mathrm{med}\left(\mathrm{VSV_i}\right) = \frac{\sum_{i=1}^{n} E_i \times PIB_i}{n}\left(\text{monetary units}\right), \tag{3.16}$$

where

i = the year of the VSV valuation.
n = the number of years of the period under analysis.
$\mathrm{E_i}$ = life expectancy in year 'i'.
$\mathrm{GDP_i}$ = GDP/capita in year 'i'.

In Table 3.5, we present some comparisons between the VSL levels obtained from international research and those obtained using the (3.15) formula.

Data collected from the OECD (life expectancy and GDP per capita) and EUROSTAT (GDP per capita) databases (OECD, 2020; EUROSTAT, 2021) have been used in this analysis.

Comparing the data in columns 5 and 6 of Table 3.5 (their ratio is presented in column 7), we can easily observe that, if we remove the extreme values, the VSV levels obtained with the (3.15) formula show a margin of variation compared to those obtained by other methodologies, ranging from −48.2% to +66.4%. We consider this margin to be acceptable as the calculated values are in the global range of VSVs obtained by different researchers, internationally, at different times and using different methodologies.

Furthermore, in our opinion, the (3.15) formula is more appropriate for estimating VSV than other methods. The explanation is very simple. If we compare columns 5 and 6 of Table 3.5 and agree that VSV is a function related to E and GDP, the results obtained for VSV using other methods of other authors are not realistic. For example, from the data in Table 3.5, taking the VSV and GDP values given by some previous methods and considering the formula (3.15) and redoing the calculation, we observe large shortcomings. Thus, the ratio of VSL to GDP shows, for example, some life

*Table 3.5* Comparisons Between VSL Dimensions Calculated Using the (3.15) Formula and Different Methods Previously Used Internationally

| *Country* | *Year* | *E*[a] *(Years)* | *GDP/capita (USD)*[b] | *VSV According to Formula 3.15 (USD)* | *VSV Calculated by Other Methods (USD)* | *Ratio Between the Results in Columns 5 and 6 of the Table* | *Rate*[c] | *Research Author* |
|---|---|---|---|---|---|---|---|---|
| *1* | *2* | *3* | *4* | *5* | *6* | *7* | *8* | *9* |
| USA | 2004 | 81,1 | 41.921,80 | 3.399.857,98 | 1.421.025 | 2,39[d] | - | Alberini |
| Italy | 2007 | 84,2 | 40.005,20 | 3.368.437,84 | 3.598.485 | 0,94 | 1,460 | Alberini |
| Poland | 2006 | 79,7 | 9.500,70 | 757.205,79 | 795.082 | 0,95 | 1,319 | Giergiczny |
| USA | 2004 | 81,1 | 41.921,80 | 3.399.857,98 | 6.247.818 | 0,54 | - | Hakes & Viscusi |
| China | 2006 | 75,5 | 2.082,20 | 157.206,1 | 115.515 | 1,36 | - | Hammitt & Zhou |
| Japan | 2007 | 86,0 | 34.033,70 | 2.926.898,2 | 1.280.220 | 2,29[d] | - | Itaoka |
| Sweden | 1996 | 81,7 | 32.587,30 | 2.662.382,41 | 4.652.973 | 0,57 | - | Johannesson & others |
| United Kingdom | 1985 | 77,6 | 8.652,20 | 671.410,72 | 5.226.967 | 0,13[d] | - | Jones-Lee & others |
| Canada | 2002 | 82,0 | 23.995,00 | 1.967.590,00 | 1.758.343 | 1,12 | - | Krupnick |
| China | 2006 | 75,5 | 2.082,20 | 157.206,10 | 562.225 | 0,28[d] | - | Krupnick |
| Austria | 2008 | 83,3 | 48.860,74 | 4.070.099,64 | 2.445.763 | 1,66 | 1,392 | Leiter & Pruckner |
| New Zealand Dollar | 2009 | 82,6 | 27.998,60 | 2.312.684,36 | 2.870.491 | 0,81 | - | Leung |
| Switzerland | 2009 | 84,6 | 71.678,30 | 6.063.984,18 | 4.362.827 | 1,39 | 1,433 | Rheinberger |
| Sweden | 2009 | 83,5 | 47.737,75 | 3.986.102,12 | 7.693.884 | 0,52 | 1,433 | Svensson |
| Italy[e] | 2007 | 84,2 | 37.716,40 | 3.175.720,88 | 3.598.485 | 0,88 | - | Alberini |
| Poland[e] | 2006 | 79,7 | 8.999,70 | 717.276,09 | 795.082 | 0,90 | - | Giergiczny |
| Austria[e] | 2008 | 83,3 | 51.386,40 | 4.280.487,12 | 2.445.763 | 1,75 | - | Leiter & Pruckner |
| Switzerland[e] | 2009 | 84,6 | 69.672,00 | 5.894.251,20 | 4.362.827 | 1,35 | - | Rheinberger |
| Sweden[e] | 2009 | 83,5 | 46.207,10 | 3.858.292,85 | 7.693.884 | 0,50 | - | Svensson |

[a] We take higher values of E (corresponding to female gender).

[b] For Europe, GDP per capita data come from https://wits.worldbank.org/CountryProfile/en/country/by-country/startyear/LTST/endyear/LTST/indicator/NY-GDP-PCAP-KD

[c] Rate = exchange rate between USD and EURO for the period 2006–2009 is based on X-RATES at http://www.x-rates.com/historical/?from=USD&amount=1&date=2016-03-30

[d] Values we consider extreme and not taken into account in the analysis.

[e] GDP per capita data are from the OECD database at http://data.worldbank.org/indicator/NY.GDP.PCAP.CD

expectancy values of 161 years for Sweden in 2009, 270 years for China in 2006 and 150 years for the USA in 2004. This is obviously not the reality. The (3.15) formula is therefore more accurate than many of the other methods used to calculate VSL.

For the function presented in (3.14), it is assumed that every person in a country has the same VSL during the same period of analysis. This theory is different from that of Jones-Lee, who emphasises that the VSV level and individual age are related (Jones-Lee et al., 1993). In this theory, Jones Lee considers the reference value (maximum value) of VSV to be for an individual aged 40 years. In his opinion, the lowest VSVs are for younger people (under 18 years) and older people (over 65 years).

Another controversial issue is whether VSV is the same for every individual regardless of place of birth or place of residence.

In other words, should an African have a different VSV than a European or an Asian?

Surely, in our view, the value of the life of every individual on earth, regardless of nationality, colour, religion, political beliefs, economic status, health status, etc., should be the same. However, unfortunately, there are many distortions related to economic activities, age, place of living and others.

It is also obvious that, given the (3.15) formula described above, VSV differs from country to country. This is due to changes in GDP per capita. Highly developed countries will have statistically more 'valuable' inhabitants compared to the inhabitants of poor countries. However, in our opinion, life expectancy corresponds fairly closely to the level of GDP/capita of the countries we investigate.

According to this idea, if we want to make a uniform VSL for a larger area (region, continent) or for the whole world, we can use arithmetic averages of the formulas (3.15) and (3.16) for all countries in these areas (see formula (3.17)).

The average level for the VSL in a period and for the union of some countries will be based on the formula:

$$\text{Med}\left(\text{VSV}_{\text{UT}}\right)/\,\text{m} = \left\{\sum_{j=1}^{m}\sum_{i=1}^{n}\left(\frac{\text{Eij}\times\text{PIBij}}{n}\right)\right\}\left(\text{monetary units}\right) \tag{3.17}$$

where (in addition to those defined above):

$\text{UT}\ \overline{\text{VSLcu}}$ ='union of countries'.
j = country 'j.
m = the number of countries in the union of countries.
$E_{ij}$ = life expectancy for country 'j' in year 'i'.
$GDP_{ij}$ = GDP/capita for country 'j' in year 'i'.

## References

EAMA (European Automobile Manufacturers' Association), *External costs in the transport sector – A critical review of the EC internalization policy*, 2008, http://www.z-f-v.de/fileadmin/archiv/hefte---2008_1_2_3/2008-2/ZfV_2008_Heft_2_02_Baum%20Gei%DFler%20Schneider%20B%FChne%20-%20External%20Costs%20of%20Transport%20%96%20A%20critical%20review%20of%20the%20EC%20Internalisation%20Policy.pdf

EC, *EU energy and transport in figures, statistical pocketbook 2020*, ISBN 978-92-76-17565-0, 2020.

EC Delft, *Handbook on estimation of external costs in the transport sector*, 2008, https://ec.europa.eu/transport/sites/transport/files/themes/sustainable/doc/2008_costs_handbook.pdf

EC Delft, *Handbook on the external costs of transport Version 2019*, 2019, https://op.europa.eu/en/publication-detail/-/publication/9781f65f-8448-11ea-bf12-01aa75ed71a1

EUROSTAT, Main GDP aggregates per capita, 2021, https://appsso.eurostat.ec.europa.eu/nui/submitViewTableAction.do, accessed April 2021

Jones-Lee M. W., Loames G., et al., *The value of preventing non-fatal injuries: Findings of a willingness to pay national sample survey*, Working Paper SRC/2, Transport and Road Research Laboratory, Crowthorne, 1993.

Lamure C. and Lambert J., *Impact des transports terrestres sur l'environnement*, INRETS, Lyon, ISBN: 2-85782-383-5, 1993.

Langdon F. J., Monetary evaluation of nuisance from road-traffic noise: An exploratory study, *Environment and Planning*, vol. 10/1978, 1978, doi:10.1068/a101015

Maddison D., Pearce D., Johansson O., Calthrop E., Litman T., Verhoef E., *The true costs of road transportation*, Earthscan Publications Ltd., London, ISBN: 1-85383-268-5, 1996.

Mattauch M. and Leonard C., *Thematic guidelines. Moving in the right direction, European mobility week, Brussels*, European Info Point, EUROCITIES, Belgium, 2012, https://mobilityweek.eu/fileadmin/user_upload/materials/participation_resources/2012/Thematic_guidelines_2012_FINAL.pdf

Miola A., Paccagnan V, Turvani M., Andreoni V., Massarutto A. and Perujo A., *Review of the measurement of external costs of transportation in theory and practice*, Institute for Environment and Sustainability, Italy, doi:10.2788/77724, 2008.

OECD, *Health Status: Life expectancy*, 2020, http://stats.oecd.org/, accessed March 2023

Pearce D. W. and Markandya A., *Environmental policy benefits: Monetary valuation*, OECD, Paris, ISBN: 978-9264131828, 1989.

Pearce D. W. and Turner R. K., Economics of natural resources and the environment, *American Journal of Agricultural Economics*, vol. 73, Issue 1, 1991. doi:10.2307/1242904, https://www.researchgate.net/publication/31662420_Economics_of_natural_resources_and_the_environment_DW_Pearce_RK_Turner

Rabl A., Air pollution mortality: Harvesting and loss of life expectancy, *Journal of Toxicology and Environmental Health, Part A*, 2005, Taylor & Francis Online, https://doi.org/10.1080/15287390590936049

Rothengatter W., Do external benefits compensate for external costs of transport? *Transportation Research Part A: Policy and Practice* vol. 28, Issue 4, pp. 321–328 Elsevier, 1994, doi:10.1016/0965-8564(94)90006-X

Rueda S., Coord, *Methodology for the development of sustainable mobility plans, Agentia d'Ecologia Urbana de Barcelona-BCneologia*, 2010, http://bcnecologia.net/en/projects/methodology-development-sustainable-mobility-plans, Barcelona

Sansom, et al., *The overall UNITE methodology. Deliverable 1*, ITS, University of Leeds, 2000, http://www.its.leeds.ac.uk/projects/unite/downloads/D1-2_2.pdf
Stern N., *The stern review on the economic effects of climate change*, Wiley Online Library, 2006, https://onlinelibrary.wiley.com/doi/abs/10.1111/j.1728-4457.2006.00153.x
UIC, *Greening transportation. Reduce external costs, Paris*, ISBN 978-2-7461-2091-4, 2012, https://uic.org/IMG/pdf/external-costs2012brochure_complete_web-2.pdf
van Gent H. and Rietveld P., Road transport and the environment in Europe, *The Science of the Total Environment* Vol 129, 1993, doi:10.1016/0048-9697(93)90171-2
Walters A. A., *Noise and prices hardcover*, Oxford University Press, London, ISBN: 978-0198281979, 1975.

# 4 Interactions between overall economic development and the level of external transport costs

In implementing a sustainable mobility system, several factors need to be taken into account: transport costs, travel times, quality of infrastructure and means of transport, safety and frequency of travel. Of all these parameters, the most important is obviously cost. The cost of travel has a major influence on mobility choices, which generally changes the market share of transport modes at all levels.

## 4.1 General considerations

As stated before, the importance of externalities caused by transport varies considerably with the mode of transport under investigation, as well as with where and when the investigations are carried out. Clearly, there is a strong link between the preference for a particular mode of transport and the minimum cost criterion. In these circumstances, cost reduction must take into account two distinct levels in terms of the actions required:

- The first level concerns the educational side, supported by information and awareness-raising actions on the effects produced by a given type of transport and the need to change behaviour towards the choice of travel mode.
- Secondly, administrative measures (fiscal, legislative) to decouple transport from economic activity.

It should also be borne in mind, however, that beneficiaries are not always able to choose the mode of transport they want. There may be various reasons for this, but to a large extent, the obstacles are of a technical nature. Therefore, if we consider intermodal relations to be, on the whole, substitutable, competing, we are simply visualising a distorted relationship.

On a more realistic analysis, the modes of transport are in a state of limited competition, each mode being characterised by certain determining properties. Some of the most important ones are as follows:

- The economic area of intervention, delimited by the intervention costs of each mode.

DOI: 10.4324/9781003743064-5

- The technical area of competence, which can be characterised from several points of view (geographical, technical, managerial, etc.).

The existence of a certain stability of these properties is generally not valid for longer periods. Thus, in the long term, infrastructures and means of transport evolve in different ways, leading to changes in both the technical fields of competence and the economic areas of intervention. This is mainly due to changes in cost formation conditions.

Changes in the economic area, which have an impact on intermodal competition, may also be influenced by other factors, such as:

- Changes in the structure of the economy (structural changes in economic activities, downsizing, etc.).
- Improvements in the performance of the transport activity.
- The emergence of new competitors on the market.
- The adoption of legislative or fiscal measures aimed at encouraging or discouraging certain modes of transport along certain lines of action (spatial, temporal, social, environmental, etc.).

However, an analysis of the evolution of transport market sharing in the long run shows that it has a cyclical aspect, alternately giving each mode of transport the upper hand (Cuncev, 1993).

For example, the flexibility of road transport makes it gain from the rail market. It reacts by reorganising its activities, improving its operating technology (quality of service), gradually regaining market segments and gaining others that traditionally belonged to road transport. It is then the turn of road transport to modernise, for example, by developing new logistical services and gaining market segments, and the cycle starts again.

As already mentioned, the transport sector produces a wide range of harmful effects on the environment and human health. However, each mode of transport has its own particularities in this context, with the extent of the damage varying according to a number of specific characteristics (energy consumption, infrastructure, vehicles, traffic intensity, etc.). We also pointed out earlier that transport also produces numerous benefits (positive externalities).

However, the values of costs and external benefits do not cancel each other out. On the basis of this principle, it is possible to highlight trends towards more efficient activities. Thus, in transport, and especially in road transport, an increase in economic efficiency can also be achieved by reducing the values of activity-specific characteristics (e.g. travel distance).

Such a result (increased efficiency) can also be achieved by levying a tax equal in size to the marginal external cost, also known as the Pigou tax (named after the person who first developed this theory). The introduction of this tax can lead to the internalisation of externalities, resulting in a social benefit.

In general, the benefits to society from transport result from the use of a transport network(s) (Verhoef et al., 2001). The decision in the process of determining the levels of transport supply and demand is, in fact, made on two distinct levels:

1 The first decision-making level is usually dominated by groups 'authorised' in the field. Central, regional or local authorities are mainly responsible for these actions, but the existence of private offers of transport infrastructure and its operation has become increasingly evident recently.

  An economic justification for this position is that transport infrastructure, especially when used at levels below that at which congestion can occur, is non-competitive in terms of consumption (use) and can be considered 'non-excludable' (especially for roads) on the basis of cost considerations. These are, in fact, the two major criteria that distinguish public goods from private goods.

  However, since these two criteria only apply with limited validity, most analysts agree that transport infrastructure is in fact a 'quasi-collective' good.
2 The second decision-making level is occupied by individual agents. In the case of public transport, the operator acts as an intermediary, offering certain services, depending on the availability of infrastructure and the demand of individual agents.

Clearly, the benefits of infrastructure provision and use are closely linked: without use, there are no benefits of infrastructure provision (except perhaps 'optional value'), and the total benefits of infrastructure can thus be considered as the total net benefits of use (the net value of the costs of use) over the lifetime of the infrastructure.

In this context, different points of view can be considered for the economic assessment of transport costs and benefits.

One of these is the 'traditional' economic valuation, exclusively according to the principle of allocative efficiency. From this perspective, it can be seen that the benefits of transport infrastructure exhibit, as explained above, to a considerable extent, non-rivalry of consumption (especially at low levels/lack of congestion) and non-excludability (especially for road transport in view of the cost of exclusion).

In view of these characteristics, the provision of transport infrastructure should mainly be the responsibility of local, regional, national or EU (multinational) public authorities. Therefore, as stated above, the total benefits of infrastructure investment can be considered as the net benefits of infrastructure use.

Two major obstacles in estimating these benefits – in addition to the more common problems associated with the evolution and uncertainty of infrastructure supply – are related to the state of the infrastructure network and the derivative nature of transport demand.

This highlights that the distinction between generative and distributional effects is relevant. Therefore, the impact of an investment – which can be both positive and negative – can usually be found in a much wider area of influence than that revealed by the direct relationship (given by the cost of use, time period, etc.). In the case of a more detailed analysis, information on the evolution of infrastructure systems in the economic area under analysis should also be taken into account, where the assessment of the benefits of such investments is envisaged.

However, although the benefits of infrastructure investment are often reaped in other (non-transport) economic sectors, these benefits are not Pareto-relevant externalities: 'there is no *a priori* reason to believe that market forces would not be able to bring about the benefits from infrastructure improvements' (Verhoef et al., 2001). On the cost side, however, externalities prevent the market from achieving allocative efficiency.

In fact, *optimal Pigouvian taxes* can be derived on the basis of marginal external costs at the individual level (which means that 'intra-sectoral' externalities such as congestion should indeed be taken into account). Despite the economic attractiveness of the allocative efficiency criterion, considerations of *equitable impact* and *social utility* are often at least as important in the design of infrastructure policies.

From this perspective, transport turns out to be a rather complicated sector for economic analysis.

Looking at the benefits of infrastructure investment, it can be seen that, unlike many other types of public investment, which often involve only 'winners' and 'losers', improvements in transport infrastructure can produce positive effects for some groups and negative effects for others, because the incidence of gains and losses for different interest groups generally varies over time and space.

Accessibility is both an absolute and a relative concept. Equity considerations (spatial and sectoral) are therefore likely to be determinants of the viability of infrastructure policies, and investments that can be justified on efficiency alone may often be unacceptable to certain groups in society (e.g. road construction at the expense of infrastructure for other modes of transport).

In terms of regulating negative externalities caused by road transport, it turns out that the most effective policies (based on the Pigouvian principle) are usually the least popular. Therefore, regulators often have to make trade-offs between efficiency and social utility.

Another source of tension between efficiency and equity considerations relates to the different policy conclusions that can be drawn depending on the point of view adopted. Unfortunately, there often does not seem to be a clear link between what is efficient and what is socially equitable.

Therefore, in the process of economic analysis, it must be accepted that social utility and equity implications may prevail in transport policy decisions.

## 4.2 Developments and comparisons between overall economic development and the dimension of transport-specific external costs for selected EU member states and Romania

The environmental impacts of each mode of transport (road, rail, air, maritime) are caused by freight and passenger traffic and mainly concern:

- Infrastructure construction and maintenance.
- Manufacture, maintenance and disposal of all types of means of transport.
- The transport activity itself.

It is obvious that the level of demand for transport at the individual level is affected by the size of income and the level of GDP in the country for which the analysis is made. In this context, it is possible to establish a causal relationship between individual transport consumption and the GDP of different countries.

Of the total energy consumption at the EU level in 2018, the transport sector consumed 30.5%, individual households and services 40.7%, industry 25.8% and agriculture 3.0% (EC, 2020, p. 123). The share increased very little in the following period so that by 2022, the whole transport sector consumed about 31% of the 902.2 Mtoe (EC, 2024). Within the sector, for the 27 EU Member States, in 2018, rail mode consumed 2% of the total (1.8% in 2022), road 93% (93.8% in 2022), inland navigation and pipeline transport 2% (2.3% in 2022) and inland air transport 2% (2.1% in 2022). An almost constant evolution over the last five years.

Concomitant with these energy consumptions, transport has caused numerous and major negative environmental impacts, as we have already underlined.

In the EU alone, the total amount of greenhouse gases in 2022 amounted to 1,043.7 million tonnes of $CO_2$ equivalent (EC, 2024), down, however, by about 4.7% from the highest value recorded in the last 20 years (1,095.7 million tonnes of $CO_2$ equivalent in 2018, representing 28.1% of the total for the EU economy and up by 1% compared to 2017).

Internationally, for the year 2018, total greenhouse gas emissions for the OECD countries amounted to 3,893.1 million tonnes $CO_2$ equivalent, a decrease of about 2.1% compared to the previous year.

Noise emissions have also increased significantly over the last decades as a result of the continuous intensification of human activities and, consequently, of levels of urbanisation, mobility and industrialisation. Noise emissions and their effects are different depending on the particularities of transport modes. The noise level of road transport results from the superposition of engine noise, rolling noise (contact of tyres with the road surface) and other recurrent noises, while the main noise emissions of rail transport result from the contact of wheels on the tracks, the operation of engines, the aerodynamic effect and the vibrations of railway structures above the ground. The main source of aircraft noise is aircraft engines, with noise levels sometimes exceeding 120 dB(A). Noise is estimated to become stressful for humans when the noise level exceeds the limit of 55.65 dB(A) (Moliner et al., 2013).

During 2010, approximately 44.3% of the residents of 25 European Union countries plus Norway and Switzerland were exposed to noise levels exceeding 55 dB(A), which, although officially tolerable, is unpleasant and undesirable. These noises were caused by road transport (35.8%), rail transport (5.4%) and air transport (3.1%). The situation was and is dramatic in urban areas. Here, road traffic congestion mainly led to a situation where 11% of the population was exposed to a noise level of/above 70 dB(A). Because of this, for EU countries, it has been estimated that the willingness to pay to avoid noise exposure increased by almost 0.11% of GDP/inhabitant for each dB(A) unit when the noise level exceeds 55 dB(A) (Profillidis and Botzoris, 2010).

Transportation infrastructure occupies space that could have other uses in urban and non-urban areas. Awareness of this effect is more evident in densely populated countries such as the Netherlands and Belgium. If the carrying capacity of all modes of transport is compared with their land occupancy, then railways have a clear advantage, as the space allocated to a passenger by car is 22 times larger compared to that allocated to rail passengers and to a bus passenger 1.7 times larger compared to that travelling by rail (Profillidis and Botzoris, 2010).

Traffic congestion is a significant effect of transport activity, especially when infrastructure capacity has already reached its limits. The cost of congestion is the sum of lost passenger time and increased operating costs due to slower travel speeds. The basic differentiation of congestion from the other transport-specific negative externalities is that while all other categories of effects are felt by society as a whole, including people not directly involved in transport activities, congestion and its effects are considered to be felt only by transport participants. The users of different modes of transport, even if they cause significant delays in some cases, do not impose major external costs on the rest of society. However, delays in transportation do imply additional production costs for certain industries and, in some cases, for changing routes or transport or production schedules. At the 2011 level, the total annual cost of congestion was estimated for EU countries plus Norway and Switzerland to be around €200 billion, which represented 1.6% of the combined GDP of these countries (Profillidis et al., 2014).

The impact of transportation accidents (and road traffic in particular) in terms of injuries, disabilities and deaths is a global social and public health challenge. In absolute numbers, the number of deaths worldwide is well over half a million cases and rising every year. In highly motorised countries, at least 5% of people involved in crashes are killed or seriously injured each year, and 50% are hospitalised due to injury. It is estimated that life expectancy in these countries is reduced by at least six months due to road accidents. In the same context, the economic effects of accidents for all modes of transport amounted to about 2% of the GDP of EU countries in 2010 (van Essen et al., 2011).

Table 4.1 shows some examples of the dynamics of the degree of motorisation and of certain road accidents for the EU-27 and some of the EU countries for the period 1990–2018 (EC, 2020).

Table 4.1 shows a clear upward trend in the level of motorisation for the period considered and the countries considered. Unsurprisingly, due to the paradigmatic changes in the economies of the former socialist bloc countries (the case of Romania, Poland and Bulgaria in the previous analysis), the

*Table 4.1* Comparisons Between Trends in Car Ownership and the Number of Accidents Between 1990 and 2022 for Some EU Member States[a]

| *Country* | *Degree of Motorisation (No. of Passenger Cars/1000 Inhabitants)* | | | | *Number of Road Accidents (Thousands)* | | | |
|---|---|---|---|---|---|---|---|---|
| | *1990* | *2022* | *2018/1990 (%)* | *Trend* | *1990* | *2018* | *2018/1990 (%)* | *Trend* |
| EU-27 | 339 | 563 | 166.0 | ↑ | 71.77 | 20.65 | 28.8 | ↓↓↓ |
| Romania | 56 | 413 | 737.5 | ↑↑↑↑↑ | 3.78 | 1.63 | 43.1 | ↓ |
| Germany | 461 | 578 | 123.4 | ↑ | 11.05 | 2.79 | 25.2 | ↓↓↓ |
| Italy | 483 | 682 | 141.2 | ↑ | 7.15 | 3.16 | 44.2 | ↓ |
| Great Britain | 361 | 491 | 136.0 | ↑ | 5.40 | 1.75 | 32.4 | ↓↓ |
| Spain | 309 | 553 | 178.9 | ↑ | 9.03 | 1.75 | 19.4 | ↓↓↓↓ |
| France | 464 | 570 | 122.8 | ↑ | 11.22 | 3.26 | 29.0 | ↓↓ |
| Poland | 138 | 584 | 423.2 | ↑↑↑↑ | 7.33 | 1.89 | 25.8 | ↓↓↓ |
| Bulgaria | 152 | 449 | 295.4 | ↑↑ | 1.57 | 0.53 | 33.8 | ↓↓ |

*Source:* data and processing based on EC (2020, pages 88 and 110) and EC (2024, pages 88 and 105).

Legend: ↑ – increasing trend; ↓ – decreasing trend.

[a] The table shows the countries with EU membership in 2018.

number of passenger cars in these countries increased much more sharply than the EU average during the period under analysis. This development is mainly due to two economic factors. Firstly, the increase in the income of the inhabitants of these countries (reflected in the GDP/inhabitant) has created greater opportunities for purchasing cars. Secondly, the very low level of car ownership in these countries, coupled with the increase in the supply of cars for purchase (new or second-hand), has contributed to the situation shown in Table 4.1.

However, looking at the number of road accidents between 1990 and 2022, a strong downward trend can be observed throughout the EU. Only in Romania and Italy was the trend less marked. In this context, the most relevant case is Spain. During the period under review, the number of road accidents fell by more than 80%, while the level of motorisation almost doubled. It is clear that there is no causal relationship between the two indicators so the external costs specific to accidents are also not directly related to the number of vehicles on the road and those directly involved in accidents.

Also, as we have already pointed out, the increase in the number of cars (mainly) per 1000 inhabitants may point to an increase in the 'appetite' to purchase this type of vehicle (increased demand) against the background of rising population income (as evidenced by the level of GDP/inhabitant) and an improvement in supply (lower purchase prices, for example).

Road accidents, air pollution and noise are closely linked to common factors such as traffic intensity, travel speed, vehicle engine power and traffic composition (passenger cars, heavy goods vehicles, etc.). Policies aimed at preventing accidents, air pollution and noise are based on the same principles: managing travel demand, improving both the safety and environmental performance of vehicles and enhancing, encouraging and promoting the use of modes of transport that are more efficient, in terms of traffic safety and environmental protection (such as railways and public passenger transport). However, it has not been shown that a measure taken to improve traffic safety will automatically have the same environmental impact and vice versa.

Some examples of this divergence (Chapman, 2007; Proost and van Dender, 2012; Profillidis et al., 2014):

- The use of electric or hybrid vehicles proves to be beneficial in terms of reducing air and noise pollution, but the quietness of an electric vehicle's travel will increase the degree of unsafety for pedestrians and cyclists, as they may become aware, too late, of the presence of these vehicles.
- The construction of noise barriers can reduce noise levels in areas close to traffic but could have a negative impact if the visibility of drivers or pedestrians is affected.

As mentioned above, the assessment and quantification in monetary units of external cost levels specific to transport have been the subject of extensive and laborious scientific research work carried out internationally. Unfortunately, the number and frequency of their occurrence are not large, which makes it difficult to obtain analysis data for longer periods of time.

At the European level, only a few important analyses have been carried out in the last four decades, and in this context, the evolution of the dimension of external costs, quantified in monetary terms, can highlight the effects of transport on economic processes, human health and the level of environmental degradation.

Thus, in 1995, total external costs (excluding congestion-specific costs) amounted to 530 billion euros, representing 7.8% of the total GDP of the EU-17 (Infras/IWW, 2000). Accidents were the largest cost category, with 29% of the total. Air pollution and climate change costs amounted to 48%. When broken down by mode of transport, road transport was the most important, accounting for about 92% of the total cost, followed by air transport with 6%. Rail (2%) and shipping (0.5%) were of minor importance. It was also emphasised that two-thirds of the costs were caused by passenger transport and one-third by freight.

In 2003, a study carried out by a team of researchers from the *Central European Initiative* (OECD, 2003) revealed the total level of external costs in 1995 for the countries of Eastern and Central Europe, which amounted to about 40 billion euros per year, representing about 14% of the GDP of these countries. Since the EURO has been in force only since 1999, for the year 1995, in the present study, the values expressed in USD have been transformed into EURO on the basis of the official calculator of the European Commission. In these circumstances, for the year 1995, the relation 1 EURO = 1.3069 USD was used, resulting in the arithmetic average of the monthly values found in that year.

For the year 2000, at the EU-17 level, the assessment of the level of external costs caused by transport was carried out in 2004 and presented in a report under the aegis of the International Union of Railways (UIC, 2004).

The situation in the countries of the former socialist bloc, for the same year 2000, the level of total external costs was estimated by research teams coordinated by the Swiss company INFRAS (INFRAS, 2004).[1]

An analysis aimed at assessing the levels of external costs caused by transportation was also carried out in 2011 by a team of researchers led by Huib van Essen (van Essen et al., 2011). The findings showed that at the EU-25 level (plus Norway and Switzerland), the total external costs of transport (less congestion) amounted to about 4% of the total GDP of these countries in 2008. In value terms, this amounted to more than € 500 billion (van Essen et al., 2011). Accidents were the largest cost category, accounting for 43% of total costs. Air pollution represented 10.4% and climate change accounted for 29% of total costs. Costs due to noise pollution amounted to 9.6% of total costs, and specific costs for damage to the natural environment, landscape and undesirable urban effects amounted to 1.0% of total costs.

The most recent assessment of the external costs caused by transport was carried out by the European Commission in 2019, with the data in the study being for EU countries at the 2016 level (EC Delft, 2019). From the values presented, it emerged that the total level of external costs caused by transport in 2016 was €841.1 billion, representing 5.7% of the GDP of the EU-28 countries. Of this amount, road transport was responsible for more than 97.5%.

In an effort to highlight possible links between economic development and the evolution of the level of external costs caused by transport, the following analysis will elaborate on aspects and data presented in the studies briefly outlined above.

More specifically, the data will highlight the situations in 1995, 2000, 2008 and 2016. Also, in order to have a balanced analysis, we have chosen to highlight the existing situations in the European Union, taken as a separate entity, and some representative countries for the western part of the UNION (France, Germany, Spain and Sweden) and from the former socialist bloc, which have gradually joined the EU (Czech Republic, Romania, Poland and Hungary). The choice of these countries was random, with no particular selection criteria.

In this situation, the evolution of the main indicators, specific to general economic activity, transport activity and external costs caused by transport, on the environment and human health, for the period 1995–2016 and for the countries mentioned above is presented in Table 4.2.

It should be noted that in the years 1995 and 2000, not all the countries of the former socialist bloc under analysis were part of the EU, but for the accuracy and continuity of the data series in the proposed analysis, data from international studies and EUROSTAT or country archives have been included in the table.

From the analysis of the data presented in Table 4.2, some important points can be highlighted:

- Indices of external costs caused by transport have, in general, similar trends to those of GDP and GVA Transport, with differences in intensity (see Figure 4.1).
- The increase in the level of external costs caused by transport is lower than GVA transport and GDP, with some exceptions (Romania and Spain).
- In the case of road transport, the increase in the degree of motorisation and the increase in the distance travelled by goods have a greater influence on the increase in specific external costs (Figure 4.2).
- In relation to GDP, the external costs caused by road transport (the highest in this economic sector) in the countries analysed have, for the most part, shown a downward trend, with the exception of Romania, where they have shown a positive trend.
- For the period and countries analysed, Romania recorded the highest increases in the indicators considered. The exception is the value of total goods transported by road, where Poland is the leader.
- There is a difference in the intensity of the evolution of the indicators analysed for the two groups of countries considered (Western countries with old EU status and former socialist bloc countries).

*Table 4.2* Evolution of Some Indicators Specific to the General Economic Development and Transport for the EU and some EU Member States, 1995–2016

| *Indicator* | *U.M.* | *Year* | *Country* | | | | | | | | |
|---|---|---|---|---|---|---|---|---|---|---|---|
| | | | *EU* | *EC* | *FR* | *GE* | *PL* | *RO* | *SP* | *SE* | *UN* |
| Population | Mil. | 1995[a] | 445,9 | 10,4 | 57,8 | 81,5 | 38,6 | 22,7 | 39,3 | 8,8 | 10,3 |
| | loc. | 2000[a] | 450,4 | 10,3 | 58,8 | 82,2 | 38,7 | 22,5 | 40,0 | 8,9 | 10,2 |
| | | 2008[c] | 495,6 | 10,4 | 61,9 | 82,2 | 38,1 | 21,5 | 45,3 | 9,2 | 10,0 |
| | | 2016[b] | 510,2 | 10,6 | 66,6 | 82,2 | 37,9 | 19,8 | 46,4 | 9,9 | 9,8 |
| | % | **2016/1995** | **114,4** | **101,9** | **115,2** | **100,9** | **98,2** | **87,2** | **118,1** | **112,5** | **95,1** |
| GDP | bn. | 1995[g] | 7364,0 | 46,0 | 1224,7 | 1977,6 | 108,8 | 28,6 | 470,2 | 204,3 | 35,5 |
| | EURO | 2000[g] | 9667,2 | 67,0 | 1478,6 | 2109,1 | 186,7 | 40,6 | 647,9 | 285,2 | 51,2 |
| | | 2008[g] | 13079,0 | 162,0 | 1992,4 | 2546,5 | 366,0 | 146,6 | 1109,5 | 354,9 | 108,2 |
| | | 2016[g] | 14856,3 | 177,4 | 2234,1 | 3134,7 | 427,1 | 170,1 | 1113,8 | 466,3 | 116,1 |
| | % | **2016/1995** | **201,7** | **385,7** | **182,4** | **158,5** | **392,6** | **594,8** | **236,9** | **228,2** | **327,0** |
| GDP/LOU | Thousand | 1995 | 15,2[h] | 4,4 | 21,2 | 24,3 | 2,8 | 1,3 | 12,0 | 23,2 | 3,4 |
| | EURO | 2000 | 19,8[h] | 6,5 | 25,1 | 25,7 | 4,8 | 1,8 | 16,2 | 32,0 | 5,0 |
| | | 2008 | 26,1[h] | 15,6 | 32,2 | 31,0 | 9,6 | 6,8 | 24,5 | 38,6 | 10,8 |
| | | 2016 | 29,3[h] | 16,7 | 33,5 | 38,1 | 11,3 | 8,6 | 24,0 | 47,1 | 11,8 |
| | % | **2016/1995** | **192,8** | **379,5** | **158,2** | **156,9** | **402,5** | **660,8** | **200,0** | **203,0** | **348,4** |
| Degree of motorisation | Car/1000 seats | 1995[d] | 377,0 | 295,0 | 468,0 | 495,0 | 195,0 | 97,0 | 360,0 | 411,0 | 218,0 |
| | | 2000[d] | 414,0 | 336,0 | 489,0 | 475,0 | 261,0 | 124,0 | 429,0 | 450,0 | 232,0 |
| | | 2008[e] | 470,0 | 423,0 | 498,0 | 504,0 | 422,0 | 187,0 | 483,0 | 462,0 | 305,0 |
| | | 2016[f] | 507,0 | 502,0 | 479,0 | 555,0 | 571,0 | 279,0 | 492,0 | 477,0 | 338,0 |
| | % | **2016/1995** | **134,5** | **170,2** | **102,4** | **112,1** | **292,8** | **287,6** | **136,7** | **116,1** | **155,0** |
| External costs (Total) | Billion | 1995 | 530,0 | 6,9[l] | 79,1[m] | 124,9[m] | 12,6[l] | 3,1[l] | 27,6[m] | 9,9[m] | 4,5[l] |
| | EURO/year | 2000 | 650,3[m] | 13,4[p] | 87,5[m] | 149,1[m] | 29,1[p] | 9,7[p] | 58,2[m] | 13,7[m] | 8,5[p] |
| | | 2008[n] | 501,1 | 8,9 | 63,8 | 108,5 | 22,7 | 7,9 | 44,5 | 9,3 | 7,1 |
| | | 2016[o] | 841,1 | 14,0 | 111,0 | 172,3 | 41,5 | 21,8 | 65,1 | 15,8 | 11,5 |
| | % | **2016/1995** | **158,7** | **202,9** | **140,3** | **138,0** | **329,4** | **703,2** | **235,9** | **159,6** | **255,6** |
| External costs (Road) | Billion | 1995 | 487,6 | 6,3 | 72,8 | 114,9 | 11,6 | 2,9 | 25,4 | 9,1 | 4,1 |
| | EURO/year | 2000 | 544,4[m] | 8,2[p] | 74,2[m] | 129,1[m] | 16,5[p] | 5,2[p] | 48,3[m] | 11,5[m] | 5,2[p] |
| | | 2008[n] | 463,4 | 8,2 | 60,2 | 101,5 | 21,4 | 7,2 | 39,8 | 8,4 | 6,5 |
| | | 2016[o] | 820,4 | 13,6 | 109,1 | 165,7 | 40,2 | 21,2 | 64,3 | 15,3 | 11,1 |
| | % | **2016/1995** | **168,3** | **215,9** | **149,9** | **144,2** | **346,6** | **731,0** | **253,1** | **168,1** | **270,7** |

(*Continued*)

*Table 4.2* (Continued)

| *Indicator* | *U.M.* | *Year* | *Country* | | | | | | | | |
|---|---|---|---|---|---|---|---|---|---|---|---|
| | | | *EU* | *EC* | *FR* | *GE* | *PL* | *RO* | *SP* | *SE* | *UN* |
| External costs Total/ GDP | % | 1995 | - | 15,0 | 6,5 | 6,3 | 11,6 | 10,8 | 5,9 | 4,8 | 12,7 |
| | | 2000 | - | 20,0 | 5,9 | 7,1 | 15,6 | 23,9 | 9,0 | 4,8 | 16,6 |
| | | 2008 | - | 5,5 | 3,2 | 4,3 | 6,2 | 5,4 | 4,0 | 2,6 | 6,6 |
| | | 2016 | 5,7 | 7,9 | 5,0 | 5,5 | 9,7 | 12,8 | 5,8 | 3,4 | 9,9 |
| | | **2016/1995** | - | **52,7** | **76,9** | **87,3** | **83,6** | **118,5** | **98,3** | **70,8** | **78,0** |
| External costs Road/ GDP | % | 1995 | - | 13,7 | 5,9 | 5,8 | 10,7 | 10,1 | 5,4 | 4,5 | 11,5 |
| | | 2000 | - | 12,2 | 5,0 | 6,1 | 8,8 | 12,8 | 7,5 | 4,0 | 10,2 |
| | | 2008 | - | 5,1 | 3,0 | 4,0 | 5,8 | 4,9 | 3,6 | 2,4 | 6,0 |
| | | 2016 | 5,5 | 7,7 | 4,9 | 5,3 | 9,4 | 12,5 | 5,7 | 3,3 | 9,6 |
| | | **2016/1995** | - | **56,2** | **83,1** | **91,4** | **87,9** | **123,8** | **105,6** | **73,3** | **83,5** |
| Travel | Billion tonnes × km | 1995[i] | 1623,9 | 54,2 | 233,1 | 372,3 | 120,3 | 40,7 | 112,6 | 51,0 | 23,4 |
| Total freight transport | | 2000[i] | 1865,5 | 60,2 | 270,8 | 429,9 | 130,2 | 33,3 | 160,9 | 55,1 | 28,8 |
| | | 2008[j] | 2465,7 | 66,3 | 255,8 | 521,3 | 217,3 | 80,3 | 253,5 | 65,5 | 45,0 |
| | | 2016[k] | 2390,1 | 65,9 | 196,7 | 486,3 | 341,5 | 74,9 | 227,5 | 64,1 | 52,5 |
| | % | **2018/1995** | **147,2** | **121,6** | **84,4** | **130,6** | **283,9** | **184,0** | **202,0** | **125,7** | **224,4** |
| Track | Billion tonnes × km | 1995[i] | 1127,2 | 31,3 | 178,2 | 237,8 | 51,2 | 19,7 | 101,6 | 31,6 | 13,8 |
| Road freight transport | | 2000[i] | 1343,9 | 37,3 | 204,0 | 280,7 | 75,0 | 14,3 | 148,7 | 35,6 | 19,1 |
| | | 2008[j] | 1877,7 | 50,9 | 206,3 | 341,5 | 164,9 | 56,4 | 243,0 | 42,4 | 35,8 |
| | | 2016[k] | 1831,0 | 50,3 | 155,8 | 315,8 | 290,7 | 48,2 | 155,8 | 42,7 | 40,0 |
| | % | **2018/1995** | **162,4** | **160,7** | **87,4** | **132,8** | **567,8** | **244,7** | **153,3** | **135,1** | **289,9** |
| Travel | Billion passengers × km | 1995[i] | 3596,5 | 62,5 | 688,0 | 886,3 | 137,3 | 58,9 | 267,0 | 94,4 | 53,9 |
| Total passenger transport | | 2000[i] | 3999,0 | 71,2 | 762,6 | 906,7 | 154,2 | 62,6 | 322,7 | 111,9 | 55,9 |
| | | 2008[j] | 5134,2 | 79,2 | 805,2 | 934,1 | 293,7 | 77,4 | 363,1 | 109,4 | 50,3 |
| | | 2016[k] | 5279,4 | 81,0 | 830,3 | 1042,6 | 223,0 | 100,6 | 356,6 | 127,3 | 64,4 |
| | % | **2016/1995** | **146,8** | **129,6** | **120,7** | **117,6** | **162,4** | **170,8** | **133,6** | **134,9** | **119,5** |
| Travel | Billion passengers × km | 1995[i] | 3283,8 | 54,5 | 641,2 | 815,3 | 110,7 | 40,0 | 250,4 | 87,6 | 45,4 |
| Passenger-road transport | | 2000[i] | 3660,4 | 63,9 | 687,7 | 831,3 | 130,1 | 51,0 | 302,6 | 103,7 | 46,2 |
| | | 2008[j] | 4725,0 | 72,4 | 720,2 | 852,3 | 273,5 | 70,5 | 339,1 | 98,4 | 42,0 |
| | | 2016[k] | 4829,3 | 72,3 | 742,5 | 946,8 | 203,8 | 95,6 | 329,9 | 114,5 | 56,7 |
| | % | **2016/1995** | **147,1** | **132,7** | **115,8** | **116,1** | **184,1** | **239,0** | **131,7** | **130,7** | **124,9** |

| | | | | | | | | | | | |
|---|---|---|---|---|---|---|---|---|---|---|---|
| Gross value added – transport and storage | bn. EURO | 1995[q] | 310,6 | 2,9 | 46,3 | 71,2 | 4,3 | 1,4 | 22,9 | 10,8 | 1,9 |
| | | 2000[q] | 415,6 | 4,4 | 58,5 | 78,7 | 8,7 | 2,3 | 28,0 | 14,7 | 2,5 |
| | | 2008[q] | 570,6 | 9,6 | 82,0 | 106,3 | 18,7 | 10,3 | 42,3 | 18,5 | 5,2 |
| | | 2016[q] | 645,1 | 9,2 | 91,9 | 123,8 | 24,3 | 10,8 | 46,7 | 21,8 | 6,5 |
| | % | **2016/1995** | **207,7** | **317,2** | **198,5** | **173,9** | **565,1** | **771,4** | **203,9** | **201,9** | **342,1** |
| Adjusted value added (VAAj) | bn. EURO | 1995 | - | -4,0 | -32,8 | -53,7 | -8,3 | -1,7 | -4,7 | 0,9 | -2,6 |
| | | 2000 | - | -9,0 | -29,0 | -70,4 | -20,4 | -7,4 | -30,2 | 1,0 | -6,0 |
| | | 2008 | - | 0,7 | 18,2 | -2,2 | -4,0 | 2,4 | -2,2 | 9,2 | -1,9 |
| | | 2016 | -196,0 | -4,8 | -19,1 | -48,5 | -17,2 | -11,0 | -18,4 | 6,0 | -5,0 |
| | % | **2016/1995** | - | **120,0** | **58,2** | **90,3** | **207,2** | **647,1** | **391,5** | **666,7** | **192,3** |

*Sources:* data and processing data after: a – (EC, 2006, p. 55); b – (EUROSTAT, 2021b); c – (EC, 2009, p. 11); d – (EC, 2020, p. 88); e – (EC, 2010, p. 155); f – (EC, 2016, p. 87); g – (EUROSTAT, 2021a); h – (EUROSTAT, 2021c); i – (EC, 2020, pp. 38–55); j – (EC, 2010, pp. 109–123); k – (EC, 2018, pp. 36–53), l – (OECD, 2003); m – (UIC, 2004); n – (EC Delft, 2011); o – (EC, 2019); p – (INFRAS, 2004); q – (EUROSTAT, 2021d).

Legend:

CE = Czech Republic; FR = France; GE = Germany; GE = Germany; PL = Poland; RO = Romania; SP = Spain; SE = Sweden; UN = Hungary.

For 1995 and 2000, the *Population* EU-25 data are shown.

EU-27 data are presented for the 2008 population.

EU-28 data are presented for the population of 2016.

For motor depletion corresponding to the year 1995, 2000 data are presented EU-27.

EU-28 data are presented for the motorisation rate corresponding to the years 2008 and 2016.

For GDP and GDP/place, according to official Eurostat statistics, EU-28 data are presented, even if in the years 1995, 2000 and 2008, the EU did not have 28 members.

For freight transport travel (total and road) and passenger travel, the data are for the EU-27, except for 2016, when they are for the EU-28.

For the total freight and passenger transportation journey, data for air and maritime modes are not included.

For road external costs, at the 1995 level, their share of 92% of total external costs has been taken into account.

For total external costs and road external costs, for the years 1995 and 2000, data are presented for the EU-17; for the year 2008, data are presented for the EU-27; and for the year 2016, data are presented for the EU-28.

x – According to CAEN classification, Revision 2.

For GVA (transportation and warehousing), EU data are for EU-28.

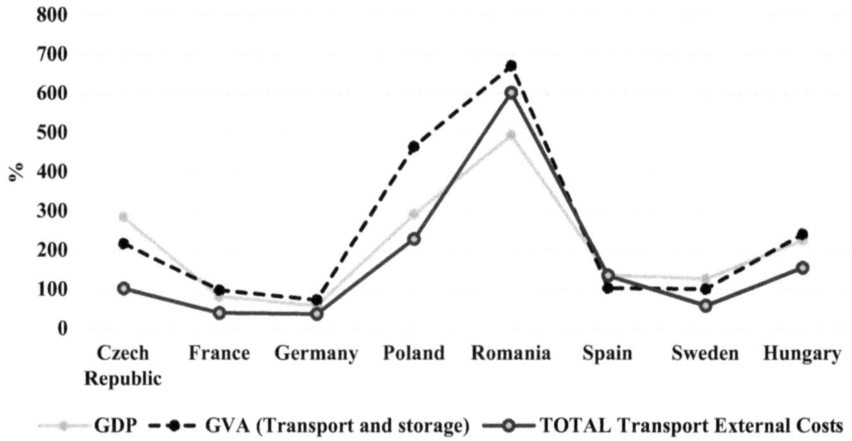


*Figure 4.1* Comparisons between developments in some indicators specific to general economic activity and transport for Romania and other EU countries in 2016, compared to 1995. (By author.)

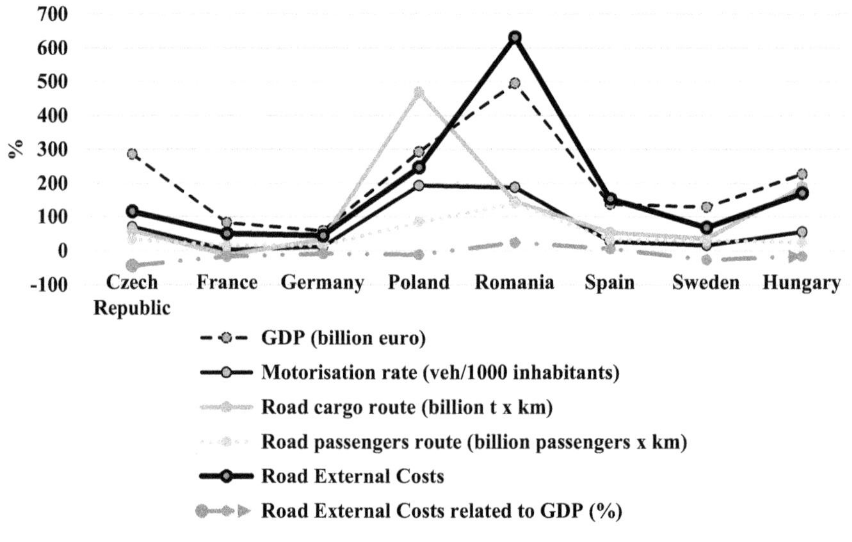


*Figure 4.2* Developments in some indicators' characteristics of general road transport and economic activities, for 2016 compared to 1995, in the case of Romania and other EU representative countries.

(*Source:* author.)

Taking into account the fact that the GVA in transport is the measure of the value realised from the activities carried out in the whole transport sector (realised income) and that the dimension of external costs caused by these activities represents expenditures not fully covered by the providers of transport activities (creating the GVA per sector), we define the Adjusted Value

Added ($VAA_j$) as the real value of the activities carried out in the transport sector, adjusted by the monetary dimension of the negative externalities produced by them.

To calculate the VAAj, we have the following formula:

$$VAA_j = GVAT - TEC(\text{euro}), \tag{4.1}$$

where GVAT is the gross value added in the transport sector, and TEC is the total external costs produced by the transport sector.The values calculated for the period 1995–2016 have been listed in Table 4.2.

From the analysed data, the opposite trend of the evolution of the value of external costs and the $VAA_j$ can be observed. Thus, it can be observed that, to a large extent, with a decrease in the $VAA_j$, the level of external costs is increasing.

This clearly highlights the fact that an increase in the level of external costs causes a decrease in the efficiency of the transport activity, the real value of which is different from the one usually reported by GVA. Moreover, there is a clear differentiation between the directions of evolution of GVA and $VAA_j$, which justifies the need to include the value of external costs in the indicator that measures the size of the transport activity. At the same time, the negative values of the $VAA_j$ mean that without the 'indirect subsidisation' of road transport activity, mainly through the non-internalisation of external costs, the efficiency, profit and market position of this mode of transport are misperceived. While looking at the GVA, the activity seems to be on the rise, the $VAA_j$ gives us a different, even contrary, perspective.

In the case of Romania, although GDP has increased substantially, transportation activity has not kept pace with this trend. Moreover, the level of external costs exceeded the level of activity (GVA) for most of the period analysed, which led to negative values for $VAA_j$.

For a clearer highlighting of these aspects, Figure 4.3a–c presents the evolutions of the indicators in, comparatively, at the level of 1995, 2000, 2008 and 2016 in some significant Eastern European countries under analysis.

For the period under analysis, 1995–2016, the increase in motorisation in Romania (Table 4.1) supported the intensification of road passenger and freight transport activities. The increase in the total number of journeys for these types of transport thus amounted to more than 200% during the period in question, while the degree of motorisation increased almost threefold. This can be explained by the growth in transport capacity, with an increasing supply. The demand for transport has also increased in line with the increase in the general level of economic development, which for Romania during the same period of analysis was about 6.6 times, taking into account the evolution of GDP per capita. The extremely favourable evolution of this indicator was one of the prerequisites and also the support for the development of the entire transport activity so that the GVA of this sector grew 7.7 times between 1995 and 2016. A similar development (growth of GDP/place and GVA in transport (or GVAT), with similar values) has been observed, according to the data

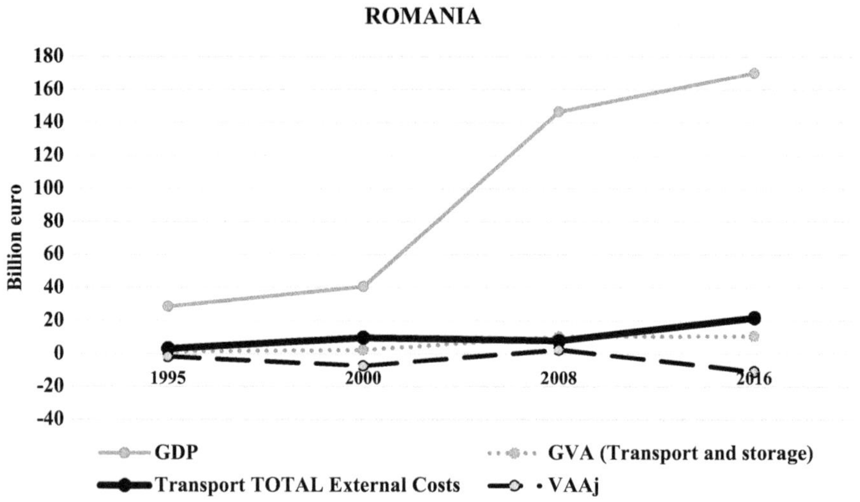


a)

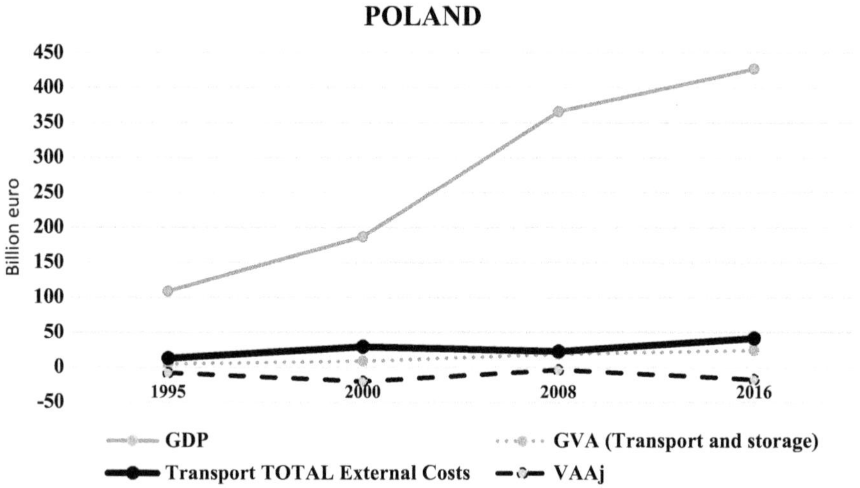


b)

*Figure 4.3* Developments in indicators specific to general economic activity and transport for some EU countries between 1995 and 2016. (By author.)

*(Continued)*

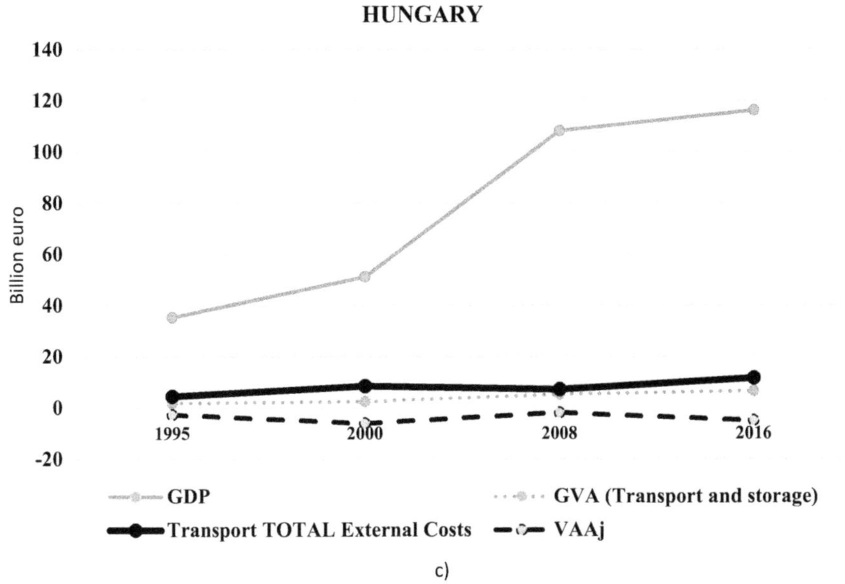


*Figure 4.3* (Continued)

analysis in Table 4.2, both at the EU level and in most of the countries investigated. An exception is the case of Poland and Hungary, where GVA growth was much higher than that of GDP/inhabitant.

Returning to the situation in Romania, for the period under analysis, the increase in the level of GVA in transport caused a similar evolution of the level of external costs specific to the road sector and of the negative value of the $VAA_j$. Also, the total level of external costs caused by transport was about 703% higher in 2016 than in 1995, which is very significant in the context of the almost sixfold increase in national GDP over the same period. However, in Romania's case, the concomitant upward trend in GDP levels and total external costs is quite clearly linked. In this context, the ratio between the level of external costs specific to transport and GDP increased by almost 1.2 times between 1995 and 2016.

Romania's situation is, however, different from that of the other countries analysed, where the positive evolution of GDP did not coincide with the increase in the share of total external costs caused by transport in this indicator. It is interesting to note that in all cases, including Romania's, the percentage increase in the share of external costs in GDP caused by road transport was higher than that recorded for the entire transport sector. This highlights the fact that road transport has remained the largest contributor to the emergence of negative externalities, and, unfortunately, the trend remains negative.

In view of all these aspects, we can state that the level of external costs remains high and is constantly rising, which calls for the swift adoption of firmer and more effective measures to internalise them.

From this perspective, the process requires regular analysis in order to adopt concrete and effective long-term policies and strategies. In this process, it is very important to provide a detailed presentation of all aspects, causes and effects for the actors involved in these activities (the general public, transport operators, decision-makers, etc.).

The need to adopt an economic (transport-oriented) strategy that takes into account the use of total social costs (thus including external costs) is an important current issue and requires an analysis of all influencing factors and 'actors' involved. It is obvious that there are different perspectives and even interest in this type of analysis. Therefore, an effective and comprehensive strategy must take into account the views and demands of all market groups.

As mentioned above, in the last decades, there has been a growing interest in expanding the analysis of the role and importance of external costs in right-sizing transport activities. The laborious, costly and even not uniformly agreed methodologies among specialists worldwide have led to the situation where estimates of the value of these external costs are very rarely made.

Estimates of total costs were often necessary in the work undertaken, since tariff regulations based on marginal costs often required adjustments that were based on total social costs.

Moreover, the findings of numerous studies have shown that external cost values have limited usability. Therefore, it may be more important to have generally accepted methodological guidelines rather than a constant assessment of the values of these types of costs.

Pricing policy proposals in the transport sector should be made in line with the pricing principles of the White Paper. Thus, for example, air transport (air traffic management and airport charges) may in the near future have similar information needs as rail and road freight. For the specific cases of urban road and interurban passenger car charging, there may be a greater direct interest in external cost estimates. Therefore, the development of methodological approaches and additional information related to this area of analysis is important.

In the context of the above, Table 4.3 summarises the influences of transport policies on economic development in general, health, environmental protection and energy security in particular.

As can also be seen from Table 4.3, the transport sector unfortunately has a significant negative impact on nature and the built environment. Fossil fuel combustion associated with transportation leads to emissions of pollutants that cause damage to human health and sensitive ecosystems. Transport can also contribute to the degradation of urban environments, with a marked reduction in the quality of life and economic efficiency.

Policy decisions to significantly reduce $CO_2$ emissions and the climate change they cause lead to effective action across all sectors of the economy, including transport.

In addition, the need for a sustainable mobility plan for transport requires the implementation of appropriate legislative instruments. In this approach, the

*Table 4.3* Links between Transport, Economic Development, Environment, Health and Energy Security

| *Areas of Analysis* | *Areas of Influence* | *Types of Supporting Activities* | *Reducing Road Transport Demand/Diversifying Mobility Options* | | | | | | | *Improving Vehicle Performance* | | *Promoting the Use of Clean Fuels* | | |
|---|---|---|---|---|---|---|---|---|---|---|---|---|---|---|
| | | | *Promoting Compact Development and Mixed Land Use (Infrastructure and Natural Habitats, etc.)* | *Building/ Modernising Non-motorised Transport Infrastructure* | *Internalising External Costs* | *Appropriate Regulation and Charging of Parking Spaces* | *Promotion of Public Passenger Transport* | *Improve and Diversify the Mobility Offer* | *Use of Intelligent Transport Systems* | *Effective Measures to Reduce Polluting Emissions* | *Investing in Research, Development, Innovation* | *Adoption of Supporting Legislation* | *Investments to Support the Production of Environmentally Friendly Vehicles* | *Information Campaigns* |
| Economic development | Functionality and efficiency | Reducing transportation costs and time | | | | | | | | | | | | |
| | Operations | Improving infrastructure quality | | | | | | | | | | | | |
| | General economic development | Creation of new jobs, reducing losses caused by traffic congestion/ congestion | | | | | | | | | | | | |
| Urban development and equity | Accessibility | Easy access to varied mobility opportunities. Reduced travel times. | | | | | | | | | | | | |
| | Affordability | Reduced transportation costs per household | | | | | | | | | | | | |
| | Acceptability | Adoption of transport policies after public consultation | | | | | | | | | | | | |
| Health and environment | Traffic safety | Reducing accidents | | | | | | | | | | | | |
| | Universal access | Promoting solutions for access to transport services for all categories of people | | | | | | | | | | | | |

(*Continued*)

*Table 4.3* (Continued)

| *Areas of Analysis* | *Areas of Influence* | *Types of Supporting Activities* | *Reducing Road Transport Demand/Diversifying Mobility Options* | | | | | | | *Improving Vehicle Performance* | | *Promoting the Use of Clean Fuels* | | |
|---|---|---|---|---|---|---|---|---|---|---|---|---|---|---|
| | | | *Promoting Compact Development and Mixed Land Use (Infrastructure and Natural Habitats, etc.)* | *Building/ Modernising Non-motorised Transport Infrastructure* | *Internalising External Costs* | *Appropriate Regulation and Charging of Parking Spaces* | *Promotion of Public Passenger Transport* | *Improve and Diversify the Mobility Offer* | *Use of Intelligent Transport Systems* | *Effective Measures to Reduce Polluting Emissions* | *Investing in Research, Development, Innovation* | *Adoption of Supporting Legislation* | *Investments to Support the Production of Environmentally Friendly Vehicles* | *Information Campaigns* |
| | Physical activities | Improving health through sport | [black] | [black] | [white] | [white] | [white] | [white] | [white] | [grey] | [grey] | [grey] | [grey] | [grey] |
| | Air pollution | Reducing air pollution | [black] | [black] | [black] | [black] | [black] | [black] | [black] | [black] | [black] | [white] | [black] | [white] |
| | Noise | Reducing noise pollution | [black] | [black] | [black] | [black] | [black] | [black] | [black] | [white] | [white] | [white] | [white] | [white] |
| | Impact on ecosystems | Protection of ecosystems. Reduced use of space for infrastructure | [black] | [white] | [white] | [white] | [white] | [white] | [white] | [grey] | [white] | [grey] | [grey] | [white] |
| Greenhouse gas (GHG) emissions | GHG level | Reduction of GHG pollution | [black] | [black] | [black] | [black] | [black] | [black] | [white] | [black] | [black] | [white] | [white] | [black] |
| Energy security | Diversification of energy sources | Increased use of alternative fuels | [black] | [black] | [white] | [white] | [black] | [black] | [white] | [grey] | [white] | [black] | [black] | [black] |
| | Resilience | Increased security from threats and hazards | [black] | [black] | [white] | [white] | [black] | [black] | [black] | [black] | [black] | [black] | [black] | [black] |
| | Independence from conventional (fossil) fuels | Elimination of fossil fuel use | [black] | [black] | [black] | [black] | [black] | [black] | [black] | [black] | [black] | [black] | [black] | [black] |

*Source:* processing based on Profillidis et al. (2014).

Legend: BLACK – Essential: academic agreement, positive examples of implementation; GREY – Uncertain: academic disagreement, lack of substantive evidence; WHITE – Limited/potentially opposite: academic agreement and evidence of limited or contrary effect.

general objective should be that each mode of transport pays for its external costs, one of the most commonly used procedures being known as internalisation of external costs (details of this technique will be given in a future study).

Clearly, the use of alternative methods and tools to determine the total social costs of transport has different implications for transport accounts and the government budget. Introducing external costs into the calculation of transport costs will, unfortunately, lead to a reduction in social welfare levels by making transport services 'more expensive' and thus reducing the attractiveness or even accessibility of such services for certain segments of the population. This clearly indicates that much attention should be paid to pricing policies for transport services, as the process of internalising external costs cannot be justified from a social justice (equity) perspective.

According to Meyeres, the use of marginal social costs generally increases social welfare. However, the magnitude of the social gain depends on the relative importance of the various externalities, the presence of a budget constraint and the flexibility of this constraint (Mayeres et al., 2004). However, not all groups are affected equally by marginal social cost prices. The impact on equity depends on how the budget is balanced. When society becomes less and less unequal, the instrument of 'income reorientation', which is more beneficial for lower income groups, will be preferred. In this context, it can be argued that income reorientation instruments play an important role in increasing the acceptability of policies aiming at transport price dimensioning that also include the level of external costs.

## 4.3 Influences and recalibrations of specific (energy, material, etc.) transport consumption in urban areas

Global consumption of resources and raw materials has tripled in the last 50 years, per capita consumption is almost twice the equilibrium threshold, and in the last six years, the world economy has exploited and used more than in the entire 20th century, creating major natural imbalances and unsustainable prospects for several regions, species and industries (Deloitte, 2023).

However, the pace of the transition to the circular economy remains outpaced by the annual volume of resource exploitation, including vulnerable or virgin resources. The global economy is now only 7.2% circular, down from an estimated 9.1% in 2018. This is a percentage decrease, not an absolute decrease, which shows that more and less responsible exploitation and consumption are taking place, and policies to green industries are failing to keep pace.

From an environmental perspective, alarming situations have emerged: over the last 30 years, a total area of more than 420 million hectares has been deforested globally, while almost half of the planet's soils are seriously degraded, oceans are more than 30% above the permitted limits of acidity, 85% of fish stocks are in collapse and wildlife has declined by 70%. In addition, over 70% of global carbon emissions come from the extraction and processing of natural resources in traditional, linear economy (take, make, waste) processes.

From this perspective, the shift away from fossil fuels, through the use of which transport is responsible for almost half of carbon emissions, with the prospect of exceeding 60% by 2050, is not only a major requirement but also a 'zero' priority.

In this context, meeting as many of the transport sector's energy needs as possible from sustainable sources can support the sector's transition to zero net emissions by 2050 – in line with the EU's climate change policy framework objectives. However, the plan to reconfigure the transport supply and, in particular, the vehicle fleet needs to be carefully considered, as this depends not only on political will but also, and above all, on the new technologies that can be applied and the financial strength of the 'players involved' (transport companies, private users, public authorities, etc.).

Transport accounts for about 29% of world energy demand and about 61.5% of annual oil consumption (Rodrique, 2022). The impacts of transport on energy consumption are diverse, including those specific to the construction, maintenance and upgrading of infrastructure and all related installations (buildings, supply and storage facilities, etc.). Energy consumption has a strong correlation with the level of development, with transport accounting for between 20% and 25% of the energy consumed in developed economies.

In 2020, against the backdrop of the health crisis, the largest decrease in energy consumption worldwide was recorded in the transport sector (down by 14%). Lower consumption rates were observed in air, rail and public transport, also as a direct consequence of the health crisis.

Between 2010 and 2020, energy intensity fell fastest in the freight transport sector by about 2.2% per year (IBIRD, 2023). This is steeper than the 0.4% annual decrease observed between 2000 and 2010.

New vehicle sales also declined, resulting in an overall vehicle stock that was older and less efficient. However, sales of electric vehicles performed well in 2020, with about three units sold. This represented 5% of overall vehicle sales.

In statistical terms, by 2022, according to the RFIDTIRES portal, there were 1.49 billion vehicles in operation worldwide, of which 1.11 billion were passenger vehicles and 380 million freight/commercial vehicles (RFIDTIRES, 2023). To understand these figures more accurately, we can recall that the number of motor vehicles worldwide was one billion units in 2009, which represents an increase of almost 50% of the world's vehicle fleet in just 13 years. This was more than three times faster than population growth over the same period. Thus, in 2009, the Earth had 6.872 billion people, and by the end of 2022, it had reached eight billion (Wall-Street, 2023).

Unfortunately, despite all the efforts in recent years to promote environmentally friendly vehicles such as battery electric, plug-in hybrids and fuel cell electric vehicles, more than 98% of the total number of vehicles in operation in 2023 were still powered by gasoline or diesel.

The 1.49 billion cars emit the equivalent of 6.9 billion tonnes of $CO_2$ (financially estimated at around $17 trillion) into the atmosphere each year, equivalent to burning all the coal in a train that stretches some 558,000 km long enough to wrap the Earth 13.9 times around the equator.[2]

Of the total number of cars in use worldwide by 2020, 390 million were in China (a 56.7-fold increase from 1990), 294 million in the USA and 283 million in Europe, with the three areas accounting for 62.9% of the world's vehicle fleet (ACEA, 2022).

In the EU, despite strong growth in sales in recent years, green passenger cars still account for only 5.3% of the total fleet. Electric cars and plug-in hybrids[3] represent only 0.5% and 0.6% of the fleet, respectively.

Light commercial diesel vehicles are still dominant in all EU countries except Greece. Thus, 91.2% of the EU van fleet runs on diesel, and only 0.4% of EU vans are battery electric.

Some 96.3% of all goods vehicles in the EU run on diesel, while gasoline powers less than 1% of the fleet; 0.24% of trucks on EU roads have a zero-emission powertrain, up from 0.04% in 2019.

Diesel buses account for 93.5% of the EU fleet, with only 0.9% electric and 1.4% hybrid electric. Significant electric bus fleets can be found in the Netherlands (12.4%) and Luxembourg (6.6%).

It is likely that these increases in recent years would not have reached the same level had it not been for the massive $CO_2$ emission reduction requirements in the industry. 'Fit for 55'[4] (adopted in April 2023) and the 26th Climate Change Conference (held in Glasgow, Scotland, in 2021) were, in this context, key points in the policy to support clean cars.

As one of the very few viable options for reducing carbon emissions, alongside electrification, and amid diminishing access to conventional fuels, hydrogen is enjoying growing market demand, with the EU estimating that by 2030, every €1 billion invested in hydrogen projects will create 10,000 new direct and indirect jobs (Deloitte, 2023). Of the 600 hydrogen projects announced or operational globally in 2022, the largest 25 account for about 70% of total current capacity, indicating an already diverse market with many small-scale, local projects that can cover one-off demands and a few major, regionally focused projects concentrated in Europe, the Middle East, the USA and Australia.

Hydrogen is optimised for industrial processes that take place at high temperatures, has a higher energy density than batteries, so provides more power for heavy machinery and equipment, and has the ability to store energy from intermittent renewable sources such as solar and wind, the study shows. Countries such as Japan, South Korea and the USA are considering using hydrogen to generate electricity or produce synthetic fuels.

Green hydrogen is the 'cleanest' of all the hydrogen options, and the potential and interest in adopting it varies from industry to industry. For example, the chemical and petrochemical industries already use grey hydrogen, so they would need minimal investment to convert some of their infrastructure to make some products cleaner and cheaper. Heavy industry, aviation and heavy road transport are encouraged by specific regulations to turn to green hydrogen with a horizon of 2030, while for the other transport categories, decarbonisation plans remain uncertain and are expected to materialise post-2030. As for cars and buildings (heating), they benefit from the electric alternative, so are less prone to transition to hydrogen.

The infrastructure costs for companies can decrease by up to 95% in the hydrogen hub scenario compared to the individual investment scenario (Deloitte, 2023). In addition to cost optimisation, this solution also brings benefits such as systematisation of the green transition, diversification of the labour market and development of new business models, for example, synergistic cement, methanol and steel production chains.

## Notes

1 It should be noted, however, that in this study the values of total external costs, for passengers and freight, refer only to cars and railroads, as statistical data for freight and passenger journeys for the ship and air modes are missing. In these circumstances, the unit values for the analysis in this study (for the Czech Republic, Poland, Romania and Hungary) will be, at the level of the year 2000: a) External costs caused by ROAD transport: Passenger – 76.0 euro/1000 pkm; Freight – 87.8 euro/1000 tkm; b) TOTAL external costs caused by transport: Passenger – 98,9 euro/1000 pkm; Freight – 105.7-euro 1000 tkm.

2 In making this assessment, the following was taken into account: a car using conventional fuel (gasoline or diesel) produces about 4.6 tons of carbon dioxide per year; burning one ton of coal produces 1.9 tons of carbon dioxide; each 15-m-long coal wagon holds about 100 tons of coal; and the circumference of the earth at the equator is 40,075 km, that is, 40,081 km (RFIDTIRES, 2023).

3 The Plug-in Hybrid Electric Vehicle, abbreviated 'PHEV', is a highly appreciated automotive technology for two very clear reasons: 1) the car is equipped with two propulsion engines, one thermal (gasoline, diesel) and one electric (battery powered); 2) the battery is of large capacity (perhaps even more than 10 kWh), but no more than a 100% electric vehicle, that is, it charges both while braking and individually at a household outlet or charging station. There are therefore two major advantages. First, you have lower fuel consumption than a conventional car with a combustion engine. Second, you get a longer range than an electric car, of course, also with the internal combustion engine (Grosoiu, 2023).

4 According to the EC, 'Fit for 55' is a set of proposals to review and update EU legislation and to implement new initiatives to ensure that EU policies are in line with the climate targets agreed by the Council and the European Parliament, namely, to reduce EU pollutant emissions by 55% by 2035 and fully by 2050 (EC, 2023).

## References

ACEA, *Vehicle in use in Europe, report 2022*, 2022, https://www.acea.auto/publication/report-vehicles-in-use-europe-2022/

Chapman, L., Transportation and climate change: A review. *Journal of Transport Geography*, vol. 15, issue 5, pp. 354–367, 2007, https://www.sciencedirect.com/science/article/abs/pii/S0966692306001207

Cuncev I., *Reflecţii privind competiţia rutiero-fero-viară, Revista căilor ferate române nr. 2* (Reflections on road-rail competition, Romanian Railways Review No. 2), Bucharest, 1993, https://www.researchgate.net/publication/303315099_Reflectii_privind_competitia_Cuncevrutiero-feroviara

Deloitte, *Hydrogen. Making it happen*, https://ro.register-deloittece.com/forms/registration.html?docid=354&utm_source=CP&utm_medium=RO&utm_campaign=deloitte-romania-hydrogen-making-happen, 2023

EC, *EU transport in figures-2009*, ISBN 978-92-79-79-10728-3, 2009, https://transport.ec.europa.eu/facts-funding/studies-data/eu-transport-figures-statistical-pocketbook/statistical-pocketbook-2009_en

EC, *EU energy and transport in figures, statistical pocketbook 2010*, ISBN 978-92-79-79-13815-7, 2010.

EC, *EU energy and transport in figures, statistical pocketbook 2016*, ISBN 978-92-79-79-73952-1, 2016.

EC, *Mobility and transport: Transport in the European Union. Current trends and issues*, 2018, link: https://ec.europa.eu/transport/sites/transport/files/2018-transport-in-the-eu-current-trends-and-issues.pdf

EC, *Handbook on the external costs of transport version 2019*, Luxembourg, ISBN 978-92-79-79-96917-1, 2019.

EC, *EU energy and transport in figures, statistical pocketbook 2020*, ISBN 978-92-76-17565-0, 2020.

EC, *EU transport in figures – Statistical pocketbook 2023*, ISBN 978-92-68-68-06324-8, 2023.

EC, *EU transport in figures-2024*, ISBN 978-92-68-19812-4, 2024, https://transport.ec.europa.eu/facts-funding/studies-data/eu-transport-figures-statistical-pocketbook/statistical-pocketbook-2024_en

EC (European Commission), *Population Statistics*, ISBN: 92-79-0164-3, 2006

EC Delft, *External costs of transport in Europe*, 2011, https://cedelft.eu/wp-content/uploads/sites/2/2021/03/CE_Delft_4K83_Handbook_on_the_external_costs_of_transport_Final.pdf

EC Delft, *Handbook on the external costs of transport Version 2019*, 2019, https://op.europa.eu/en/publication-detail/-/publication/9781f65f-8448-11ea-bf12-01aa75ed71a1

EUROSTAT, *Main GDP aggregates per capita*, 2021a, https://appsso.eurostat.ec.europa.eu/nui/submitViewTableAction.do

EUROSTAT, *Population on 1 January*, 2021b, https://ec.europa.eu/eurostat/databrowser/view/tps00001/default/table?lang=en

EUROSTAT, *GDP and main components*, 2021c, https://appsso.eurostat.ec.europa.eu/nui/submitViewTableAction.do

EUROSTAT, *National accounts aggregate by industry, transportation and storage*, 2021d, https://ec.europa.eu/eurostat/databrowser/view/NAMA_10_A64__custom_883756/default/table?lang=en

Grosoiu G., *What plug-in hybrid means and what models are on the market*, https://www.autovit.ro/blog/ce-inseamna-plug-in-hybrid-si-ce-modele-sunt-pe-piata/#Ce%20%C3%AEnseamn%C4%83%20plug-in%20HybridRFIDTIRES, 2023, How many cars are in the World, https://www.rfidtires.com/how-many-cars-world.html

IBIRD, *The energy progress report, 2023*, 2023, https://iea.blob.core.windows.net/assets/9b89065a-ccb4-404c-a53e-084982768baf/SDG7-Report2023-FullReport.pdf

INFRAS, *External costs of transport*, ISBN: 978-2-7461-0891-7, 2004

INFRAS/IWW, *External costs of transport. Accident, environmental and congestion costs in Western Europe*, 2000, https://www.infras.ch/en/projects/external-costs-of-transport-accident-environmental-and-congestion-costs-of-transport-in-western-europe/

Mayeres I., Proost S., Van Dender K., Wickart M., Suter S. and Van Nieuwkoop R., *Testing alternative integration frameworks – What are the effects of alternative pricing policies?* European Union Contract: 1999-AM.11157, UINTE – ITS, University of Leeds, 2004.

Moliner, E., Rosario, V. and Vicente, F., A fair method for the calculation of the external costs of road traffic noise according to the Eurovignette Directive. *Transportation Research: Transport and Environment*, vol. 24, pp. 52–61, 2013, https://www.sciencedirect.com/science/article/abs/pii/S1361920913000862?via%3Dihub

OECD (Organisation for Economic Co-operation and Development), External costs of transport in Central and Eastern Europe, https://rosap.ntl.bts.gov/view/dot/34169, 2003

Profillidis, V. and Botzoris, G., Internalization of external costs and impact on the demand of various transport models, *Proceeding of the 5th International Conference on the Research on Transportation*, pp. 148–161, Volos, Greece, 2010.

Profillidis V. A., Botzoris G. N. and Galanis A. T., Environmental effects and externalities from the transport sector and sustainable transportation planning – A review, *International Journal of Energy Economics and Policy*, vol. 4, no. 4, ISSN: 2146-455553, 2014, www.econjournals.com

Proost, S. and van Dender, K., Energy and environment challenges in the transportation sector. *Economics of Transportation*, vol. 1, issue 1-2, pp. 77–87, 2012, https://www.sciencedirect.com/science/article/abs/pii/S221201221200010X

RFIDTIRES, *How many cars are in the world?*, 2023, https://www.rfidtires.com/how-many-cars-world.html

Rodrique J. P., *Transportation and energy*, 2022, https://transportgeography.org/contents/chapter4/transportation-and-energy/

UIC (Union of Railways), *External costs of transport*, ISBN No 2-7461-0891-7, 2004, https://transportpublic.org/images/pdf/20041000-infras-externalcosts.pdf

van Essen H., Schroten A., Otten M., Sutter D., Schreyer C., Zandonella R., Maibach M. and Doll C., *External costs of transport in Europe*, Delft, Netherlands: CE Delft, INFRAS and Frannhofer, 2011, https://ce.nl/wp-content/uploads/2021/03/CE_Delft_4215_External_Costs_of_Transport_in_Europe_def_1343133550.pdf

Verhoef E. T., Nijkamp P., Rietveld P. and Lakshmanan T. R., *Benefits and costs of transport classification, Methodologies and policies*, doi:10.1007/PL00013618, 2001, https://www.sciencedirect.com/science/article/pii/S1056819023004840

Wall-Street, 2023, https://www.wall-street.ro/articol/Social/291545/care-era-populatia-planetei-in-anul-in-care-te-ai-nascut-si-unde-se-situeaza-romania-ca-numar-de-locuitori.html#gref

# 5 Prospects for sustainable transport development

The sustainable future of transport is closely linked to the implementation of efficient solutions that discourage environmental and health damage. Analyses carried out internationally over the last decades show that there are two major drivers of unsustainable transport development.

The first is transport demand, which is overwhelmingly oriented towards road transport.

The second is the supply of transport in large urban agglomerations, which is also overwhelmingly oriented towards supporting polluting mobility (mainly by road transport).

In contrast to these realities, sustainable transport development must be based on the adoption of solutions that meet the growing demand for mobility, diversify the transport offer and reduce negative environmental and health impacts.

## 5.1 Solutions for a real sustainable mobility system

In our opinion, three courses of action can be used to achieve sustainable mobility:

- Reducing transportation demand.
- Predominant use of green transport systems (vehicles, infrastructure, logistics, etc.).
- Developing a policy using a mix of the above two types of action.

Taking all these elements into account, it is obvious that using a combination of simultaneous actions to reduce transport demand and the use of green transport systems is the ideal solution.

But what would these two elements of economic policy entail? Reducing demand for transport can only be achieved by actions to reduce its specific components.

However, the general microeconomic theory of demand is not applicable to transport in its established form. This is mainly due to the obvious differences between transport and any other consumption activity.

DOI: 10.4324/9781003743064-6

Thus, the demand for transport is a derived demand, which means that transport is an intermediate consumption in relation to a final consumption at the destination of a different nature. Second, the quality of the service offered is of particular importance in transportation demand (e.g. speed and safety of travel). Third, there is a much more important random component in the mode choice process than in other forms of consumption (Kanafi, 1983).

Transportation demand can vary under two concurrent circumstances: the volume of transportation (the number of passengers or the amount of freight) varies, or the distance over which this volume is transported varies.

Demand, in terms of volume, structure and level of consumption requirements, changes from one period to another and is therefore dynamic. The main factors on which the dynamics of demand depend are needs, income and price.

The law of demand expresses the relationship between demand and price in which demand moves inversely to price. The notion of price elasticity lies at the heart of the demand for transportation and refers to the variation in demand in response to a variation in cost. The price elasticity of transport demand can influence the strategic behaviour of economic actors.

For example, container shipping lines face highly inelastic demand due to the combined effect of the lack of close substitutes and the small impact of transportation rates on total costs (Rodrigue, Notteboom, 2020). The only alternative transport mode in intercontinental transportation of high-value intercontinental freight is air transport, but this market segment has much lower freight capacity and much higher prices.

For most shipments, the total transportation price represents only a small part of the total value of the shipment, usually less than 5%. Since container lines cannot influence the size of the final market, they try to increase their market share in the short term by reducing prices.

As such, shipping lines can reduce freight rates without substantially affecting the underlying demand for container shipping. The only additional demand may come from low-value products, which will only be shipped abroad if shipping rates are very low (e.g. the waste paper and scrap metal market).

These temporary markets tend to disappear once the transportation rate exceeds a threshold level, not allowing a profit on trading these products abroad. The relatively inelastic nature of demand for shipping services is the central problem for the poor financial performance of container shipping lines.

Shipping lines have developed an intense cost focus and have negotiated long-term contracts with large shippers to secure cargo that will only be shipped abroad if freight rates are very low (e.g. the waste paper and scrap metal market). These temporary markets tend to disappear once the transportation rate exceeds a threshold level, not allowing a profit on trading these products abroad.

The relatively inelastic nature of demand for shipping services is the central problem for the poor financial performance of container shipping lines. Shipping lines have developed an intense cost focus and have negotiated long-term contracts with large shippers to secure cargo to be shipped abroad only if freight rates are very low (e.g. the waste paper and scrap metal market).

The most important factors influencing transportation demand are as follows:

- From demographic perspectives:
  - Number of people (residents, employees and visitors).
  - Employment rate.
  - Wealth/income.
  - Age/life cycle.
  - Lifestyle.
  - Preferences.
- From the perspectives of economic activities:
  - Number of jobs.
  - Type of economic activity.
  - Freight transportation.
  - Tourism.
- From the perspective of transportation options:
  - Walking.
  - Cycling.
  - Public transport.
  - Transport in partnership (ridesharing).
  - Taxi.
  - Telecommuting.
  - Transportation services for delivery to a predefined address.
- From a land-use perspective:
  - Density.
  - Potential for use as transportation infrastructure.
  - Connectivity.
  - Design of infrastructure networks.
  - Proximity to centres generating transport flows (economic, educational nodes, etc.).
- In line with demand management:
  - Prioritisation of the use of transport networks.
  - Price reforms.
  - Promotion campaigns.
  - Detailed and comprehensive information to users of transport services.
  - Creating conditions for fair intermodal competition.
- From a price perspective:
  - Level of fuel prices and related taxes.
  - Level of transport insurance (correct setting of the Statistical Value of Life indicator).

- Internalisation of external costs.
- Labour costs.
- Level of total social costs, expressed by the formula:

$$C_g = P + vt + aC + bS = cE \tag{5.1}$$

where

$P$ = transportation fare.
$t$ = transportation time.
$v$ = individual monetary value attributed to time.
$C$ = comfort conditions indicator.
$S$ = safety conditions indicator.
$E$ = indicator of external effects on environmental and health factors (noise, pollution, vibration, etc.).
$a, b, c$ = parameters corresponding to comfort, safety and external effects.

In conclusion, we can say that transport demand has the following characteristics:

- It is derived.
- It is different from the notion of traffic.
- It is strongly influenced by the quality of service as a supply parameter.
- The elasticity of transport demand must also be analysed in terms of the generalised cost, which contains terms 'particularising' (individual unit monetary value) and indicators of comfort, safety and environmental conditions. Thus, the generalised cost formula is as follows:

Therefore, in the case of transportation, we consider demand to be a function of the form:

$$C_t = f(P_t, CA_t) \tag{5.2}$$

where

$CA_t$ = transport-specific characteristics.
Or more concretely

$$CA_t = f\left[(V_t, D_t, T_t) / P_t\right] \tag{5.3}$$

where

Ct = transportation demand.
$P_t$ = transportation price.

$D_t$ = transportation distance.
$T_t$ = transportation time.
$V_t$ = transport volume (goods and passengers).

It is obvious that transport demand decreases with increasing prices, and increases with increasing specific transport factors (volume and distance).

Moreover, statistically, at the international level, transport demand is measured by the size of the freight and passenger journeys.

From all these considerations, it is clear that solutions must be found to reduce transport demand (in order to reduce the natural degradation of environmental factors) by reducing the distances or volumes transported.

Among the less popular methods would be to increase transport prices. Unfortunately, this is happening today for reasons totally different from environmental principles. More specifically, the price of transport increases if more environmentally friendly vehicles are used (an example is the higher cost of transporting goods by rail over the same distance, compared with road), which is contrary to the objective of reducing the negative effects on the environment and human health. This situation must be remedied, and an immediate solution would be to make it compulsory for all transport operators (road, rail, air and sea) to pay all the social costs of transport, including external costs (which are largely subsidised indirectly, especially in the case of road transport).

The use of environmentally friendly transport systems has been gaining momentum recently, especially for road vehicles. In this case, electric vehicles are becoming more widespread, but unfortunately, only part of the problem is being solved globally. Thus, while pollution levels in congested areas will decrease significantly, the level of pollution at the sources of electricity generation (the main source of battery electric vehicles) will increase.

In this context too, the electric vehicle solution must be considered as a good one for areas with high road traffic intensity, but in order to keep the environmental balance, it is necessary to accelerate the shift of a large share of transport demand to more environmentally friendly modes (multimodal, rail, etc.).

In the future, according to VIRTA estimates, there are three possible scenarios (VIRTA, 2022):

- The state policy scenario suggests that, by 2030, the global stock of electric vehicles (excluding two/three-wheelers) will reach almost 200 million vehicles and account for about 10% of the global vehicle fleet.
- Under a more ambitious scenario, 270 million electric vehicles will be roaming by 2030, and the share of electric vehicles in the global vehicle stock will reach 14%.
- By 2050, the scenario anticipates that the global electric vehicle stock will reach 350 million electric vehicles and its share will increase to 20% by 2030.

The next two years will be crucial for Europe to secure its leadership. We expect to reach the first milestone of 14 million electric vehicles in 2025. Thereafter, low estimates are 33 million electric vehicles by 2030, while high estimates are 40 million electric vehicles travelling in Europe at the same time.

From 2035, all new cars sold in Europe are expected to be electric.

Among the European concepts to be taken into account in developing a sustainable transport development strategy, the most important are the following:

- **Sustainable development of** the public and private transport system, a concept which encompasses both the resource, waste, pollution, etc., aspects and the social dimension of transport development – mobility of the population.
- **Developing competition**:
  - Competition on the market, between modes of transport, between groups of operators or between individual operators.
  - Competition for the market through competitive bidding for concessions to use individual routes or the entire network.
- **Integration of services** at the modal level, intermodal interface, integration with other public services and integration into the socio-economic environment.
- **Stimulating the dissemination of information** and knowledge on transport services, using best practices and advanced technologies.
- **Establishing quality criteria** for transport modes.
- **Establishing an appropriate policy and legislative framework** to promote the increased use of transport systems.
- **Efficient use of financial instruments** so as to optimise the economic contribution of the transport system for local, regional, national and international benefit.
- **Social equity**, given that transport is interdependent with the whole field of land use, socio-economic activity planning.
- **Environmental protection**.

The European Commission proposed to the Council of Europe at the Gothenburg Conference a series of measures to improve transport systems in line with the principles of sustainable development.

It is worth noting that for the first time in the European Union's political strategies, transport and land use are treated under the same group of priority issues.

This means that the emphasis in the development of European transport will be on **intermodal rather than on door-to-door transport**.[1]

There is, of course, a case for this option, given the density of transport networks throughout Europe and the territorial limitations on their further development, as well as the high proportion of short-haul transport. This, however, leads to another set of problems, posed by the increase in the volume of marginal services (primarily transhipment and storage services for goods transport) and the increase in transport times.

While in land and inland waterway transport, this option can be considered optimal or optimisable, in air and maritime transport, it is questionable in terms of economic efficiency. It is primarily a question of the impact that the development of intermodality will have on air and sea shipbuilding: this option implies an increase in the production of small and medium-sized vessels and relatively short routes. Sea and air transport have a global component that no longer restricts transport routes to the European area, which would imply promoting the construction of large-capacity ships capable of long voyages.

The following goals and tasks result from the document adopted in Gothenburg, with a view to developing transport on a competitive and sustainable basis:

Main objectives

- Decoupling the development of transport from that of GDP in order to reduce congestion and other negative effects of transport.
- To induce a shift in transport from road to rail and water transport and from private to public passenger transport, so that the share of road transport is no higher than that in 1998.
- Promoting a balanced development of regions by reducing disparities in economic activities and maintaining the viability of urban and rural communities, as recommended by the European Spatial Development Perspective (ESDP)

As an EU member country, Romania will have to continue the process of harmonisation of the national strategic framework with the main components of the strategy on sustainable transport policies.

For the period up to 2030, the overall objective of Romania's development strategy is the sustainable development of the economy and the improvement of the quality of life, based on the balanced development of the transport system, ensuring modern and sustainable transport infrastructure and services.

Achieving this objective will contribute directly to ensuring the sustainable development of the transport sector, increasing Romania's accessibility, ensuring the inter-modality of the transport system, promoting the balanced development of all modes of transport and increasing the quality and efficiency of services.

The specific objectives envisaged to achieve the general objective are as follows:

- Modernisation and development of the transport network of European interest and national interest, increasing safety conditions and quality of services.
- Full liberalisation of the internal transport market.
- Stimulating economic development and competitiveness.
- Strengthening social and territorial cohesion at regional and national levels.
- Environmental compatibility.

Improving public services in road passenger transport aims to achieve a minimum commercial speed of 65 km/h between origin and destination (with maximum waiting times of one hour for transhipments) by 2030, while rail should offer commercial speeds on long-distance routes at least equal to private road transport by 2030.

In freight transport, services will be improved by modifying the contractual framework, focusing on correcting existing imbalances, creating a greater degree of computerisation between the different modes and clarifying the rights and obligations of each party in intermodal services.

*Energy efficiency* aims at improving the efficiency of the transportation system in order to significantly reduce the energy dependency of the Romanian economy. To this end, over the next few years, specific energy consumption must be reduced by at least 10% in relation to the number of passenger-km and tonne-km volume, compared to the levels of 2000, while achieving a progressive reduction of up to 20% by 2030. For these purposes, the introduction of alternative, environmental-friendly fuels in the transport sector is envisaged, in compensation for conventional fuels, in line with the objectives of the European Union. At the same time, the aim is to increase the fleet of electric vehicles.

Reducing the negative impacts of transport in relation to the environment will focus on two strands: reducing the overall impacts of transport (especially climate change) and improving environmental quality in the natural and urban environment.

A real solution to reduce energy consumption and imports of oil products is undoubtedly to stimulate the long-term development of multimodal transport. This is in view of the fact that multimodal transport leads to the efficient use of modes that have spare capacity (rail, inland waterways and maritime transport) with beneficial effects on energy consumption and pollutant emissions.

Recent studies have highlighted the net energy benefits of combined transport compared to road, for example (Dragu, 2009):

- Up to 29% reduction in energy consumption for origin-to-destination travel.
- Up to 30% reduction in energy consumption per kilometre.
- Reduction of up to 55% of $CO_2$ emissions for origin-to-destination transportation.
- Up to 60% reduction of $CO_2$ emissions per kilometre.

The difference in costs between road transport and the other modes of transport is mainly because the negative effects produced by this mode of transport are not taxed at their real value.

The creation of an efficient multimodal transport system is an important element of Romania's policy in this field, because of the benefits it can bring. The achievement of the multimodal transport system is, however, inextricably linked to the quality of the infrastructure and the degree of interoperability between the various modes of transport. Therefore, the efficiency of the multimodal transport network is directly linked to the investment resources that can be attracted for the modernisation and development of infrastructure, equipment and multimodal transfer points.

From the super-centralised market of the socialist society, to the one supported by the rules of the liberal and, later, neoliberal economy, Romanian transport has been in a continuous effervescence. The encouraged modes of transport, in a certain period, reached an accelerated decline, in another, the infrastructure deteriorated but, here and there, it also adapted to the new requirements and, overall, we can say that the evolution was an acceptable one, from an economic perspective, but not from an environmental perspective.

## 5.2 Landmarks in the development of the Romanian transport market from a sustainable perspective

In Romania, the strategic framework for sustainable transport policy has been aligned (conceptually) with the European policy on sustainable transport outlined in the Transport White Paper (and its updates) and has supported, in its actions, the overall objective of ensuring sustainable transport.

The focus has been on achieving a structural reform of the transport system in its entirety by restructuring, reorganising and strengthening the institutional and administrative capacity of the line ministry and its institutions/units.

Competition in the transport market has been encouraged, and direct state participation in the market has been reduced through the privatisation of state-owned transport companies.

The changes brought about by liberalisation, and the transition to a market economy, have radically altered both the organisation of specialised transport units and the ownership structure. The number of private businesses and joint ventures has increased, with the number of private transport operators in the transport and related sectors reaching about 98% of all operators. The number of active, authorised and licensed economic operators operating in the transport sphere was about 51,700 as of 2019, of which more than 51,000 were in land and pipeline transport, 261 in naval transport, 61 in air transport and 3,523 in ancillary activities, with a total number of employed personnel of more than 327,000 employees (NIS, 2022).

As a result of the structural changes in the economy and the transition to a market economy, the evolution of freight and passenger transport has fluctuated from a sharp decrease to a slight increase.

The various modes of transport have thus 'settled' in the market, with, contrary to the measures set out in the White Paper, an aggressive growth trend in road (freight and passenger) and air (passenger) transport and an uneven development of the other modes of transport, with the objective of a balanced shift to environmentally friendly modes of transport far from being achieved.

The share of road transport in the structure of the freight transport market in 2020 was 67.7% by volume and 68.0% by freight transport by volume, comparable to the EU-27 (76.5%) but higher than the White Paper (44%).

The modal evolution in the specific transport markets (freight volume and freight journey) changed substantially between 1990 and 2020:

- Air and inland waterway transport have seen very large increases in both freight volume and freight journeys.
- However, compared to the total freight volume (over 2.1 billion tonnes in 1990 and 470 million tonnes in 2022), inland waterway transport remains relatively small (0.5% in 1990 and 5.9% in 2022).
- The same trend is observed for the movement of goods by sea, with the exception that in 2022, as a result of the more than 15-fold increase in the value of this indicator, sea transport will have moved into second place in the modal hierarchy of the national market, ahead of rail transport.
- Moreover, the most unstable trend is observed in rail freight. In terms of the total volume of freight, rail freight has increased by more than 11.4% over the period under analysis, while freight journeys have fallen by almost 50%. This can be explained by the shift in rail freight from medium and long distances to shorter distances, which are, in principle, also less efficient in terms of transport costs.
- A completely opposite trend is observed for road freight transport, where the volume of goods transported in 2022 is almost four times higher than in 1990, against the background of a decline in the volume of goods transported. This situation is mainly caused by the goods transported by trucks (TIRs), which transport goods over long distances, with small volumes of goods, but cover areas and transport segments that cannot be transported by other modes (no railways, waterways, airports, etc.).

Market demand for freight transport has generally increased by 2–3% above the annual GDP growth rate, mainly due to developments in the European economy and 'globalised' production systems, and the high share of low-value-added goods in the national economy, which does not indicate a dissociation and orientation towards sustainable transport.

In fact, total energy consumption has continuously increased, reaching 31% higher at the EU level in 2006 than in 1995, and in 2022, it will reach 279.9 Mtoe[2] (EC, 2024). Compared to 2006, when the figure was 370.1 Mtoe (EC, 2009), there is, however, a significant decrease of almost 25%. This development is largely due to the policy emphasis of recent years in which the European Commission has imposed strict rules on Member States to improve efficiency and reduce transport-specific energy consumption and to reallocate significant shares of the transport market from energy-intensive and polluting modes to the more environmentally friendly ones.

Road transport grew by 1.58%, while other transport modes decreased by between 1.38% and 3.08%. By 2022, transportation in Romania had a final energy consumption level of 7.4 Mtoe, equivalent to about 31% of national energy consumption (EC, 2024). Out of the amount consumed by the transport sector, more than 95% was accounted for by road transport (goods and passengers).

Pollutant emissions from the transport sector have steadily increased compared to 1990, largely due to the road mode.

As an example, if in 1990, transport in Romania was responsible for about 13.3 million tonnes $CO_2$ equivalent, in 2022, this value increased by more than 61%, reaching 21.5 million tonnes $CO_2$ equivalent (EC, 2022a). Of this, 73.2% was due to road transportation.

Road traffic noise has also increased in recent years, with an estimated 10–15% of the population exposed to high noise levels (between 65 and 75 dB).

The policy of renewing and modifying the fleet structure (developed by privately owned transport operators) has created the conditions for a significant decrease in pollutant emissions (especially NOx), comparable to the limits of the National Emission Ceilings Directive 2001/81/EC.

The gradual introduction of the EURO-2 (1998), EURO-3 (2002), EURO-4 (2008), EURO-5 (2011) and EURO-6 (2015) standards, unleaded gasoline and low sulphur diesel, have led to significant decreases in some pollutant emissions from motor vehicles (in particular, sulphur dioxide, volatile organic compounds, benzene, lead and carbon monoxide).

However, the increase in the number of vehicles, and hence in road traffic, has led to the undesirable situation where the number of accidents and fatalities in accidents has remained high.

Moreover, although in 2022 Romania reported only 1,633 road fatalities, almost 57% fewer than in 1990 (EC, 2024), it remained one of the most dangerous countries in the European Community in terms of this type of event. In 2022, Romania would rank first in the EU in terms of the number of road accident fatalities per million inhabitants (85 people), per 10 billion passenger × km (156 people) or per million vehicles (232 people). By comparison, the EU average in the same year 2022 was, for the three indicators presented above, 42 persons, 51 persons and 76 persons. Romania's values are therefore two, three and even four times higher than the EU average, which is unacceptable.

The conclusion is that the number of vehicles in traffic has increased, the infrastructure has not 'kept pace' with this increase, and some drivers have 'adapted' to the new conditions by flouting traffic rules and real traffic conditions, which has led to the negative trend shown. From this perspective, sustainable transport has corrective tasks, based on offering efficient alternatives from an economic and environmental perspective.

At present, although Romania's transport infrastructure has undergone some development/modernisation, both partially and in certain modes, it has not managed to achieve the expected competitiveness parameters, lagging behind both the EU Member States and the objectives set out in the White Paper and the European Sustainable Development Strategy. The most serious, however, has been the fact that, over the last three decades, the quality and size of the transport infrastructure network have not kept pace with national needs, which has not been conducive to Romania's better and more intensive socio-economic development.

The inadequate capacity of the infrastructure network has reduced the degree of coping with the increase in freight and passenger traffic in some areas (especially the entrances to major cities), at certain times of the year (especially summer and winter in mountainous areas, weekends, etc.) or in the event of accidental events (snow, ice, accidents), which has led to an increase in the number of congestion and automatically to a worsening of transport quality parameters and failure to achieve the objectives imposed by the White Paper directives.

The changes in urban development policy have led to the growth of residential and commercial areas and the expansion of the intra-urban space, especially along the national road network, integrating national roads into the street network of localities for tens of kilometres, with direct consequences on the operating and safety parameters that national roads must provide. The same situation is becoming increasingly apparent at airports, most of which have been/will be integrated into residential areas.

This situation, together with the capacity limitations of the urban road network systems (including parking spaces), which are inadequate for the explosive growth in the number of road vehicles, has turned national roads and urban settlements into veritable *parking lots* and has increased the level of pollutant emissions and traffic noise.

Access to western European corridors, as well as to eastern and southern European corridors, is hindered and limited by the traffic capacity and poor quality of certain elements of Romania's transport infrastructure, which is in a position to disrupt the free movement of goods and persons and to reduce international goods/passenger traffic transiting Romania.

The delays, for objective reasons (securing funding, drawing up technical documentation, contracting works and execution of works), in carrying out transport network modernisation projects have increased the number of networks that have reached the limit of technical and operational exploitation, increased the volume of project realisation/implementation costs and generated the risk of not achieving the White Paper objectives on the development of the trans-European transport network, in particular the TEN-T corridors through Romania.

The situation of the national transport network was the main factor that led to the diminished possibilities to capitalise on the opportunities offered by Romania's geo-political position, namely the following:

- Location at the crossroads of the roads linking East and West Europe and South and North Europe, as well as on the transit axes linking Europe and Asia.
- The Danube and the opening to the Black Sea.

In the context of limited financial resources, progress in the development of the national transport network has been minimal, with notable results being achieved more in areas such as the modernisation and partial rehabilitation of existing national infrastructure.

The signing of Romania's Accession Treaty to the EU on 25 April 2005 (EC, 2022b) accelerated the decision-making process to support the national sustainable transport development process. In this context, the first official act that inserted the notion of sustainable transport was the January 2007 Strategic Environmental Assessment for the Sectoral Operational Program – Transport Infrastructure 2007–2013 (POS-T) carried out by the Romanian authorities in 2006. In the evaluation document, from the very beginning, it is emphasised that it is necessary to: 'complement and modify the overall and specific objectives of the POST by emphasising the objective of developing sustainable transport' (SEA-POST, 2007). After Romania's effective accession to the EU on 1 January 2007, transport legislation has had to adapt permanently to EU requirements, and sustainable transport has played an increasingly important role in this context.

Romania, as a member state of the United Nations Organization (UN) and the European Union (EU), has expressed its adherence to the 17 Sustainable Development Goals (SDGs) of the 2030 Agenda, adopted by UN General Assembly Resolution A/RES/70/1, during the UN Summit on Sustainable Development in September 2015. The EU adopted the EU Council Conclusions of 20 June 2017 through the document 'A sustainable future for Europe: the EU response to the 2030 Agenda for Sustainable Development', which set out the main directions for action by EU Member States to implement the 2030 Agenda for Sustainable Development. In this new context, in 2018, the National Strategy for Romania's SUSTAINABLE DEVELOPMENT 2030 – NSDSDR 2030 (RG, 2018) was drafted, in which an extremely low place was given to transport. More specifically, the extremely brief situation of the national transport infrastructure was presented, emphasising in the opening of the chapter dedicated to transport that:

> The transport sector is considered a priority in the context of Romania's development plans, given its interdependent relations with the other branches of the national economy, the value of the services offered to the population and its considerable impact on the environment. Major shortcomings include sustainability, economic efficiency, safety, environmental impact, contribution to the development of other branches of the economy, contribution to ensuring internal and external connectivity and securing funding for investment.

## 5.3 Supporting elements for a sustainable transport development strategy in Europe

Ensuring sustainable transport requires a combination of several types of policies, including supportive and corrective ones, ensuring the involvement of an increasing number of actors, representatives of the transport sector, public administration, and citizens.

The negative effects of individual actions can only be mitigated by applying a mix of measures that are also acceptable to the majority of citizens.

Of the many measures needed, we mention those which are directly or indirectly related to road transport:

1. The use of technological progress (measures to improve energy efficiency, new standards/regulations for engines and fuels, use of new technologies and alternative fuels).
2. The implementation of new market-based instruments (taxes/charges based on environmental impact or congestion, fiscal incentives, the adoption of an 'Emissions Trading Scheme (ETS)' taking into account the specific characteristics of the different modes of transport).
3. Optimising the use of means of transport and specific infrastructure to encourage businesses and citizens to change harmful habits (overuse of private cars for daily journeys, reduction in unjustified demand for mobility, etc.).

From this perspective, EU Member States should focus their actions on key areas of the system as follows:

1. Congested metropolitan and urban areas, where the majority of journeys are made, mainly by road.
2. Major European inter-urban corridors, where most intra-EU and international trade is concentrated.
3. Environmentally sensitive areas (Alpine region, protected areas, etc.).

It is equally important to use economic incentives to promote fuel-efficient and low-emission vehicles, such as tax exemptions or reductions (annual taxes, compulsory insurance, road user charges, etc.); reforming the system of vehicle excise duties according to the level of pollutant emissions and energy efficiency; introducing incentives to take the most polluting vehicles off the road and to purchase new, low-emission cars.

The Council and the Member States are to accelerate their investment in infrastructure and intelligent transport systems, aiming in particular at the following:

1. The completion, as soon as possible, of priority projects in the Trans-European Networks sector, which are of key importance for freight transport logistics and for a sustainable European transport policy.
2. Addressing the problems of congestion in urban areas and along corridors crossing sensitive areas.
3. Improving intermodal transport.

It is also considered necessary to develop a generally applicable, transparent and comprehensible model for assessing the external costs of all modes of transport to serve in future as a basis for calculating infrastructure charges.

This model should be accompanied, in line with the Eurovignette Directive, by an analysis of the effects of the internalisation of external costs for all modes of transport and a strategy for the progressive application of this model to all modes of transport;

As regards the problem of urban agglomerations, which are responsible for 40% of $CO_2$emissions and 70% of other pollutant emissions from motor vehicles, a more ambitious approach must be taken, in compliance with the principle of subsidiarity, by implementing a strategy of cooperation and coordination at the European level.

An effective urban mobility policy must take into account both passenger and freight transport and therefore be based on as holistic an approach as possible, able to bring together the optimal solutions for each problem. Urban areas continue to offer a good economic potential for modal shift policies in favour of public transport, walking and cycling and a new approach to urban logistics. It is considered particularly important in this context to focus on technological innovation (wider use of intelligent transport systems, ITS), better use of existing infrastructure, in particular through demand management measures (use of differentiated charging – congestion charging and road pricing), innovative solutions to optimise the integration of urban freight flows and, lastly, new solutions to optimise the use of private cars, such as car sharing, car-pooling and home-working arrangements.

There is a pressing need to reduce the external costs caused by transport systems (activities and infrastructure). The scale and diversity of the external costs caused by transport make it necessary to take a firm stance in support of all actions to reduce them. Of course, reducing external costs can be achieved through two major types of action.

The first is aimed at reducing transport demand and hence transport activity. The adoption of measures that lead to a reduction in transport demand, without being imposed (forced), needs massive support from the legislature (authorities and decision makers) and the economic environment as a whole. More specifically, there is the possibility of reducing business travel where economic activities and legislation allow it. For example, employees who are able to work from home no longer have to travel to the company's head office, but can solve the specific problems of their work via the communications system (e.g. the Internet). Similarly, elementary education (pre-school, school) can be provided remotely, to avoid pupils having to travel to educational establishments. The examples go on. However, things appear to be more complicated for the transport of goods, where it is obvious that it is not possible to teleport goods for the time being, and it is therefore necessary to reorganise the movement of goods by adopting solutions that reduce the number of goods transported to the same destination. The best example of this is multimodal transportation. Instead of numerous trucks transporting goods on congested roads, trains, ships or aeroplanes can be used for such routes between the main hubs. From those points, obviously, for the time being, the solutions only offer transportation by car.

The second type of action aims at internalising external costs. For this type of policy, all those measures are envisaged which reduce the effects borne by society as a result of specific transport activities. As mentioned above, the measures that can be adopted to achieve this aim are fiscal, administrative, legislative and technological.

Clearly, an effective and coherent strategy must simultaneously adopt measures for each type of externality, incorporating them into a single system designed to achieve the proposed objectives. Therefore, to begin with, it is necessary to know which policies are specific to each type of externality.

In the case of air pollution, economic theory emphasises the need to introduce corrective measures directly at the source, aimed at counteracting negative effects. Obviously, measures can be both regulatory and fiscal.

Regulatory policy is currently based on actions aimed at lowering the level of polluting emissions by setting certain maximum permissible limits. At the same time, the introduction of new types of environmentally friendly (clean and energy-efficient) vehicles and fuels should be encouraged.

Fiscal policy aims to achieve two objectives simultaneously: to stay within the maximum permissible limits and to avoid market failures in the specific (transport) market.

The quantity of pollutants emitted depends, however, on numerous characteristics such as the technical parameters of the vehicles (engine power and type, etc.), the quality of the fuels or the way the means of transport are used (driving style, for example). From this perspective, an effective policy must adopt the solutions which entail the lowest costs and, at the same time, spread the marginal cost of each action in the fairest possible way, while taking account of existing situations.

The introduction of a system of charges proportional to the level of emissions (e.g. a charge based on distance travelled and travel characteristics) is the most 'tempting' economic instrument that can be used to internalise the external costs of air pollution. The level of taxes imposed on the sources of pollution should be equal to the marginal value of the damage caused. At the same time, the number of fiscal instruments should be differentiated according to the number of pollutants due to each source, or more precisely, by the level of induced external costs. For example, differentiation of taxation according to the quality of the fuels used will result in encouraging the introduction and accelerating the use of alternative, non-polluting fuels. In Europe, the tax system (excise duties) imposed on gasoline has evolved in favour of unleaded. In 1986, for example, the difference in price between normal (leaded) gasoline and unleaded gasoline was 1%, whereas in 1993, the difference between unleaded and leaded gasoline was 53%.

The exclusive adoption of a taxation system based exclusively on the level of polluting emissions does not, unfortunately, encourage the use of alternative systems designed to reduce the negative effects (the use of catalytic converters in cars, for example). However, recent studies show that the costs of improving fuel quality or vehicle characteristics (from an eco-energy perspective) are lower than the costs borne by society as a result of air pollution.

Here, too, there are a number of peculiarities. Where the equipment used to reduce pollution is expensive, a differentiated system of vehicle purchase taxes should be introduced to encourage the purchase of environmentally friendly vehicles. Such policies have been adopted in many European countries, such as Germany and the Netherlands.

Applying such tax measures is advantageous and relatively simple to administer. The drawback, however, is that, once the taxes have been paid, the haulier is no longer given an incentive to reduce pollution.

Another difficult problem is the non-uniformity of pollution levels along the routes, which is due to differences in both traffic patterns and sources of pollution. In this context, it has been observed that, in general, a relatively small percentage of vehicles (especially old ones) are very polluting compared to the total. Alberini shows, for example, that in the USA, although the share of in-service vehicles manufactured before 1980 is relatively small (less than 18% of the total fleet, covering 8% of the total vehicle miles travelled), they still cause more than 40% of the damage from emissions of unburned hydrocarbons (HCT) and 25% of the damage from NOx (Fistung, 1999). The only way to remedy this situation is to replace these vehicles with more environmentally friendly ones. However, carrying out a cost–benefit analysis for such a measure is very difficult for several reasons. First of all, it is difficult to adopt a function that would show the links between the age of vehicles and the level of pollution they cause. There are also situations in which 'older' vehicles pollute less than 'newer' ones, due to their maintenance or because they are less used.

From this point of view, any method adopted must address the problems in general, and not just target old vehicles. In this way, the introduction of cost-effective economic instruments will encourage the 'natural selection' of polluting vehicles over others.

Inadequate maintenance of vehicles or poor (environmentally friendly) fuel quality can also be causes of high pollution. Betts shows that only 6% of diesel vehicles need repair work to reduce air pollution, compared with 30% of gasoline engines equipped with catalytic converters and 80% of normal gasoline engines without catalytic converters (Betts et al., 1992).

In Great Britain, in 1998, 10% of the entire car fleet was responsible for about 44% of the level of air pollution, while in Romania, in the same year, more than 80% of the car fleet in operation did not comply with the maximum permissible limits for air pollution.

This is why it is necessary to introduce measures making it compulsory to carry out routine vehicle maintenance and to measure whether vehicles fall within the limits considered permissible in terms of environmental pollution.

At the same time, other measures should also be taken to reduce fuel consumption and, consequently, the level of pollution, which should be aimed primarily at improving traffic conditions (improving the quality of infrastructure, traffic flow, etc.).

Reducing pollution therefore involves both economic and technical solutions and changes in consumer behaviour. From this perspective, the most economically efficient method is to introduce a tax system where the level of taxes imposed is proportional to the degree of pollution per litre of fuel consumed. In this way, users will have to adapt (technically, behaviourally) to the new situation, with the higher overall costs forcing them to opt for environmentally efficient solutions.

With regard to reducing noise pollution, it appears to be quite difficult to introduce a set of economic measures that are both efficient and equitable. Theoretically, in the case of the adoption of fiscal policy elements, the level of taxation should be related to the value of the marginal effect produced by vehicles. This effect depends on where, when and for how long. It is therefore necessary to adopt, at the same time, a policy regulating the way in which traffic is monitored, highlighting the noise levels reached. This is very costly and therefore unlikely to be achieved any time soon. Rather, speed reduction measures can be adopted in certain places and at certain times, as the level of noise pollution is, as mentioned earlier, strongly influenced by, among other things, the speed of vehicles.

Certain types of fiscal instruments are currently being used in Europe and are imposed:

- New vehicles, depending on the level of noise produced during operation.
- Vehicle manufacturers, which are also differentiated according to the noise level in which production falls.
- Vehicle owners, with the size of the tax differentiated according to the noise level of the vehicle in operation and tested in special laboratories;
- Users who own 'noisy' vehicles when they travel in environmentally 'sensitive' areas.
- All those involved (manufacturers, users, etc.) to encourage the production and/or purchase of quiet vehicles; in the Netherlands and Germany, for example, such a method is being used successfully. The reduction of excise duties by 7.5% and 5%, for certain types of vehicles with a noise level 6 dB(A) and 3 dB(A) below the maximum permissible limit, respectively, has meant that in the Netherlands alone, over a period of a few years, 60% of all heavy goods vehicles have been within the 5 dB(A) limit below the maximum limit. Obviously, the fiscal instruments listed can be used to reduce pollution from road, rail and air transport. In the future, in the case of rail transport, it will also be possible to introduce certain charges for use of the network, depending on the technical characteristics of the trains or the conditions in the areas of travel. For air transport, take-off/landing charges may be levied, differentiated according to the level of noise pollution.

With regard to mitigating the negative effects on infrastructure, the aim must be to establish a fair link between causes and effects for each user (a fair tax system must be based on the marginal external cost of the tax). Marginal external cost charging is important for the efficiency of transport systems because it forces users to reduce their costs (lower external costs will correspond to lower charges).

At the same time, the amount of revenue raised from charging must fully cover infrastructure costs. This is partly because of the need to recover the costs incurred by the owners of the networks (in most cases, the infrastructure is publicly owned) and partly because of the desire to maintain an acceptable level of investment in the sector. Any mechanism used will, however, require actions to be taken in full transparency in order to better understand the approaches and avoid undesirable conflicts.

The introduction of a system of charging for infrastructure use generally gives rise to some negative problems as well as benefits. Some of these relate to the analysis of the efficiency of the allocation of the space on which the infrastructure is built. According to some authors, the solution to such problems lies in the change of ownership (the privatisation process in the case of public property). As highlighted above, reducing externalities can be done in two ways. The first is the introduction of fiscal policy elements aiming at internalising external costs, while the second is the creation of a regulatory framework for the ownership of a public good.

Theoretically, as long as there are a large number of sellers and buyers of infrastructure space (mainly roads), the costs imposed by infrastructure operators will be high enough to cover construction and maintenance costs. The space allocated for a road, for example, will be traded like any other asset on a market, but the route chosen will be the one for which the user will obtain the greatest own benefits (reduced length, low tolls, etc.).

The privatisation of infrastructure will certainly lead to some maintenance and repair issues being resolved quickly; otherwise, users (hauliers) will have the option of using other routes. This would, however, lead to a loss of revenue for the infrastructure owner (through non-collection of user charges), which is obviously not desirable, as the owner would be tempted to maintain the network to the required quality standards. Another benefit of privatisation is that it will also relieve the state budget of the costs of maintaining and repairing the networks. In this case, the state will only have the role of monitoring the efficiency and fairness of measures to internalise external costs.

However, there are also economists who take a different view from the above. Rothengatter, for example, is keen to point out that networks such as roads and railways are natural monopolies and therefore any attempt to privatise them will not lead to the creation of a real market, as there are no counter bidders (Rothengatter, 1994). Moreover, many construction costs become sunk. What characterises transport infrastructure is the fact that the needs it directly meets are linked both to the mobility of the activity of economic agents producing or providing services and to the mobility of people.

However, the benefits accruing to producers from more efficient transport activity determine how they react to changes in transport supply. At the same time, the effects of infrastructure investment depend on the conditions in the national economy, the size of the savings in transport costs being determined by the level of economic activity, but also by the general constraints imposed by the macroeconomic framework.

Is infrastructure charging an efficient solution?

As presented above, by introducing an optimal charging system, society (made up of all individuals) will gain certain benefits. In this case, there will be a transfer of revenue from the hauliers to society. If, on the other hand, the revenues accruing from the introduction of these measures are not also directed in some proportion to those who pay them (in the form of compensation, etc.), the only beneficiary of the benefits of transport activity will be society, to the detriment of transport. Therefore, the success of introducing a tax system for infrastructure use will depend on the redistribution of the effects among those involved (hauliers, individuals in society, etc.).

Most charges imposed to recover the costs of infrastructure damage are based on the principle of differentiating their value according to the distance travelled (charge per kilometre travelled) and traffic values (degree of congestion). The simplest way of reducing traffic intensity and hence the likelihood of congestion in urban areas is to increase parking charges and reduce the number of parking spaces.

However, as has been observed, this method is only very effective for low-income users, while for other categories, the effects are minor (high-income private users, traders, professional hauliers, etc.). Moreover, the forecasts for Europe, in terms of the number of situations where congestion occurs, show only a very rapid increase. This entails huge additional costs.

An important characteristic of congestion is the variation in its intensity over time and space, which must be taken into account in any policy aimed at reducing specific negative situations.

A simplistic method of achieving these objectives has been presented above. There are, however, more complex techniques which, in principle, aim to achieve a system (mechanism) of permanent monitoring of flows. The systems currently applied differ only in terms of the technologies used or tax policy (permanent taxes with constant value, permanent taxes with different values, etc.).

What is very important remains the way and the degree to which these taxes are levied among the various users. Normally, each of them should be obliged to pay a charge equal to the amount of degradation caused to the network. At the same time, it should also be taken into account that the level of degradation also depends on the climate, the quality of the road and the technical characteristics of the vehicles (weight/axle, for example). Increasing the axle weight of vehicles can lead to an almost exponential increase in the amount of damage to infrastructure. Therefore, heavy vehicles with a low number of axles cause much more damage than passenger cars, considering that both types of vehicles travel the same distance. It is therefore necessary to differentiate the level of charging according not only to the type and characteristics of the vehicles, but also to the distances travelled and the quality of the infrastructure. From this perspective, a tax levied on the price of fuel is not effective, as the level of damage caused is not proportional to the volume of fuel consumed. The most correct way is to impose a charge per kilometre driven. However, the size of this charge should be modelled according to the characteristics outlined above. Moreover, the condition (quality) of a road has a direct influence on both the level of fuel consumption and on vehicle maintenance costs.

It is extremely difficult to calculate the relationship between road quality and each component of the operating cost, due to differences in fleet structure and the degree of wear and tear of the means of transport. However, according to some studies, the operating cost, depending on the quality of a road, ranges (USD/km) from 0.045 to 0.053 for passenger cars, 0.292 to 0.462 for buses and from 0.323 to 0.507 for heavy trucks. The savings that can be obtained, for example, for changing a road in a lowland area from the worst to the best technical condition are USD 0.006/km for cars, USD 0.07/km for buses and USD 0.059/km for heavy trucks.

It can therefore be concluded that adopting a strategy that includes the use of fiscal instruments, together with regulatory measures, can lead to savings compared to the level of expenditure incurred in maintaining and developing competitive networks.

Clearly, a just policy on the development of a sustainable transport system must contain specific elements at both national and local levels.

1 At the national level, depending on the constraints imposed by the level of economic development, fiscal (charges, taxes), compensatory or informational measures may be adopted. In these steps, regulatory actions are of great importance, as they are aimed at creating the legislative basis for the future development of the process. The adoption of coherent and stimulating legislation is the key to success.

However, fiscal policy measures are also very important, as they must aim both to reduce the proliferation of negative effects caused by transport activity (e.g. chemical and noise pollution) and to compensate for factors not directly involved in the activity. To a large extent, this can be achieved by internalising external costs. The following model can be used to achieve this effect in the case of road transport:

a An annual tax corresponding to the fixed infrastructure costs.
b Transportation charges corresponding to variable costs, for example:

  i Road maintenance tax.
  ii Traffic surveillance and accident tax.
  iii Energy and $CO_2$ emission charges.
  iv Charges for air pollution by NOx and volatile organic compounds.
  v Noise pollution charge.

The annual charges should be differentiated according to the characteristics of each type of vehicle. Transportation taxes should be included in the price of fuel. In addition, for heavy-duty vehicles, an additional charge should be introduced, generically referred to as a charge per kilometre.

Since the internalisation of external costs will increase the price of transport by polluting vehicles, the need and purpose of these steps must be clearly explained to the public. It is also necessary to continuously adapt the process to changing circumstances.

The entire transport modernisation process must be supported from the revenue generated by the collection of these taxes, which can be estimated at many hundreds of millions of dollars annually, in the spirit of sustainable development principles, with a significant percentage of this revenue being directed towards the introduction of social measures for those affected by the negative phenomena caused by transport activity. We do not agree that the funds collected through internalisation should be paid into the overall state budget, as this is both uneconomical because it generates inflation, and inconsistent with the principle of sustainable development, which aims to reduce polluting and energy-intensive activities while compensating those affected.

In addition to these actions, consideration should also be given at the national level to the adoption of measures aimed at the following:

a Encouraging fewer polluting modes of transport.

In this context, it is worth mentioning the support for combined road–rail, road–river, road–river–river or road–rail–sea freight transport. From this point of view, Romania has a clear advantage in terms of both rail and waterway–maritime infrastructure, which allows such an approach, and the existence of properly equipped intermodal hubs.

b Rehabilitation of existing infrastructure in order to ensure the quality of transport activity. From this perspective, we are in favour of measures to modernise existing road infrastructure rather than implementing anti-environmental programmes for the construction of new infrastructure, which is not even economically justified.

c Encouraging the use of 'clean' fuels by imposing much lower prices for them than for other types.

d Improving emission monitoring systems so that measures can be taken in good time to keep emissions within the permitted limits.

e Adopting measures to reduce the effects of noise pollution, vibrations, etc.

2 At the local level, fiscal measures can be introduced to support national measures, but the main actions to be taken are of an administrative nature. In summary, these types of action should aim at the following:

a Promoting less or non-polluting transport, such as public transport, walking or cycling.

b The adoption of regulations restricting travel speeds on certain stretches, car access to certain areas (historic, central, etc.), the use of clean fuels for government vehicles, etc.

c Adoption of fiscal measures at the local level, when those at the national level are insufficient. It should be noted that it is essential to avoid imposing several taxes for the same type of effect (e.g. in the case of internalising the external costs of road accidents, a fixed annual tax and a variable annual tax levied on fuel prices should not be introduced at the same time).

It is clear that, overall, the total external costs exceed the total external costs resulting from user charges, the difference being borne by society.

In the case of rail and maritime transport, the degree of recovery of external costs is even lower than in the case of road transport, because of the particular problems encountered in assessing the costs of infrastructure use.

For all modes, however, in order to increase the degree of recovery, measures to internalise these external costs must be adopted. As indicated above, this can be achieved either through regulatory policy or by using specific economic instruments. In the case of an overall strategy, the balance between adopting one or the other has to be very carefully considered on a case-by-case basis (road, rail, etc.).

## Notes

1 In the United States, for example, reducing transportation times by promoting *door-to-door* transportation is a priority, and this applies to all modes of transportation.
2 Mtoe = megaton oil equivalent.

## References

Betts W., Floys S. and Kvinge F., *The influence of diesel fuel properties on particulate emissions in European cars*, Society for Automotive Engineers Technical Paper 922190, Warrendale, Pennsylvania, USA, 1992.

Dragu V., "Tansporturile multimodale – soluţii eficientă pentru economisirea resurselor şi Minimizarea efectele externe negative (Multimodal transport – efficient solutions for saving resources, and minimizing negative externalities)", in *Buletinul AGIR nr.4/2009*, Bucharest, Romania, 2009, https://www.buletinulagir.agir.ro/numar_revista.php?id=28

EC, *EU Transport in figures-2009*, ISBN 978-92-79-79-10728-3, 2009, https://transport.ec.europa.eu/facts-funding/studies-data/eu-transport-figures-statistical-pocketbook/statistical-pocketbook-2009_en

EC, *EU transport in figures*, ISBN 978-92-76-76-53698-7, 2022a.

EC, *Short history of relations between Romania and the European Union*, 2022b, https://romania.representation.ec.europa.eu/despre-noi/scurt-istoric-al-relatiilor-dintre-romania-si-uniunea-europeana_ro

EC, *EU transport in figures-2024*, ISBN 978-92-68-19812-4, 2024, https://transport.ec.europa.eu/facts-funding/studies-data/eu-transport-figures-statistical-pocketbook/statistical-pocketbook-2024_en

Fistung D., *Teorie economică, ecologie, legislatie (Transportation. Economic theory, ecology, legislation)*, ALL BECK Publishing House, Bucharest, ISBN 973-94435-03-3, 1999.

Kanafi A., *Transportation Demand Analysis, Mc Graw Hill*, New York, ISBN: 978-0-0-0-70332-71-3, 1983.

NIS (National Institute of Statistics - Institutul Național de Statistică), 2022, *Anuarul Statistic al României 2021 (Statistical Yearbook of Romania 2021)*, 2022 https://insse.ro/cms/sites/default/files/field/publicatii/anuarul_statistic_al_romaniei_carte_0.pdf

RG (Romanian Government), *Strategia Națională pentru Dezvoltarea Durabilă a României 2030 (National Strategy for Sustainable Development of Romania 2030)*, 2018, https://sgg.gov.ro/1/wp-content/uploads/2018/10/SNDD-2030-_-varianta-dup%C4%83-Comitet-interministerial-4-octombrie-2018.pdf

Rodrigue J.P., Notteboom T., The Provision and Demand of Transportation Services, The Geography of Transport Systems, Routledge, 2020, https://transportgeography.org/contents/chapter3/provision-and-demand-of-transportation/; doi.org/10.4324/9780429346323

Rothengatter W., Do external benefits compensate for external costs of transport? *Transportation Research Part A: Policy and Practice*, vol. 28, Issue 4, pp. 321–328 Elsevier, 1994, doi:10.1016/0965-8564(94)90006-X

SEA-POST, *Environmental report (strategic environmental assessment) – Sectorial operational program transport infrastructure - Romania*, EuropeAid/121373/D/SV/RO, 2007, https://www.fonduri-ue.ro/files/programe/INFRASTRUCTURA/POST/Introducere_SEA_vers_2.pdf

VIRTA, *The global electric vehicle market overview in 2022: Statistics & forecasts*, 2022, https://www.virta.global/en/global-electric-vehicle-market

# 6 Sustainable mobility in the 'building green cities' process

The evolution of the Earth's population has been the result of a spectacular and long-lasting economic and social transformation. The quantitative evolution of the world is synthesised in the number of its population, just as the evolutionary details are highlighted by the same individualised values of temporal, spatial or administrative and partial or structured sub-populations.

## 6.1 Milestones in the formation and development of human communities

Throughout human evolution, as the number of humans increased, certain individuals found common reasons (food procurement, defence, reproduction, etc.) to come together and form groups, cohabiting communities. A small number of these settlements grew into what we today call localities (fairs, villages, towns, cities, etc). This type of growth corresponded to the changing organisation of work in each historical period.

The first cities appeared thousands of years ago in areas where the land was fertile, such as the cities founded in the historical region known as Mesopotamia around 7500 BC, which included Eridu, Uruk and Ur (National Geographic, 2023). These cities were among the many communities between the Euphrates and Tigris rivers (the so-called Fertile Crescent). The first city-states also emerged here. The ancient city-states had a king as early as the 4th millennium BC, who was also commander of the army, supreme judge and priest. Some of these city-states later became empires. Among the earliest such settlements were following: Babylon, Akkad, Tyre or Lagash.

Cities were also formed along the Nile River in Egypt, the Indus River valley on the Indian subcontinent and the Yellow (or Huang) River in China, as people began to cultivate crops and settle in communities. The agricultural production, in these fertile areas, meant that people could give up a nomadic lifestyle as hunters and gatherers to take advantage of food and other produce, some of which even existed in surplus. Settlement along waterways also provided a much-needed transportation system that facilitated trade and cultural exchange.

DOI: 10.4324/9781003743064-7

Over the following millennia, cities continued to grow in number, size and stature. Some of the world's largest cities have grown steadily for hundreds of years, while others seem to have flourished overnight.

Some of the world's once most populous cities, such as those of Mesopotamia, no longer exist, while others have experienced massive population declines. For example, in the 1st century BC, Rome (Italy), topped 1 million people, making it the largest city in Europe; its population fell to just 20,000 during the Middle Ages, which was only a 'temporary blip' in time. But these are exceptions; most of the world's cities continue to grow, and some are experiencing unprecedented growth. In fact, Rome's population was over 2.3 million in 2022 in the city alone, as the metropolitan area alone recorded almost 4 billion inhabitants (Zhuji World, 2023).

Throughout history, people have been drawn to cities as centres of commerce, culture, education and economic opportunity, but the resulting urban population growth has not always been steady. Indeed, urbanisation is a relatively recent phenomenon; until recent years, the vast majority of people lived in rural areas. For example, more than 90% of the global population lived in rural areas in Europe in 1800. The US Census Bureau indicates that more than 94% of the US population lived in rural areas in 1800; over a hundred years later, that number had dropped to 60%, at the expense of cities.

One of the main reasons for the growth of cities was the Industrial Revolution, which began in England in the mid-18th century and then spread to the United States and other parts of Europe. The Industrial Revolution helped to drive the growth of factories, creating additional demand for workers in urban areas.

Over the next century, millions of people in the United States and England moved from farms to cities. As other parts of the world industrialised, they too became more urbanised.

Thanks to the Industrial Revolution, London, for example, grew from a population of 1 million in 1800 to over 6 million a century later. Within a few decades, the so-called 'Second Industrial Revolution' spurred urbanisation in the United States, spurring the New York City metropolitan area to become the largest city in the world by 1950, with a population of 12.5 million.

Over the last few hundred years, the world's population has grown significantly and economies have industrialised; as a result, many more people have moved to cities. This process is known as *urbanisation*.

Even after the rise of cities, however, the vast majority of people lived and worked in rural areas. Only when large-scale industrialisation began in the 18th century did cities really start to grow. Almost half of all people now live in urban areas. They are attracted by jobs in manufacturing and the professions, as well as increased opportunities for education and entertainment.

Urbanisation is often discussed in reference to countries that are currently in the process of industrialisation and urbanisation, but all industrialised nations have experienced urbanisation at some point in their history. Moreover, urbanisation is increasing across the globe.

One effect of this huge increase in the population living in urban areas is the growth of the megacity, which is a city that has over 10 million inhabitants. Now there are cities with even more than that. Tokyo, Japan, for example, has almost 40 million inhabitants.

Another effect of urbanisation has been the spatial spread of agglomeration. Urban sprawl is when a city's population is dispersed over an increasingly large geographical area. This movement from higher-density urban cores to lower-density suburbs means that as cities expand, they often begin to occupy significant areas of land previously used for agriculture. This sprawl also increases the need for utilities such as transportation infrastructure.

As we move into the 21st century, the global population will certainly continue to grow. Urban areas will continue to grow with the population. This continued growth presents complex challenges as we prepare for the cities of the future. How we choose to manage urbanisation will have consequences for our world for many years to come.

Urban population growth has also been fuelled by new technologies, in particular technologies that have enabled cities to thrive. Innovations in steel enabled the development of skyscrapers, which allowed for greater population density. And, of course, the invention of the elevator made it possible for skyscrapers to take people up to their apartments and offices in the sky.

In the case of Romania, there was a predominantly downward trend in the rural population from 1930 to 2022 (almost a century), while, against the background of general demographic growth, the urban population grew steadily until 1992, after which the trend changed.

In fact, after 1992, the total resident population of Romania also started to decrease, the main cause of this evolution being the phenomenon of 'economic emigration'.

What is interesting is the opposite current trend of population migration within the country. Thus, if in the socialist period, mainly due to accelerated industrialisation, the workforce was forced to move to urban areas. In recent decades, this phenomenon has been reversed, with many urban dwellers moving voluntarily to rural areas.

Although most Romanian cities have experienced reductions in their residential population in recent years, their surface area has expanded outwards, especially in peri-urban areas.

Between 1992 and 2002, for example, Bucharest added to its urban size an area the size of Ploieşti. Similarly, the cities of Brasov and Cluj-Napoca expanded their urban area by about 24%, while Craiova saw an increase of 27%.

But development in peri-urban areas has been chaotic. Some specialists in the field have called this phenomenon 'parcel urbanism', highlighting how each developer has interpreted the planning rules according to personal needs (Fistung and Popescu, 2018).

One of the most negative side effects of *parcel urbanism*, however, has been the fragmentation of peri-urban land and the emergence of isolated, low-density residential areas.

Looking at the global picture, more than half of the world's population now lives in urban areas, and around two-thirds of people will live in cities by 2050.

Much of this growth is due to natural increase – that is, the births of people already living in cities – as well as the continued migration of people from rural to urban areas in search of new opportunities.

Urbanisation in recent years has fuelled the growth of ever-larger cities and a new type of city: the megacity, which is defined as having a population of 10 million or more. The metropolitan areas of New York City and Tokyo, Japan, became the world's first megacities in the 1950s; by 2018, there were 37 megacities across the globe.

While the biggest cities of the Industrial Revolution were in North America and Europe, today Asia and Africa are experiencing the greatest urban growth due to industrialisation.

As of 2019, Tokyo, with more than 37 million people, is the world's largest urban area; Delhi, India, has nearly 30 million. Shanghai, China; Mexico City, Mexico and São Paolo, Brazil, all have metropolitan area populations of more than 20 million.

## 6.2 Structural changes in human settlements

Whereas in 1950 more than half of the world's population lived in rural areas, according to the UN, in 2007, the balance between the sizes of populations living in rural and urban areas was achieved. Since that time, the majority population of the world has been recorded as living in urban areas (UN, 2018).

In fact, according to Richie and Rover (2019), between 1950 and 2020, the number of cities in the world with a population of more than 300,000 inhabitants more than six times increased from only 306 cities in 1950 to about 1934 in 2020.

As the aforementioned authors have already pointed out, the trend towards an increase in the number of cities with more than 300,000 inhabitants worldwide will continue in 2035. In this case, the number of large cities will be about 2363, which is an increase of more than 12% compared to 2020 and almost eight times more than in 1950.

Looking to the future means looking at the growth of today's cities. The United Nations (UN) has predicted that there will be 41 megacities by 2030 (National Geographic, 2023).

India, which already has five cities with metropolitan areas of more than 10 million people, is expected to grow to seven by 2030. Some of the megacities could be in places that surprise you, like Kinshasa, the capital of the Democratic Republic of Congo. Already a megacity with a population of 11.6 million, Kinshasa's demographics and growth trends suggest it may quickly surpass some of the world's largest cities.

Just as fast-growing cities in England and North America struggled to keep up with population growth in the industrial age, cities in Africa and Asia are also struggling to service rapidly growing populations. Lagos, Nigeria, Africa's largest city, is home to 12.6 million people, about two-thirds of whom live in slums.

The population continues to grow, not through migration to the city but because of high birth rates. While there are a number of downsides to growing too fast – from traffic congestion to the growth of slums – the population in the world's megacities continues to grow.

According to British researcher Noreena Hertz, by 2050, almost 70% of the planet's population will live in cities, and one in ten people will live in cities with more than 10 million inhabitants (Hertz, 2021).

But there are other views. For example, American professor Joel Kotkin argues that: 'It is only a matter of time before today's modern metropolises, such as New York or London, will end up having the fate of Ancient Rome, whose decline lasted four centuries' (Ungureanu, 2023). In his analysis, Joel Kotkin notes that the skyscrapers in which corporations rented office space, a symbol of urban grandeur, have become as anachronistic as the cathedrals of the Middle Ages. That is because office occupancy has dropped significantly since the turn of the century, and migration to big cities has started to fall in recent years, as people have increasingly visibly moved out of metropolitan areas to smaller towns or rural areas. Moreover, pandemics and rising crime have accelerated this process.

With more than 80% of global GDP generated in cities, urbanisation can contribute to sustainable growth through increased productivity and innovation if managed well (World Bank, 2022). However, the speed and scale of urbanisation bring challenges, such as meeting the accelerated demand for affordable housing, viable infrastructure, including transport systems, basic services and jobs, especially for the nearly one billion urban poor who live in informal settlements to be close to opportunities. Rising conflicts contribute to the pressure on cities, as 50% of forcibly displaced people live in urban areas.

Once a city is built, its physical form and land-use patterns can be locked in for generations, leading to unsustainable sprawl. The expansion of urban land consumption is outpacing population growth by up to 50%, which is expected to add 1.2 million $km^2$ of new urban area worldwide by 2030. Such expansion puts pressure on land and natural resources, resulting in undesirable outcomes; cities account for two-thirds of global energy consumption and account for more than 70% of greenhouse gas emissions.

Cities play an increasingly important role in tackling climate change, as their exposure to climate risk and disasters increases as they grow. Since 1985, about 76,400 $km^2$ of newly urbanised land has been added in locations with flood depths of more than 0.5 m during severe floods – this corresponds to about 50 times the area of the Greater London area. Globally, 1.81 billion people (1 in 4 people) live in areas at high risk of flooding. Exposure is particularly high in the densely populated and rapidly urbanising river plains and coasts of developing countries, where 89% of the world's flood-prone people live.

In urban areas, human behaviour undergoes many important changes. Of these, we need only mention the increase in travel speed. Thus, according to Noreena Hertz (2021), the walking speed in cities is on average 10% higher than in the 1990s, especially in the countries of the Far East. In the study conducted in 32 cities around the world, it was observed that in 2007, compared to 1990, walking speed increased by 30% in Singapore and by about 20% in Guangzhou (China). It has also been shown that travel speed increases with the prosperity of the city. In the richest cities, people walk several times faster than in less prosperous ones. There are many reasons for this phenomenon, mainly related to the time taken to cover increasingly long distances, possible congestion or bottlenecks, which can lead to wasted time and money (the famous capitalist saying 'time is money'), etc.

Cities are also on the front line in the fight against epidemics. The COVID-19 pandemic was a massive challenge for cities and their citizens, rich and poor alike. Its impact and the measures taken to control the spread of the virus have had a disproportionate impact on the poor, marginalised and vulnerable, revealing fault lines in the economic structure of cities, their preparedness for such a crisis – in particular the state of public health and their service delivery systems.

In normal times, there may be many attributes that cities strive to compete and excel on a global scale, including liability, competitiveness and sustainability; but on any given day, and especially in a time of crisis, a city must perform well for all its citizens. Inclusion has thus come to the fore in the recent pandemic.

If much data show the slow but steady demographic decline of big cities, what might the urban future look like?

Joel Kotkin argues that the suburbs and suburban regions will increasingly gather companies and employees (Ungureanu, 2023). The American professor also proposes a solution by developing so-called 'garden cities'. The garden city is a concept theorised by British urban planner Ebenezer Howard in 1898 as an alternative to polluted industrial cities. The concept involves public control of land, the presence of a green (possibly agricultural) belt around the city (to provide food and reduce the levels of air pollutants entering the city), a relatively low density of buildings, the presence of public facilities such as parks, shopping and cultural centres in the city centre, and control of the actions of economic entrepreneurs in the urban space. American professor Joel Kotkin notes that such *garden cities* lend themselves to remote working and shorter commutes, making them ideal for energy efficiency and carbon reduction as well as for reducing transportation demand and encouraging sustainable mobility. In Professor Kotkin's view, such a direction of development could prevent some cities from going into total decline, even if they will no longer be the unrivalled centres of population, economic power and innovation.

There is currently no universal definition of what 'urban' means. The UN reports figures based on nationally defined urban shares. But the problem is that countries adopt very different definitions of urbanisation. Not only do the thresholds between urban and rural vary, but the types of metrics used also differ. Some countries use minimum population thresholds, others use population density, infrastructure development, type of employment or simply the population of predefined cities.

Worldwide, the notion of urban area is adopted differently, according to one or more criteria.

In Romania, the definition of urban areas and localities has undergone several legislative changes over time. The latest version of this field is given by LAW no. 351 of 6 July 2001 (RP, 2001), on the approval of the National Spatial Plan – Section IV – Network of localities, updated by Laws no. 106 of 3 June 2010 and no. 264 of 7 December 2011.

According to this document, the national network of settlements is composed of urban and rural settlements, prioritised by rank. The urban locality is defined as a 'locality in which the majority of labour resources are employed in non-agricultural activities with a diversified level of endowment and equipment, exerting a constant and significant socio-economic influence on the surrounding area'.

According to the latest statistical data, as of July 2022, there were 319 municipalities and cities in Romania (NIS, 2023, p. 65).

Out of these, only 201 municipalities have more than 10,000 inhabitants, which, according to the first criterion of Law 351/2001, only these can be classified as urban municipalities; the other cities, in our opinion, do not meet the legal requirements.

Until now, in Romania, changes in the status of localities have been in the sense of increasing their rank from commune to town or from town to municipality. To the contrary, downgrading a locality to a lower rank would entail, among other things, a reduction in the number of local councillors and the salaries of town hall employees, as well as a reduction in local taxes and duties, which seems difficult to achieve in practice.

## 6.3 Challenges of the future green cities

Building cities that are sustainable, or 'green' in the latest international sense, requires intense policy coordination and quite significant investment orientations in these endeavours. National and local authorities have an important role to play in acting now to shape the future of their development and create opportunities for all citizens. The history of environmentally friendly cities goes back a long way, but terms such as sustainable city or green city are more recent.

In 1987, the American researcher Richard Register launched the term 'eco-city' (*ecocity*) in his book 'Ecocity Berkeley: Building Cities for a Healthy Future'. This concept has been strongly influenced by numerous other trends for the future development of human society, such as appropriate technology, community economic development, social ecology, green movement, bioregionalism and finally sustainable development, which have given an interdisciplinary dimension to this new type of concept (Peace et al., 2016).

The origin of the term 'Eco-city' is also linked to the foundation of urban ecology, an interdisciplinary field of study that analyses the relationships between people and the environment, including cities and urban landscapes.

In the literature, the terms 'Eco-city' or 'sustainable city' are often used synonymously with 'green city'. The term is associated with 'sustainable development' as an extension of this concept, but within the activities of a city. Thus, the adjective 'green' not only refers to environmental issues but also addresses the specific issue of social and economic integration in urban development processes.

In line with these particularities, the 'green city' has, since the concept was first introduced, to date, had numerous definitions.

Among the most important is that of the United Nations Environment Program (UNEP), which states that:

> Green cities are defined as those that are 'environmentally friendly'. Greening cities requires some, or preferably all, of the following actions: (1) monitoring diseases and their influence on overall health; (2) reducing chemical and physical hazards; (3) developing high-quality urban environments for all; (4) minimising environmental cost transfers to areas outside the city and (5) ensuring progress towards sustainable consumption.
>
> (Peace et al., 2016)

Taking into consideration the Lewis (2015) and Breuste (2020) definitions about the green city we propose a new version of this concept: *A 'green city' is that urban area which adopts the political, economic and social measures necessary to achieve a high level of environmental quality and which contributes, in itself, to human well-being without compromising the development in harmony with the environment of future generations.*

Various indicators have been used by some European and international institutions to index 'green cities'. These have been used to produce the most important green city rankings, such as the European Green City Index, the European Green Capital Award and Urban Ecosystem Europe.

In this context, quantitative indicators of the dimensions of 'environmental quality' and 'human well-being' can not only provide a measure of 'green city' performance, but at the same time also assess the effectiveness of 'social and political action'.

According to Peace, 13 categories of indicators play an extremely important role in assessing the 'greenness' of cities (Peace et al., 2016):

- Amount of $CO_2$ in the atmosphere.
- Air quality.
- Energy consumption.
- Building construction.
- Transportation system.
- Water quality.

- Level and management of waste.
- Green space size and occupancy.
- Level of noise pollution.
- Health and safety.
- Level of education.
- Equity.
- Degree of participation in community action.

The 13 categories are divided into two broad and well-known groups: 1) quantitative indicators and 2) qualitative indicators.

1 The quantitative indicators are used to describe three major dimensions of the Green City, grouped into: a) Environmental quality; b) Well-being and c) Political-administrative activity

   a From an environmental quality perspective, the specific transportation indicators are as follows:

      i Share of buses with low emissions of pollutants (air and noise).
      ii Degree of use of alternative transport (non-motorised).
      iii For all trips of less than 5 km, the proportion of trips made by car, public transport, bicycle, on foot or other alternatives.
      iv Number of passengers using public transport.
      v Number of cars registered.
      vi Number of electric vehicles owned by local authorities (and share in total registered vehicles in the city).

   These indicators can assess the daily vectors of commuting and the consequences of individual mobility on the urban environment. They can also support technological innovation by assessing the proportion of low-emission buses and electric vehicles owned by the municipality, thus also measuring the efforts of public administrations.

   b In terms of well-being, the indicators correlated with transport activity are as follows:

      i Size of the non-motorised transport network (bicycle paths, pedestrian paths, etc.).
      ii Proportion of population living within 300 m of a public transport service per hour (or more frequent).
      iii Size of tram and metro networks.
      iv Availability of cycle paths and cycle lanes.
      v Number of charging points available for privately owned electric cars.

   c Among the indicators of policy-administrative actions with an impact on transport activity, the following should be mentioned:

      i Legislative initiatives to support sustainable transportation.
      ii Procurement plans for clean vehicles for public transport.

iii Sizing and approval of alternative transport networks (bicycle networks, pedestrian paths).
iv Support for the creation of protective (forest) zones to reduce levels of chemical and noise pollution.

2 For the transport activity with a view to classifying a city in the 'green' category, the qualitative indicators concern:

a Energy efficiency initiatives for buildings.
b Promoting sustainable transport.

In Peace's opinion, the use of the presented indicators can cover the analysis of the transport activity in achieving the three major dimensions necessary in the process of classifying a city as 'green', but they are not sufficient to measure and monitor the performance of the 'green city'.

Therefore, in order to overcome the measurement problem, a monitoring system needs to be in place to indirectly assess and compare qualitative aspects. More precisely, the possibility to measure and monitor over time the dimensions of 'environmental quality' and 'human well-being' through quantitative indicators is linked to the effectiveness of the city's political and societal actions. The assessment of this relationship is very useful for policy makers, as it allows checking and correcting not only local, but also national and international policy measures, minimising the environmental impact of urbanisation.

For the European continent, the *Green City Agreement* was recently launched at the European Week of Regions and Cities, committing more than 20 signatory cities to improve air quality, reduce pollution and promote green urban areas. The role of cities in the development of the continent is obvious, and the way in which this involvement and collaboration should take place takes shape through such projects and agreements.

The European Union aims to reduce emissions by 55% by 2030, yet many of Europe's urban centres have taken matters into their own hands and come up with innovative initiatives. Helsinki, for example, aims to become carbon-neutral by cutting greenhouse gas emissions by 80% by 2035. Stockholm aims to phase out fossil fuels by 2040, and Copenhagen illustrates the importance of a transition in energy supply. Other European cities are offering innovative local solutions that reduce the effects of climate change and tackle related issues. Madrid's new approach to water management, Berlin's *sponge city* concept, Utrecht's green corridors and Paris' proposed rain plan are just a few notable urban initiatives.

According to worldwide analysis, there are several cities considered to be very close to the green city concept. These include Copenhagen (Denmark), Amsterdam (Netherlands), Frankfurt (Germany), Vancouver (Canada), Frankfurt (Germany) and Portland in the United States (DGB, 2023).

The city of Copenhagen, Denmark, is recognised for its bike-friendly streets and commitment to carbon neutrality. Copenhagen has long been considered one of the greenest cities in the world. City authorities have set a target to be carbon neutral by 2025. The process has already made significant progress by increasing the use of renewable energy sources and implementing green transportation options such as bicycles and electric cars.

Amsterdam, the iconic city of the Netherlands, is another city that supports sustainability and green transport. The city's famous network of bike lanes and waterways makes it easy for residents and visitors to get around without relying on cars. In fact, there's a well-enforced 'Park and Ride' policy where it is desirable for car owners to leave their vehicles at the edge of the city in parking lots and travel around the city using green transportation. Amsterdam has also placed a strong emphasis on preserving and restoring its natural ecosystems.

Another example of a green city is Frankfurt, Germany: In recent years, Frankfurt has made significant investments in green technology and infrastructure, earning it a place among the world's top green cities. The city's public transportation system is one of the greenest in Europe, and it also has a number of green buildings, such as the European Central Bank's headquarters, which is designed to be energy efficient.

Canada's Vancouver is also considered to be 'green'. Known for its stunning natural scenery and commitment to protecting the environment, Vancouver is one of the greenest cities in North America. The city has a goal to become carbon neutral by 2050 and has implemented a number of initiatives to help achieve this goal, such as expanding its network of bike lanes.

Also in North America, but in the United States, is the 'green' city of Portland. It has a long-standing reputation as a green city because it has placed a strong emphasis on sustainable transportation and has implemented a number of initiatives to promote the use of public transport, cycling and walking. In addition, Portland has taken effective measures to increase the amount of green space in the city and protect its natural ecosystems.

As is well known, integrating green space into urban planning can lead to more sustainable and efficient infrastructure development. For example, green corridors and plant alleyways can provide connectivity between urban areas and improve pedestrian and cycling circulation. People can travel by alternative transportation from different parts of the city much faster and more efficiently.

At the same time, vegetation in green spaces helps to reduce the heat island effect in cities, which is the phenomenon whereby urban temperatures are higher than those in the surrounding rural areas. This helps to maintain a more pleasant climate and reduce the risk of extreme heat waves in cities.

Despite the current widespread crisis, city administrations have seized the moment and proposed positive changes. Milan has stimulated a city-wide expansion of cycling and walking. Florence has increased the capacity and safety of its tram system. Budapest has actively promoted its multitude of lesser-known green areas, trying to avoid congestion. Moreover, Amsterdam and Brussels promote the doughnut economy, an economic model for sustainable development.

At the root of our environmental crisis is the simple reality that we are asking more of our planet than it can provide and producing more waste and emissions than it can cope with. If we are to remedy this crisis and secure a decent future for ourselves and the natural environment, this must change. We must never forget that we are not protecting nature; nature is protecting us.

## 6.4 The sustainable integrated city – New concept supporting sustainable transport

Based on the opinion of Rocco et al. (Peace et al., 2016), we believe that we can group the influencing factors of *green city* development into the following categories:

- Environment, including:
  - Level of $CO_2$ emissions.
  - Air quality.
  - Water quality.
  - Level of noise pollution.
- Economic activities, including:
  - Industry.
  - Services.
  - Trade.
- Social, including:
  - Level of health.
  - Degree of life security.
  - Level of education.
  - Equity.
  - Level of community participation.
- Transportation, including:
  - Actual activity.
  - Size and structure of the transport fleet.
  - Specific infrastructure.
- Waste Management.
- Energy (production, transportation and consumption).
- Architecture, including:
  - Diversity, form, layout and condition of buildings.
  - Green spaces or areas for restoration/recreation.
  - The degree of spatial occupancy and spatial change trends in the urban area.

Within such a Green City, there will obviously be biunivocal relationships between all seven categories of influence, but from our point of view,

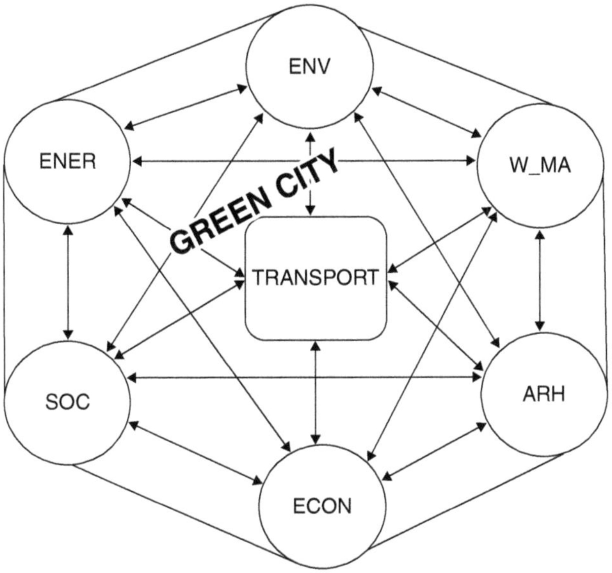


Legend: ENER - energy; SOC - Social; ECON - Economy; ARH - Architecture; W_MA - Waste Management

*Figure 6.1* Structure and influence relationships between the components of a *Green City*. (By author)

transportation occupies a central place in this structure, influencing all areas of the area's evolution (see Figure 6.1).

Of course, transportation, in turn, will also be influenced by these components of the green city structure, constantly adapting its offer according to market demand.

Looking only at the link between transport and green city development, we can prioritise a number of directions for action:

- Reducing transportation demand.
- Providing access routes for all the 'vital' needs of the community (safety, health, education, work, recreation, etc.).
- Encouraging clean transportation.
- Spatial redesign and reorientation of transportation infrastructure to a new concept of urban development defined as the 'integrated sustainable city'.

Based on these elements, we define *the Sustainable Integrated City* as a *new concept of urban development, in which the community must be served by efficient communication relationships that are able to satisfy both the mobility requirements of individuals and the protection of natural life and social and economic relations.*

In this new understanding, the concepts of sustainable development, social equity and enhanced quality of life are closely linked. All types of activities and requirements linked to the smooth functioning of community life must be ensured in such a way that the city can provide optimal living conditions.

From this perspective, there are many lines of action to be pursued:

- Spatial and infrastructural resizing of the community. In this context, the creation of *semi-autonomous* communities within larger urban areas (large cities, metropolitan areas, etc.) can be envisaged, each of which meets the requirements of an integrated sustainable city. An *integrated sustainable city-type* community should be smaller in size that can be covered on foot in no more than half an hour in all directions from the centre to the periphery. This does not mean the administrative zoning of larger localities into smaller sub-units, nor does it in any way restrict access to any other area outside the defined space. As an example, Bucharest could be resized into several integrated sustainable urban communities, which can function efficiently (economically, socially and ecologically) and independently, but which are in permanent connection with the other areas of the core city (Bucharest).
- Providing vital community functioning facilities (health, educational, cultural and recreational facilities, centres of economic activity, commercial, etc.).
- Reducing the need to travel. Smaller areas with opportunities for housing and socio-economic activities for all categories of individuals can sharply reduce the need to travel to/from centres of interest (jobs, recreation, education, health centres, etc.). Moreover, increased interest and type of 'telecommunicating' activities can support such an approach.
- Use predominantly sustainable transportation within the sustainably integrated urban community. The small size of areas can provide opportunities for much more frequent use of public transport, walking or non-motorised vehicles.
- Ensuring efficient community management and, at the same time, maintaining connections and interconnections with complementary, similar areas within a larger urban space (large and very large cities).

More evocatively, the integrated sustainable city can be likened to a molecule with many atoms, representing semi-independent hubs (neighbourhoods) in the new concept, where the inhabitants are provided with almost all the facilities for normal life.

As in the case of atoms, the neighbourhoods will not be identical and of the same size, but will be calculated according to the needs, the size of the population and the topography and size of the spaces considered.

Closely linked to this new concept of sustainable urban development, the evolution from the *garden city* to the *green city* has recently taken a new approach, that of the '15-minute city'.

Thus, following the Paris Climate Change Conference in 2015, Carlos Moreno introduced the concept of the '15-minute city' as a framework to combat greenhouse gas emissions in 2016 (Khavarian-Garmsir et al., 2023).

However, the concept only grew in popularity after 2020, when the world faced the COVID-19 pandemic. Consequently, the mayor of Paris, Anne Hidalgo, selected the concept as a centrepiece of her 2020 re-election campaign, and the Cities Climate Leadership Group (C40) recommended it as a vital strategy for post-COVID recovery. The *15-Minute City* concept aims to create self-sufficient neighbourhoods with the essential functions of living, working, shopping, commerce, healthcare, education and entertainment by decentralising urban functions and services.

Moreno envisioned dense, socially and functionally connected, mixed-use neighbourhoods based on providing for human needs that encourage walking and biking.

According to Moreno's theory, the 15-minute city has seven supporting, interconnected elements:

- *Proximity*. This was a fundamental principle of earlier concepts, such as the garden city and neighbourhood unity. The current concept aims for a proximity-oriented strategy that focuses on redistributing and relocating urban facilities and resources within neighbourhoods, allowing residents to access the six basic functions of living, working, shopping, health care, education and entertainment with a 15-minute walk or bike ride.
- *Density*. The 15-Minute City considers density as a critical dimension to ensure that residents have adequate access to daily necessities without the need to travel by private car. In this context, it has been shown that density can provide a critical mass for the operation of local services and businesses within walking distance, as well as the creation of local employment opportunities. As a result, higher density promotes the localisation of urban functions while indirectly reducing total VMT and associated carbon emissions.
- *Diversity*. This new concept aims to increase neighbourhood diversity through two mechanisms: (1) software diversity and (2) hardware diversity. Software diversity means creating communities that represent a diverse range of cultures, ethnicities, economic and social classes. Hardware diversity aims to create neighbourhoods with multiple uses, including residential, commercial and recreational. Carlos Moreno sought to transform the 15-minute city from a monocentric urban structure into a polycentric form. Based on the philosophy of chrono-urbanism, he argued that the localisation of workplaces can improve people's quality of life by preventing them from wasting time in their daily commute (reducing transportation demand) and therefore lowering carbon emissions.
- *Digitisation*. Although digitisation was not initially among the '15-minute city' principles, the capabilities offered by smart solutions during the pandemic convinced proponents of the concept to use smart city technologies to complement the initial principles. Smart technologies, such as Big Data and the Internet of Things, enable planners to access real-time and timely data, help citizens participate more actively in planning and decision-making processes and allow residents to use water and electricity resources

more efficiently. It is no less true that internet networks also greatly assist in the development of 'tele-communication' services.

- *Urban design on the scale of human needs.* The need to return to this kind of need has been one of the major lessons of the COVID-19 pandemic for cities. The size and shape of a '15-minute city' are defined based on human needs and characteristics. This concept advocates the redesign of public spaces for the benefit of citizens, rather than for cars. It promotes investment in walking and cycling trails so that city residents can fulfil their daily needs within a maximum of 15 minutes walking or cycling distance.
- *Connectivity.* Connecting neighbourhoods by public transit helps avoid isolated neighbourhoods. This ensures that neighbourhoods are seamlessly integrated into the wider urban fabric. In 15-minute city neighbourhoods, active mobility modes, such as walking and cycling, are combined with public transport, increasing the efficiency of public transport and addressing the connections between different modes of sustainable transport.
- *Flexibility.* This new concept supports the transformation of single-purpose public and semi-public spaces into multifunctional areas to maximise the use of buildings and spaces. It aims to assign multifunctional roles to public and semi-public spaces that are already being used for a specific purpose at a specific time of day or on specific days. Paris is an example where some school playgrounds become parks when school is closed and serve a distinct purpose at weekends.

To achieve social equity, the '15-minute city' concept emphasises equal access to urban services for all people and decentralises urban infrastructures and functions. It provides affordable housing, career opportunities and social services for vulnerable social groups. Moreover, different classes, including low-income and marginalised people, participate in planning procedures. The walkable environment of neighbourhoods provides public health benefits and lower crime rates. It also increases opportunities for social gatherings in public spaces and fosters a sense of community belonging. Based on the theory of chrono-urbanism, the '15-minute city' can create employment opportunities by reducing commuting time. Residents may also have more time for the essential social functions of living, healthcare, education and entertainment.

The '15-minute city' also offers several potential benefits for economic sustainability. First, the neighbourhood distribution of businesses and concentration of jobs can help communities maintain economic viability while consuming less energy and reducing transportation-related emissions. Second, a '15-minute city' allows residents to save on transportation costs that they would otherwise incur using private vehicle transportation. According to Khavarian-Garmsir et al, the transportation costs of households in walkable districts are half those of households in car-dependent districts (Khavarian-Garmsir et al., 2023). Third, because the 'city of 15 minutes' has compact neighbourhoods, the cost of maintaining and developing infrastructure in compact neighbourhoods is lower than in low-density, segregated, car-dependent areas. Lastly, the new concept provides for the development of affordable housing for low-income families and is therefore a pro-poor strategy that allows its residents to spend more on local purchases and support local businesses.

Implementing the new concept also helps the environment by reducing pollution from non-essential transportation. It decentralises urban services to reduce unnecessary travel through cities, expands walking and cycling routes, redesigns public spaces to discourage the use of private cars and develops public transport routes. The '15-minute city' gives non-car owners easier access to essential services within walking and cycling distances. Examples from New Zealand, Doha and Barcelona suggest that sustainable transport can reduce car traffic, energy consumption and carbon emissions. In addition, it supports more efficient use of urban land by reducing current pressure on suburban and agricultural land. It therefore protects agricultural and rural land from unregulated urban sprawl.

There are many similarities but also some differences between the two new concepts of the '15-minute city' and the 'integrated sustainable city'. Among the similarities is that both require a multidisciplinary approach, involving transportation planning, urban design and policy making, to create well-designed public spaces, pedestrian-friendly streets and mixed-use developments. This lifestyle change can include remote working, which reduces daily commuting and is supported by the recent widespread availability of information and communication technology (ICT).

As differences, we can point out that the '15-minute city' project has been criticised for:

- It encourages physical determinism. It thus aims to improve the current status quo through physical design and modification of the urban form, similar to classical planning movements, while social and economic concerns cannot be addressed through physical or technological measures.
- Implementation of the concept may not be feasible in practice for highly congested cities that have a sprawled urban structure displacement being dependent on cars and suffer from rigid land planning and zoning regulations.
- Cities face diverse problems that are not universal and, as a result, a '15-minute city' model based on a one-size-fits-all approach cannot be applied to all cities. Each community requires a neighbourhood development plan that takes into account its specific geography and unique issues.
- This concept pays little attention to the walking behaviours and daily needs of different social and demographic groups. For example, the walking behaviours and speeds of individuals vary by age, and therefore, the daily needs of young people may differ from older adults.
- There is some confusion about the scope of the concept. While its name implies that it is a citywide concept, its size and characteristics focus on the neighbourhood level.

- Despite Moreno's emphasis on affordable housing, ethnic diversity and proximity of housing to jobs, it is unclear how and by what means these goals will be accomplished. This issue was noted in the Portland Plan in 2012, where the location of activities and services was not operationalised due to ambiguities in providing the market range and threshold populations needed.
- The 15-minute city concept does not account for the complexity of cities. Cities are systems of systems, with subsystems that interact and affect each other. As a result, a rigid and simplistic urban structure cannot address critical concerns such as climate change, pandemics and transportation crises.
- The concept of the 15-minute city ignores critical issues such as environmental protection, biodiversity, energy efficiency, green and clean energies, as well as local vernacular language, culture, heritage and identity. As a result, a rigid and simplistic urban framework cannot address critical concerns such as climate change, pandemics and transportation crises. Finally, the 15-minute city concept ignores critical issues such as environmental protection, biodiversity, energy efficiency, green and clean energies, as well as local vernacular language, culture, heritage and identity. As a result, a rigid and simplistic urban structure cannot address critical concerns such as climate change, pandemics and transportation crises.
- The 15-minute city concept ignores critical issues such as environmental protection, biodiversity, energy efficiency, green and clean energies, as well as local vernacular language, culture, heritage and identity.

In contrast to these problems, the proposed concept of an 'integrated sustainable city' supports the reconfiguration of the envisaged spaces in such a way that the objectives can be achieved by adapting solutions to the concrete situations. In other words, it does not propose the creation of a standard model of development, but only proposes courses of action and the objectives to be pursued, with concrete tools and measures that may differ from situation to situation.

Moreover, the integrated sustainable city represents a 'greening' of current cities by developing semi-independent neighbourhoods (like atoms in a molecule) that offer the possibility of normal economic, social, health, educational, recreational, etc., living conditions without the need for multiple and long-distance journeys by polluting vehicles.

## References

Breuste J., *The green city: General concept*, 2020, https://link.springer.com/chapter/10.1007/978-3-030-37716-8_1

DGB, *The top 10 green cities in the world*, 2023, https://www.green.earth/blog/the-top-10-green-cities-in-the-world

Fistung D. and Popescu T., *The construction market in Romania after 1989*, CEIS-INCE internal study, Romanian Academy, 2018.

Hertz N., *The age of solitude*, Ed. Humanitas, ISBN 978-973-50-7275-9, 2021.

Khavarian-Garmsir A. R., Sharifi A. and Sadeghi A., *The 15-minute city: Urban planning and design efforts toward creating sustainable neighbourhoods*, 2023, https://www.researchgate.net/publication/365440746_The_15-minute_city_Urban_planning_and_design_efforts_toward_creating_sustainable_neighborhoods

Lewis, E., *Green city development tool kit*. Manila: Asian Development Bank, 2015, https://www.adb.org/sites/default/files/institutional-document/173693/green-city-dev-toolkit.pdf

National Geographic, *Cities History*, 2023, https://education.nationalgeographic.org/resource/history-cities

NIS (National Institute for Statistics), 2023, *Population and housing census, round 2021*, https://insse.ro/cms/sites/default/files/com_presa/com_pdf/rpl2021_date_provizorii_profil_teritorial_ian_2023.pdf

Peace R., Churkina G. and Rivera M., *How green is a "Green City?"*, 2016, https://www.rifs-potsdam.de/sites/default/files/files/wp_dec_2016_en_green_city.pdf

Richie H. and Rover M., *Urbanization*, 2019, https://ourworldindata.org/urbanization

RP (Romanian Parliament), LEGEA nr.365/2001, privind aprobarea Planului de amenajare a teritoriului naţional – Secţiunea a IV-a Reţeaua de localităţi (LAW no. 365/2001, regarding the approval of the National Spatial Planning Plan – Section IV Network of localities), 2001, https://lege5.ro/Gratuit/gmztgobx/localitati-urbane-lege-351-2001?dp=giytqmjvgmytqhttps://oot.rmdsz.ro/sites/default/files/Legea_351-2001_-_PATN-Reteaua_de_localitati.pdf

UN, *The world cities' in 2018*, 2018, https://www.un.org/en/development/desa/population/publications/pdf/urbanization/the_worlds_cities_in_2018_data_booklet.pdf

Ungureanu L., *The decline of big cities begins*, 2023, https://www.libertatea.ro/stiri/incepe-declinul-marilor-orase-avertismentul-unui-reputat-cercetator-asa-cum-a-prezis-h-g-wells-aici-vor-locui-elitele-servitorii-lor-si-cei-fara-copii-4490113

World Bank, *Urban development*, 2022, https://www.worldbank.org/en/topic/urbandevelopment/overview

Zhuji World, *Rome, Italy – Statistics*, 2023, https://ro.zhujiworld.com/it/1731634-rome/

# 7 Some solutions to support sustainable transport

There are many strategies and policies worldwide to support sustainable transport, some of which have been implemented, but unfortunately, many remain only at the stage of proposals, theories and debates. Some of them are, however, promising.

Many measures are being taken to reduce the level of transport demand in order to minimise the negative effects of traffic congestion or to halt the degradation of infrastructure networks.

## 7.1 Examples of international action

The concept of modernised and environmentally friendly inter-city rail networks, for example, has spread not only in Europe, where several routes have been planned and modernised, but also in Canada, China and Thailand. In these regions, rail is responsible for about 8% of all passengers traveling between cities. Passenger rail passenger activity is 75% electrified, with China and India hosting most of the existing lines as well as projected future lines (SLOCAT, 2022).

Due to growing urban congestion and the expansion of shared mobility options between 2017 and 2019, total car sales declined by 7% in Organisation for Economic Co-operation and Development (OECD) member countries and by 10% in non-OECD countries. This was made possible by the imposition of policies to support cleaner transport or to adopt traffic restrictions in places where congestion has had extremely serious consequences in terms of human health and environmental protection.

In Singapore, for example, since 1975, there has been a zone in which motor vehicles are not allowed to enter during peak hours without a permit (Fistung, 1999). Exempted from this rule are commercial vehicles or those with less than four passengers. The regulation was supplemented in 1990 with a set of measures aimed at discouraging the purchase of passenger cars. The maximum number of cars that could be marketed was set on the basis of monthly decisions. Of course, many specialists do not agree with imposing restrictions on the possibility of buying a car. However, the Singapore measure was not aimed at that, but at reducing car use. In order to reduce congestion, it is, however,

DOI: 10.4324/9781003743064-8

more appropriate to adopt measures aimed at eliminating a certain segment of users – especially those for whom the value of the trip is less than the value of the delay (increase in travel time) caused to other traffic partners. In this respect, the size of the payment agreement, equal to the marginal congestion cost imposed on other users, emphasises those whose trips are economically efficient.

Measures of this kind are currently being adopted in many countries such as France, Italy, Japan and the United States. And in cities such as Oslo, Bergen and Trondheim, there are zones within which motorised travel is only allowed after paying a fee. In Switzerland, there is a system of all-user highway access permits valid for a certain period of time (usually one year). This type of policy is easy to put into practice and does not lead to a decrease in traffic figures during peak periods. This is true because, after obtaining the permit, the user can travel at any time of the day (within the time limit of the permit) without restrictions.

In Mexico and Chile, in order to reduce the intensity of road traffic in densely populated cities, a measure has been adopted to authorise the movement of vehicles according to their registration number (one day even, one day odd). This initiative was only partly aimed at decongesting traffic, but more importantly at reducing the alarming levels of chemical pollution.

Unfortunately, even such a policy has not been particularly effective, as many Mexicans own at least two cars, one of which they register with an even number and the other with an odd number.

In counterbalance to the use and marketing of cars, motorised two-wheeled vehicles such as mopeds and motorcycles grew by 149% in India and 80% in Vietnam between 2010 and 2019, and the world's largest motorcycle fleets are in China, India, Indonesia, Pakistan and Vietnam. These modes are most prevalent in Southeast Asia and have also grown in Latin America and Sub-Saharan Africa.

In Hong Kong, due to the very rapid increase in traffic values based on the multiplication of cars in a very short period of time, an electronic road traffic monitoring system was introduced in 1985 to levy road user charges.

From a technical point of view, the system consists of a mechanism installed in each vehicle which incorporates a specific identification code, with decoders at various points in the infrastructure network designed to 'read' the traffic information. This means that at any given time, it is known how many vehicles are passing through a given point, and at the end of each month, vehicle owners are asked to pay a fee based on the number of journeys they have made. Although the system is quite expensive, a cost/benefit analysis has shown that the financial benefits obtained are about 14 times greater than the costs involved in implementing it. It was also found to be twice as cost-effective as any other type of policy aimed at discouraging car purchases.

A report on China, Europe and the United States revealed that the shared mobility market[1] in these areas was of the order of $54 billion in 2016 (Santos, 2018). According to the same report, the shared mobility market in China, Europe and the United States accounts for 0.12% of the combined GDP of these regions. The report states that in China or the United States, the shared mobility market is dominated by 'players' such as those who use their own cars for trips with a driver hired specifically for these services only (e-hailing) or companies that do not own cars but work with private car owner-drivers. The service is based on an app, provided by the company. This is the case with the well-known taxi-like public transportation services of companies such as UBER, BOLT, etc.

Some examples of measures also adopted in the United States to support sustainable transportation are briefly presented below (Leuenberger et al., 2014).

New York City's focus on improving energy-efficient transportation is an example of a sustainable transportation plan that integrates multiple systems. Goals of the plan include safer streets for students and seniors, improved pedestrian signalling and expanded educational programming. Enhanced mobility and physical infrastructure goals include new dedicated bus spaces and bike lanes and expansion of the renewable fuels initiative. Behavioural improvements include reducing city employee trips and improving citizen engagement programmes. This includes multi-system goals such as coordinating with stormwater management systems to increase permeable surfaces, partnering with a street arts programme and implementing a public environmental development programme. Today's sustainable transportation management is outcome-focused, uses data-driven change, integrates behavioural change and physical infrastructure improvements and involves citizens in multiple systems.

A similar example in Dubuque, Iowa, is the Regional Sustainability Initiative. In partnership with the University of Iowa, this initiative aims to implement measures to improve urban planning, housing, education, health and transportation. The goals of the transportation plan are citizen access and mobility with programmes such as car and bicycle sharing, improved pedestrian access, improved traffic signals, transit route redesign and decision-making based on continuously updated data (Casiello et al., 2012).

California's governor has proposed an emissions trading programme to fund his state's $68 billion high-speed rail project. In another example of creative thinking, the private sector has helped cities along rail lines recognise the economic opportunities presented by the development of rail stations. Whereas in the past, rail companies had to pay for the right to develop railway stations, planners now want to promote regeneration and commercial development by transforming deprived areas into productive real estate and quality community investment.

In Toronto, Canada, an expert panel appointed by the regional government, backed by the private sector, has recommended raising local corporate and fuel taxes to fund traffic improvement and congestion relief. This is similar to the *Crossrail*[2] project in London (UK), where work is underway on Europe's largest construction project is being funded in part by imposing an additional commercial charge and community infrastructure levy.

In addition, public authorities around the world are looking for creative solutions to sustainably pass on the cost of building infrastructure to the direct beneficiaries of the project.

Recognising the need for a coordinated regional approach to development projects, the Government of India established the Mumbai Metropolitan Region Development Authority in 1975. The organisation has a lead role in preparing infrastructure development plans, formulating policies and programmes, implementing projects and directing investments in the region. The agency owns and leases large tracts of land, and the revenue generated is used to finance infrastructure projects. The Agency's central concern at present is to improve transport links throughout the city while greening these services through massive investment and increasing the efficiency of specific activities.

Public transport is the main mode of transportation in Shanghai (China) since 1994, when limitations on car ownership were introduced to limit the growth of car traffic and alleviate congestion. New private cars cannot be driven without license plates, which are sold in monthly auctions for license plates, accessible only to locally registered residents and those who have paid social insurance taxes or individual income taxes for the past three years.

Approximately 9,500 license plates are auctioned each month, and the average price was about US$13 in 2019. Shanghai, with a residential population of 25 million, has more than four million registered cars, making it the fifth largest Chinese city by this measure. Despite the measures taken, the city remains plagued by congestion and vehicle pollution. Even though in 2018, the public transportation system handled an average of 16.05 million trips each day, of which 10.17 million (63%) were by metro and 5.76 million (36%) by bus, the traffic volume on Shanghai's expressways remained extremely high, averaging 1.215 million vehicles per day.

From all the above examples, it can be seen that sustainable transport requires certain measures to be taken, the most important of which are as follows:

- Reducing demand for non-essential transportation.
- Adopting (choosing) the use of vehicles that do not use fossil fuels.
- Shift from own cars to more environmentally friendly systems (e.g. shared mobility).
- Internalising all external costs.
- Increase the use of public transport, cycling (or other similar vehicles) or walking.
- Gain as much (political or public) support as possible for sustainable transport, technological innovations and/or laws and regulations in this field.

## 7.2 Concrete measures to support a sustainable transport system

Some broad themes and orientations in current and prospective support of achieving a real and effective system of sustainable mobility:

1 Spatial planning.
2 Traffic reduction.
3 Development of an integrated public transport system.

4 Creation and use of Intelligent Transport Systems (ITS).
5 Pre-dominant use of clean/autonomous vehicles.
6 Sharing the use of vehicles (cars, bicycles and others).
7 Vehicle and fuel taxes.
8 Parking management.

### *7.2.1 Spatial planning*

Cities need to be rethought to encourage safer and more sustainable modes of mobility. Safety should always be the priority in infrastructure planning and provision.

Sustainable transport methods focus on improving urban mobility in order to ensure rapid access and movement of people in large urban centres. Urban transfers usually involve many means of transportation, and transit can take place using different modes of transportation (Martin et al., 2019).

Urban mobility cannot be addressed by land-use planning. The way the street network is designed and developed determines which mobility options are real and how attractive some are relative to others.

If cities increase their population or density at a faster rate than they offer alternative mobility options, they become gridlocked.

Congestion is not only caused by land use. The way cities are built plays an important role in ensuring safe travel.

In this respect, cities could introduce measures to promote the densification of urban areas by limiting the development of new urban centres in peripheral areas. This reduces average travel times and increases urban vehicle use (Speirs, 2019). Another major way forward would be to redirect investment towards essential mobility and discourage optional trips.

Two land use patterns that are yielding the greatest results are mixed-use development and transit-oriented development. These two policies are mutually reinforcing and should be implemented simultaneously. Pedestrian and cycling infrastructure, together with public transport, should be prioritised over roads for cars.

By reducing the amount of infrastructure dedicated to cars and allocating more space for other transport modes, it is possible to shift the modal mix. A central advantage of mixed-use areas is that they minimise reliance on private car use.

### *7.2.2 Reducing traffic*

Traffic calming depends on two elements: lower vehicle speeds and smoother traffic flow. However, poor iteration of traffic calming can provide intermittent speed 'spikes' that only encourage drivers to accelerate quickly and brake suddenly to leave a particular area. Traffic calming should not frustrate drivers – when done well, it should not be noticed. In Malaysia, for example, most traffic calming measures have been 'implemented ad hoc without any proper standards or guidelines, but only based on the experiences of the local traffic engineer and the requests of residents' (Speirs, 2019).

In the past, in less dynamic times than the ones in which we live, the planning systems used by most organisations were mainly based on setting annual budgets and extrapolating current sales and environmental trends over five- to ten-year periods. On the basis of these forecasts (which were generally accurate because of the stability of the economic environment), managers allocated resources.

Today, however, there is no such thing as stability in any economic, environmental or political sense. The only constant is constant change.

For the punctual traffic situation in urban areas, a viable solution is the installation of an Urban Traffic Control and Management System (UTMS) using computerised technology. The effective use of this tool leads to solving traffic problems, providing gradual prioritisation to public transportation vehicles and reducing noxious emissions. UTMS cannot be realised in a single module for the entire road network, but gradual approaches are needed to eventually cover the entire urban area.

Benefits of UTMS:

- Reduced travel times.
- Reduced fuel consumption.
- Reducing the number of stops at traffic lights by improving traffic flow.
- Reducing the vehicle fleet.
- Improving energy efficiency.
- Lower vehicle operating costs.
- Reduced emissions.
- Installation of the Global Positioning System (GPS) in public transport vehicles. This is aimed at locating and positioning each vehicle and communicating traffic information by radio to the central control panel. The data are used to streamline public transportation and reduce external costs.

### *7.2.3 Development of an integrated public transport system*

The promotion of public transport is at the basis of most environmental or social projects in most localities. At comparable transport capacities, the advantages of public transport over private transport are numerous: energy and space savings, reduced chemical and noise pollution, greater safety, lower urban traffic intensity, etc.

Efforts to promote public transport are evident throughout the EU. This can be improved if the following areas are addressed: vehicle construction, improvement and modernisation of specific infrastructure, integration with other transport systems, improved quality of service, etc.

Potential users of public transport may be discouraged if numerous changes of mode are required along a route, or if the same mode is changed with long waiting times due to timetable non-synchronisation, or with additional costs (for other tickets) due to non-integration of the fare system.

Systemic integration aims at grouping all modes of public transport into one common operating environment. The principle is simple: the aim is to create a single operating network for all modes of public transport (tram, bus, trolleybus and metro).

The most important elements of the integrated system are as follows:

1 Coordination of operating timetables.
2 The existence of single ticketing systems (tickets valid for all modes).
3 Multimodal terminals.

1 By all possible measures, the aim should be to reduce the need to change modes of transport, but where this is unavoidable, waiting times when changing modes should be minimised. The timetables should be adapted so that the departures and arrivals of the different means of transport complement each other. It is preferable that a single organisation (institution) manages the whole system.
2 Often, charging systems are complex and payment methods are restrictive. For example, some tickets have to be purchased in advance and are not available on the means of transport. The integration and harmonisation of the charging system make it easier for passengers to travel, as a purchased ticket can be used anywhere along the route. The introduction of this system always leads to increased use of public transport.
3 The creation of an efficient transport system depends mainly on how different modes of transport (public systems, pedestrians and private vehicles) are integrated.

Multimodal terminals – or transfer points – can provide a quick and efficient link between different modes. The simplest form of multimodal terminal is the 'interchange space', where a private car can be replaced by public transport. The principle of such a structure is simple: to delimit an area which, through the facilities it offers, can lead to a rapid change of means of transport.

In order for a transport system to be efficient, it is essential that potential passengers are constantly informed about the status of available modes of transport. Real-time information on the rhythm and timing of activity on the different transport networks is of real importance in this respect. This data can be transmitted both via local communication systems (teletext, telephone, radio, etc.) and the media.

Quality of service is one of the main attributes on the basis of which the choice of a particular mode of transport is made. From this perspective, private transport has an advantage over public transport as it can provide door-to-door travel. This can be improved under certain conditions. For example, public transport services could be introduced, using minibuses with a fixed route but with flexibility (i.e. they can take passengers to certain destinations close to the basic route, even if this inevitably means over-distancing and then returning to the basic route). The 'maxi-taxi' system, with optional stops, even if generally suitable for passengers with reduced mobility needs, may be another example.

The following measures could be considered to improve and encourage the development of public transport:

- Adopting specially designed and reserved lanes for this type of transport can lead to the avoidance of accidents due to heavy traffic in certain conurbations.
- The introduction of a preferential flow for 'green wave' public transport systems may create conditions conducive to the use of these services by a greater number of passengers.
- Limiting access in certain areas to private transport, while maintaining public transport.
- Rationalisation of the urban planning scheme can greatly improve the quality and efficiency of public transport (e.g. industrial, commercial, residential or commercial areas or areas with a high public affluence can be designed in such a way as to allow easy access to public transport services).

Cities aspiring to a high modal share of public transport should plan urban development accordingly. An effective way to encourage public transport is to ensure that development occurs near transport nodes (Speirs, 2019).

Two key factors determine the success of public transport. Firstly, the time (efficiency) of public transport. Second, the price compared to traveling by private car. To facilitate this, all modes of public transport must be integrated.

An integrated payment system increases the efficiency of the system while at the same time reducing costs – both for users and operators.

The key to an efficient transport network that serves all parts of the city is cross-subsidisation, where crowded bus lanes subsidise buses operating in quieter areas.

The 2004 reforms imposed in the South Korean city of Seoul provide a model in this direction. Before the reforms, bus operators were paid according to ridership. So, they competed for the busiest routes, did not serve quieter communities and drove dangerously for passengers. The reforms changed the way bus operators are reimbursed based on vehicle mileage instead of rider trips (gross contracts) and fares collected for the entire system. Today, more areas of the city are served, cycling has increased (along with fare collection), and buses are driven more safely. This would not have been possible without a single transit fare.

However, without dedicated bus lanes, buses are prone to getting stuck in traffic. This is why the development of dedicated bus lanes, or, if possible, in the future, a Bus Rapid Transit (BRT) network, has started. BRT can be divided into 'full-service' and 'lite' categories. Full-service BRTs use dedicated lanes, architecturally well-defined boarding platforms and pre-boarding fare collection. An additional advantage is that such networks can be built on existing infrastructure, making them more economically viable. Dedicated lanes greatly increase the frequency and reliability of buses and make them more attractive – especially if private vehicles are trapped in their own congestion.

### *7.2.4 Development and use of intelligent transport systems (ITS)*

ITS is a concept used to describe how new information technology can be used in the transport sector. The idea with ITS is that services should provide travellers and operators with additional information to improve travel activities.

Even if telematics applications are not currently practical solutions for urban public transport, the increasing attractiveness of urban public transport calls for the evolution and expansion of the use of information systems. In the future, the demand for public passenger transport will only be able to increase if, among other things, passengers have rapid access to 'real time' data on the traffic situation and services offered by urban public transport operators. The information system needed to achieve this can be deployed on a large scale, provided that the necessary funding is made available and the appropriate technologies and solutions are adopted.

Large sums of money are currently being allocated to the development of telematics solutions to improve urban public transport operations by providing better information to passengers. This information is mainly related to:

- Routes and timetables.
- Position of means of transport and forecasts of their arrival at their destination, according to actual traffic conditions.
- Information support for passengers to choose the best public transport solutions.

Guidance and control systems are thus designed to make transport safer and more efficient in relation to passenger requirements and traffic conditions. By using computerised coordination systems and taking all the organisational and technical measures for carrying out their activities, public transport operators can intervene 'in real time' to solve unexpected transport shortcomings, such as a sharp increase in demand (introduction of additional transport vehicles) or the occurrence of bottlenecks (choice of detour routes). These systems have led to general traffic management solutions in urban areas in favour of public transport. Examples are signalling systems at intersections giving priority to public transport vehicles.

The integration of different computerised control systems is a major contribution of public transport operators to increase the benefits and competitiveness of the business, but also to meet the demands of passengers.

In addition to these monitoring systems, 'on-board' computer systems are used in public transport.

These systems oversee the operation of all subsystems on the means of transport, including passenger information, ticketing or the processing of travel data on the basis of 'smart cards'. In general, the communication between the transport supervision centre and the control systems in the means of transport is done by radio. The traffic coordinates for determining the position of the means of transport are provided by satellites (GPS).

Telematics solutions specific to urban public transport should aim to develop modern payment systems, ticketing, support operations, new and modern passenger information systems and transport prioritisation systems for surface public transport. Investments in these systems are aimed at increasing the quality and attractiveness of public transport for passengers and, at the same time, the efficiency of transport operators' activities.

In this context, there is currently a major interest among public transport operators in Central and Eastern European countries in developing telematics solutions that are as effective as possible.

The most commonly used grouping of ITS is based on system positioning, as follows (Bawangaonwala et al., 2018):

1 Advanced Traffic Management Systems (ATMS), which integrates various subsystems (such as vehicle detection, messaging, etc.) into a single articulated interface that provides real-time traffic degree data and forecasts traffic circumstances for more efficient planning and operation. Dynamic traffic control systems, highway operations management systems, incident response systems, etc., respond in real time to changing conditions.
2 Advanced Traveler Information Systems (ATIS), which provide users with travel-related information to support decision-making on routing, travel time and avoid congestion. This can be achieved by providing different information using various technologies, such as:

    a GPS enabled in vehicle direction presentation and alternative route-finding systems.
    b Road signage presentation for real-time communication of information on possible traffic congestion situations.
    c Signalling of traffic jams, accidents and alternative route information while travelling.
    d Provision of coded network maps, including road congestion levels.

3 Advanced Vehicle Control Systems (AVCS) are tools and concepts that improve drivers' control of the vehicle to make traveling safer and more efficient. For example, collision warning systems. In more advanced AVCS applications, the vehicle could automatically steer away from a collision, based on inputs from on-vehicle sensors.
4 Commercial Vehicle Operations (CVO) systems, comprising an assembly of satellite navigation systems, a small processor and a digital radio, which can be used in commercial vehicles such as trucks, vans and taxis. This system

allows constant monitoring of truck operations by the central office and ensures traceability and safety.

5 Advanced Public Transportation Systems (APTS), where information technology is applied to public transportation systems to increase operating efficiency and improve safety. This includes real-time passenger information systems, automatic vehicle location systems, bus arrival announcement systems and systems that provide priority of way to buses at signalised intersections (transit signal priority)

6 Advanced Rural Transportation Systems (ARTS) provide information about long-distance trucking and other transportation systems. Examples include automatic reporting of road and weather conditions and directional information. This type of information is useful for drivers traveling in remote or rural areas. This has been widely implemented in the United States and could be a valuable asset for countries such as India, where rural areas are widely dispersed.

In many countries, strategies and techniques have been developed and implemented, tailored to specific geographic, cultural, socio-economic and environmental specificities, to integrate the different components into a coherent system. In general, any ITS application uses a Traffic Management Centre (TMC), where data are composed, analysed and merged with other operational and control concepts to deal with complex transportation problems.

Thus, the numerous telematics solutions developed in support of urban public transportation are worth mentioning:

- The analytical system BALANCE, implemented in Munich (Germany), presents the costs, benefits and effective solutions for optimising traffic flows by controlling traffic-light intersections, taking into account the specific transport demands of private, public and pedestrian transport. The system does transport optimisation both on a macro level (for the whole city) and on a point-by-point, intersection-by-intersection basis. An 'intelligent' public transport prioritisation subsystem acts when the need to use public transport arises (traffic in the area is very high and private transport is in 'conflict' with public transport). The system has proven to be very effective, bringing specific travel cost benefits for all 'actors' involved in urban transport and large gains for public transport.
- The System of Flexible Public Transport Services for Elderly and Disabled People (SAMPO), implemented in Göteburg (Sweden), aims at coordinating a fleet of minibuses on alternative routes, flexibly chosen according to demand (with a minimum response time of 15 minutes from order). The PLANET subsystem of this network coordinates demand coverage on flexible routes and automatically records the requests and the requested travel coordinates (origin, destination, route). The system was soon extended as its users were very satisfied with the service. This drastically reduced the demand to use the much more expensive taxi network for the same type of service.

- The introduction of efficient passenger information systems provides details of the transport services on offer (routes, timetables, vehicles, etc.). Displays at stations and on public transport have always been well appreciated by the majority of passengers on public transport networks throughout Europe.

In the future, there is a need to develop those telematics solutions that are best accepted by passengers. A few examples, in this context, we consider relevant.

The EUROSCOPE ROMANSE project, developed in Southampton (UK) by a public-private consortium of organisations, provides travellers with necessary and 'real time' public transport information, covering both the pre-trip and post-trip periods. In this context, the Passenger Information and Traffic Control Centre was created. This Centre coordinates subsystems such as: Strategic Information Systems, which provide comprehensive data on the traffic in a given area and travel coordinates, presented using maps; STOPWATCH System, which provides 'real time' information on the progress of the public transportation schedule, information posted at their stations; The Travel Terminal connects with transport operators, providing information, posted in various special centres, on travel possibilities to different destinations; the Multimodal TRIPLANNER system provides information on planning a particular trip, through 24 special centres in the city and a website.

The INFOPOLIS system, developed in France, provides a detailed analysis of public transport information systems operating in Europe.

Another efficient passenger information system on public transport timetables (DELFI) has been implemented in Germany since June 2004. The Federal Ministry of Transport, Building and Housing (MFTCL) launched this project to increase the use of public transport by providing comprehensive information on the services offered by public passenger transport operators at the local level, but also regionally or over longer distances. In this way, interested travellers can obtain comprehensive information about a particular route they wish to travel, throughout the whole country, without having to consult different information systems specialised in specific transport categories (such as the Guide to Public Transport in specific cities, Train Routes, etc.).

Another telematics application was based on the use of an 'electronic ticket', the data of which were stored via a 'chip'. This method was initially used in the region of North Westphalia (Germany), where some two million traditional travellers used a dual-purpose electronic card as a season ticket. In Bremen, 'e-tickets' were purchased via a card bank, issued by a payment card, and in the future can be purchased directly by a simple bank debit for the account holder.

All these applications, aimed at increasing the attractiveness of public transport use, while reducing the marketing costs of transport operators and also increasing their revenues. By accurately determining the number of passengers for each individual route, the operator involved and the transport system used, urban public transport activity can be optimised at no extra cost and on a permanent basis.

Another example is the complex system developed in the city of Helsinki (Finland) since 1999, called HeLMi. The design of this system was based on achieving three major objectives:

1 Introduction of a priority signalling system for public transport.
2 Traveler information.
3 Control of the transport fleet.

The pilot phase of the project was known as 'Project-423', as tram line 4 and bus line 23 were the first routes to be used in the new system. The project is led by Helsinki City Transport (HKL) and the Helsinki City Traffic Planning Division. The new HELMI telematics system offers several public transport telematics functions, such as real-time passenger information, bus and tram priority at traffic signals and timetable monitoring. Although similar systems are in use in several European cities, the HELMI telematics system is one of the most sophisticated, as it relies almost exclusively on wireless communication with radio modems.

Public transport using such a telematics system has been accepted as a new tool to increase the attractiveness of bus use even in smaller cities such as Västerås and Jönköping in Sweden. In the Nordic countries, the largest telematics-based public transport system, including all city bus and tram lines, is used in Gothenburg.

In Romania, the last two decades have seen exponential growth in the IT industry, much of it dedicated to intelligent transportation solutions. For example, a large part of the software development activities at NXP Semiconductors Romania are aimed at implementing solutions for autonomous cars (e.g. ADAS Radar). Autonomous cars are vehicles that contain a large number of chips and a considerable amount of software. Also nicknamed Software Defined Vehicles, these cars benefit from a wealth of software-enabled functionality (David, 2022). Autonomous cars update like cell phone software, remotely.

### *7.2.5 Predominant use of clean/autonomous vehicles*

While still in the early stages of development and adoption, autonomous and electric vehicles (EVs) are advancing rapidly and appear poised to trigger the biggest change in the transportation industry in the last century. However, in our view, the era of EVs will be short-lived, quickly replaced by vehicles that will use less expensive and easier-to-maintain energy sources. At the moment, examples in this direction include vehicles powered by fuel cells.

EVs are much quieter, contributing very little to noise pollution. However, EVs do not address many of the problems associated with cars (Speirs, 2019). For example, they are still prone to congestion and pose a safety risk to pedestrians and cyclists.

Autonomous vehicles (AVs) can improve the mobility of people who cannot drive: children, the elderly and people with disabilities. They will also increase road flow and reduce delays caused by collisions. This type of vehicle will also reduce the cost of commuting – smoother traffic flow reduces fuel costs, fewer collisions reduce insurance premiums, shared mobility would eliminate parking fees, and time in transit could be used freely or productively – this will allow people to live farther from job to job. The result may be an increase in urban sprawl. Eventually, these vehicles could be used to run errands without passengers – to collect pickup or drop off a package.

The use of electric buses combined with rapid transit vehicles is a more energy-efficient alternative to the use of conventional private cars (Martin et al., 2019).

Electric and autonomous vehicles represent substantial improvements over traditional vehicles. They improve urban air quality and will reduce accidents and fatalities. Unfortunately, these positive outcomes do not eliminate the negative effects that can arise from the use of these new types of transportation. They can exacerbate urban sprawl, further contribute to road congestion and perpetuate class inequality as they are only accessible to those with higher incomes. Vehicle automation and electrification, however, significantly reduce the cost of travel.

Electricity is cheaper than gas and EVs have far fewer parts, making them much cheaper to maintain.

In addition to the economic decision to buy an electric vehicle, consumers have to overcome 'range anxiety' or fears that a battery cannot get them to a 'filling station'. As battery performance continues to improve and charging networks expand, this fear is likely to reduce over time. Over the past six years, the average range for EVs has extended by 56%, and some models can now drive more than 300 km on a single charge.

EU countries offer various incentives to promote the use of EVs, such as purchase subsidies, registration tax benefits, property tax benefits, business tax benefits, VAT and other financial benefits, local incentives and infrastructure incentives. In most countries, there is a clear relationship between the incentives offered and the growth in plug-in EVs. Not surprisingly, in countries where no incentives are available, such as Bulgaria, Estonia, Poland and Slovakia, there is a low propensity to buy electric cars (EC, 2018)

An interesting EV trend, which has already emerged in 2020 and continues to gain in importance, relates to public transportation and the collaborative economy (VIRTA, 2022).

Thus, electric buses have increased in popularity since 2020. The global stock of electric buses in 2021 was 670,000. One of the emerging global leaders in electric bus fleets is India, which plans to introduce 5,500 electric buses to the domestic market annually.

In EU countries, the Clean Vehicles Directive also provides for public procurement of electric buses. France, Germany, Spain and the United Kingdom are just a few of the countries experiencing a surge in sales of electric buses.

Romania has not taken a stand against this approach either. Thus, by 2021, there were around 500 electric buses in operation in Romania, running in 21 cities. Most of them, that is, 100, were purchased for the capital. The top green cities are followed by Brasov and Timisoara, with 44 electric buses each (Chisăliţă, 2021).

But not only big cities like Bucharest and Cluj have or will have public transport using electric buses. The Ministry of Public Works announced the purchase of 81 electric buses for seven smaller cities (excluding Constanţa) as follows: Alexandria (10), Brăila (11), Constanţa (21), Drobeta Turnu-Severin (3), Focşani (20), Lugoj (9) and Slobozia (7).

The Romanian government has also purchased another 123 electric buses worth €65 million for cities in the centre, north and east of the country. The Spanish group CAF, through its Polish subsidiary Solaris, won the tender launched by the Ministry of Public Works to supply zero-emission buses to six cities in the country. Iaşi, Sibiu, Sighetu Marmaţiei, Suceava, Târgu Mureş and Piteşti will thus benefit from EVs in their public transport fleets, with the buses to be delivered before the end of 2022.

As for heavy-duty trucks (HDT), manufacturers are showing their commitment to the electrification of mobility. Many truck manufacturers are striving for an all-electric future. HDT sales have grown rapidly over the past five years in the United States and Europe. However, by 2021, nearly 90% of electric trucks were registered in China.

The largest concentration of private electric two-/three-wheelers is in Asia, particularly China, accounting for 9.5 million registrations out of the total of 10 million in 2021. Vietnam and India are strong markets for two- and three-wheeled EVs. The market is also growing in Europe, where it reached a 5% market share in 2021

Rising sales of EVs have led to increased production of lithium-ion batteries for them. In 2021, demand for electric batteries more than doubled year-on-year, mainly due to increased sales of EVs in China.

China also remains the leading country in battery production, accounting for more than 75% of global production capacity. However, Europe is not lagging behind. According to Bloomberg NEF, Europe's share of global battery production could increase to 31% by 2030. By comparison, the continent's share of global battery production was only 7% in 2021.

It is clear that to support the growing demand for EVs, the different elements of the battery supply chain will need to grow, from the extraction of raw materials such as lithium or nickel to the production of EVs themselves. However, the metal supply estimates should be sufficient to meet the estimated demand set by the 2025 state policy scenario.

But, as mentioned earlier, in our view, the future will be dominated by vehicles that will be powered by cheaper and more environmentally friendly fuels using fuel cells. With this in mind, the world's first fleet of hydrogen-powered rail trains was inaugurated in Germany in mid-2022. The fleet of 14 trains, supplied by the French group Alstom in the state of Lower Saxony, runs on a 100-km route linking the cities of Cuxhaven, Bremerhaven, Bremervorde and Buxtehude, a short distance from Hamburg (Măciucă, 2022).

Alstom's Coradia iLint hydrogen-powered trains, designed in Tarbes, France, and assembled in Salzgitter, Germany, are a first in the industry, according to the source. The new train designed by Alstom combines hydrogen on board with oxygen in the ambient air, thanks to a fuel cell installed in the ceiling. This produces the electricity needed to pull the trains.

The new technology is the German authorities' preferred measure to reduce carbon dioxide emissions and replace diesel locomotives, which still operate on 20% of German rail routes. The new fleet, which costs €93 million, replacing the diesel traction on this route, will cut air pollution by about 4,400 tonnes of $CO_2$ each year. The experiment started this year is to be expanded rapidly, so that in about ten years, between 2,500 and 3,000 diesel trains in Germany could be replaced by hydrogen trains.

By 2035, some 15%–20% of the European regional rail market will be able to run on hydrogen. Hydrogen trains are particularly efficient for small regional lines, where the cost of switching to electric trains would be too high in relation to the cost-effectiveness of the links.

But it is not only rail transport that is aiming to use hydrogen mobility solutions in the future. The entire transport sector, including road and air transport, as well as heavy industry, especially steel and chemicals, is counting on this technology to reduce its $CO_2$ emissions.

Romania is also interested in acquiring hydrogen-powered trainsets to renew and modernise the current national fleet. Unfortunately, in the last 15 years in Romania, there has been no public procurement of new rolling stock for public rail passenger transport, the state-owned rail passenger transport operator has not acquired and does not have in its active fleet a sufficient number of rolling stock to serve the passenger public, and the privately owned rail passenger transport operators which have entered the rail transport market in Romania have only acquired second-hand rolling stock which is considerably old and no longer meets European quality and safety requirements. The government argues that the introduction of electric traction railcars based on hydrogen fuel cells is not a solution for the cancellation of electrification programmes in Romania, but only an environmentally friendly alternative to diesel traction on the still non-electrified lines. The public rail network in Romania has, by 2021, a length of about 10,500 km, of which 64% are non-electrified, since electric traction in the 'classic' catenary-pantograph system is not possible.

### *7.2.6 A possible sustainable transport system invented in Romania in the 1970s*

In the early 1970s, the Romanian inventor Henri Coandă proposed a revolutionary transportation system, namely 'transportation in a vacuum tube'.

Coandă dreamed his train could reach speeds of 500 km/h. In principle, it was a transportation system based on pressure and depression zones and was to be built on an air cushion. This system was based on the Coandă effect and was to be realised by capsules moving in vacuum tubes.

At that time, Coandă was proposing to build trains between Bucharest and Brasov and between Bucharest – Constanta and the City of the Future. The latter, called Delta, was a sort of Romania's Silicon Valley, located on the beach between the Danube Delta and the Sfântu Gheorghe, Sulina and Black Sea inlets. There would have been two pipelines to Brasov, one for people and one for goods, and the other route was called 'Autostrada Soarelui'[3] (Andriescu, 2013).

The first test for vacuum tube transportation was made in Bucharest in June 1971. A 300-m-long and 1 m in diameter pipeline was built there. After the successful test, the authorities set up a branch of the 'Aerotubexpress Department' in Măneciu Ungureni, in the Ciucaş mountains. Two 1,020-mm-diameter pipelines were built on a 1,300-m-long route following the Teleajăn riverbed. The propulsion station consisted of four fans. The freight containers could not reach a speed of more than 35 km/h, but they managed to follow the winding route imagined at Măneciu. The system was automated.

The first cargo shipments were realised at two mines in Baia Mare and Caraorman (Gadget Report, 2022). Later, to test the transportation of people, the team of engineers built a 1,600-m pipe.[4]

As the experiments had been successful, Henri Coandă designed the route for a passenger train to run through the pipeline on the Bucharest–Ploieşti route. The 60 km distance between the two cities was to be covered in a record time of around 7–8 minutes. In the same context, the Bucharest–Constanta route was to be covered in about 30 minutes.

In addition to passenger transport, the great scientist also wanted his invention to be applied to the transportation of goods between the port of Constanta and the steelworks at Galati.

Unfortunately, Coandă was never able to carry out his plan and died in 1972.

In the 1980s, Romanian engineers studied the possibility of building *the speed train* Coandă invented between Mangalia and Constanta.

The route would have included a glass station in the middle of the sea, where passengers could admire the marine fauna.

After 1980, more than eight years after Coandă's death, experiments were also carried out with the transportation of people in tubes. A 6-m-long steel capsule was built. It held two people at most. They built a 400-m pipeline, and the container had its own braking system. The speed was no more than 70 km/h, and the big problem was the deafening noise from the capsule. Such lines were also built in Baia Mare and the Danube Delta, but they were destroyed after 1990.

In the wake of the economic crisis of the 1980s, the project was abandoned because the authorities at the time allocated all available resources to the construction of the Bucharest metro. But after the metro was completed, the communist regime was going to support the construction of *the Coandă train*. The change of regime in 1989 would put an end to all plans to build the electric train.

However, the Romanian inventor's idea was to take shape in several projects around the world, especially in the United States.

Daryl Oster is one of those who worked on such a system of transportation of capsules through air tubes[5] – ETT (Evacuated Tube Transport), putting it into practice, on a large scale, with a patent in 1999 (Andriescu, 2013).

The Evacuated Tube Transportation system differs from maglev (magnetic levitation) because it uses closed tubes for transportation, based on the lack of air in the tunnels. This eliminates friction and could theoretically travel at very high speeds.

Researchers at China's Jiaotong University have started to develop such a train to reach speeds of 1,000 km/h, but it is not yet operational.

This new type of transportation was proposed in 1910 by American engineer Robert Goddard. His train would have run between Boston and New York in just 12 minutes.

Such trains were later proposed in the 1970s in the US by Robert Salter. He argued that the system could be easily built. The trains did not have to be fitted with engines, and station pumps would help the trains to leave.

Under these proposals, a plan was made for a Los Angeles to New York route, and the tunnels would have used a vacuum to reduce friction. The problem was that the 4,800-mph speed would have caused forces 1.4 times greater than gravitational, leading to the need for pivoting heads. The train was never built because the cost was approaching $1 trillion.

More recently, the Hyperloop project was realised by the South African entrepreneur and inventor Elon Musk (Băltărereţu, 2013).

In 2012, he announced that he wanted to build a high-speed train that would operate according to the principle thought by Coandă. The train is designed to run, in a tubular system, at speeds of up to 6500 km/h and will connect the cities of New York and Los Angeles. The 2,000-km (2,000-mile) distance between the two cities would be covered in about 30 minutes.

Unlike Coandă's inventions, the Hyperloop does not operate in a vacuum or under normal conditions, but in low-pressure tubes.

Inside the low-pressure tubes, the trains will be placed on tethers made of Inconel, an alloy also used by private space transportation company SpaceX. It can withstand pressure and heat. Air is pumped through holes inside the tethers to form a cushion of air, and there will be jet turbines at the front of the train, similar to those on Concorde. A turbo-electric compressor will transmit the pressurised air from the front to the links and cab. Magnets on the 'rails' and an electromagnetic pulse will provide the initial acceleration, and constant restarting of the engines will keep the speed at a certain level.

Hyperloop will use low-energy resources and friction will be reduced. The entire mechanism could be powered by solar energy and the system would be similar to the Maglev train, which uses the principle of repulsion between two magnets. The Hyperloop would have to travel at over 950 km/h to get from Los Angeles to San Francisco in half an hour, as originally planned. However, air will become a problem for the Hyperloop at speeds above 400 km/h. According to experts, the air inside the tubes would be 1/100 of the atmospheric pressure or 1/6 the pressure on Mars. Even at this low pressure, the air causes friction, so the 'fan' at the front of the train pulls this air in, pressurises it and sends it through the holes in the rails. This allows the train to run on a cushion of air. But as the speed increases, the air pressure will require more energy to move the vehicle forward. In addition, the friction with the air creates noise, and that would create discomfort for the other people where the Hyperloop would pass. That is why Coandă's original idea was to enclose the Hyperloop in a vacuum tube.

Apart from the technical issues surrounding the medium of transportation (air or vacuum), there are also dilemmas about the route. Over mountains or across the sea, under blocks or through space, it cannot. It cannot all be done at once. Highways could serve as 'guide lines' for Hyperloop. But at a speed of about 1,000 km/h, Hyperloop would need a turnaround 'curve' of about 65 km.

Recently, a Hyperloop project has been developed in China that has overcome the technical hurdles faced by Elon Musk. A 2-km test track has been built in Shanxi province, using a combination of low-vacuum steel–concrete tubes, artificial intelligence-controlled magnetic dampers and advanced construction techniques. This project promises speeds of up to 1000 km/h, significantly reducing travel time between cities.

Also, in 2020, Virgin Hyperloop conducted a test with two passengers. However, the feasibility and large-scale implementation of human transport still face significant challenges, including high costs, regulatory obstacles and engineering complexities related to safety, comfort and infrastructure.

### *7.2.7 Sharing the use of vehicles (cars, bicycles and others)*

Shared mobility is seen as a promising way to reduce congestion and $CO_2$ emissions, although the extent of these reductions would depend on the type of shared mobility. For example, trips made by one person or by a driver and a passenger will yield lower congestion and $CO_2$ benefits than on-demand public transport.

It has been clear for decades that compact, mixed-use cities centred on walking, cycling and public transport offer the greatest, most equitable access for everyone and at very low cost to society.

Sustainable shared mobility is achieved in two ways: by discouraging inefficient travel (low occupancy vehicles) and by encouraging efficient travel (e.g. walking, cycling, public transport). There are a variety of ways to discourage inefficient travel, including by internalising external costs. These are vital to counteract the savings generated by new mobility technologies. Efficient travel is being encouraged in a variety of ways, including planning and zoning compact, mixed-use cities centred around public transport, and building infrastructure that supports and prioritises walking, cycling and public transport. The development and evolution of ridesharing represent an opportunity. Cities can, through careful legislation, facilitate an industry that is more complementary to public transportation than the taxi industry has been to date. By partnering with public transportation, through promotions and coordination, ridesharing companies can increase not only the number of ridesharing users but also the number of people opting for public transportation (Speirs, 2019).

The 'Park and Ride' (P+R) system helps to reduce traffic volume if certain principles are followed:

- Non-centralised decisions to limit route length.
- No competition between P+R and public transport: if new P+R schemes are created on the edge of the city centre, public transport users may prefer to use their own vehicles.
- Defining management principles for P+R (e.g. level and structure of user charges, access restrictions, etc.).

Ridesharing can also fill gaps in public transport services in sparsely populated suburbs. Some cities have cancelled bus routes with the shortest routes and offered discounts for ridesharing activity for residents in these areas. For example, San Clemente (California) has partnered with *Lyft* to offer rides discounted by up to nine dollars between 6:00 a.m. and 8:00 p.m. that start and end near existing bus stops. Ridesharing has additional benefits, such as reducing dangerous behaviour. A study conducted in the area of US cities with populations over 100,000 between 2000 and 2014 estimates that ridesharing activity reduced fatal collisions caused by drunk driving by up to 11.4% (Speirs, 2019).

Shared mobility, or mobility in the collaborative economy, is also a new concept that includes a number of emerging models. The two characteristics common to all these new mobility services are that there is an element of sharing an asset (a vehicle) rather than user ownership and that they are technology-based (i.e. a digital platform). There are four main models (Santos, 2018):

- Personalised vehicle rental – peer-to-peer provision with a company as a broker. Individuals can rent their cars when not in use.
- Modern car club – short-term vehicle rentals managed and owned by a supplier. The provider owns a number of vehicles that are strategically parked in a city or certain areas of a city, either in dedicated parking lots and/or on authorised public roads.
- Uber-like service – companies do not own cars themselves, but enrol regular car owners who act as drivers. The service relies on an app, provided by the company, which also collects payment, taking a percentage of the total revenue.
- New on-demand public transportation – private cars, vans or buses and other vehicles, such as large taxis, can be rented on demand, shared by passengers going in the same direction. It is essentially a user-oriented form of public transportation. Options could include users specifying pick-up and drop-off locations and required departure and/or arrival times, which would essentially be shared taxi services.
- Taxi-bus services that use predefined stops and require advance booking.

### *7.2.8 Vehicle and fuel taxes*

Incentives to support sustainable transport can be categorised into three groups (Santos, 2018): 1) command and control; 2) financial; and 3) in-kind.

1 Command and control incentives are essentially regulations that are introduced and enforced, such as traffic restrictions.

From an economic point of view, command and control policies are and are not efficient because the objectives are not achieved at least cost. However, if the objective is to discourage drivers from using their cars and switching to shared mobility services, restricted traffic zones that restrict non-essential private cars but not shared mobility would undoubtedly be helpful.

This type of policy has been put into practice and there are several examples internationally. To reduce air pollution, authorities in Mexico City, Bogota, Athens, Athens, Santiago and São Paolo, among others, have restricted traffic according to vehicle registration number and day of the week. Officials in Paris and Madrid adopted such a measure in December 2016, albeit only temporarily, to combat unusually high air pollution. When it was introduced permanently, the policy had little success, because many car owners bought a second vehicle so they could also have a complementary number plate to the one on their original vehicle and thus be able to drive on any given day. Unfortunately, in the majority of cases, the second vehicle was purchased 'second-hand' and was even more polluting than the first one they owned.

Another example of a command-and-control policy is the restriction of the circulation of vehicles in certain areas, such as in several Italian cities, including Rome, Florence, Pisa and Milan. These cities have created restricted traffic zones in their historic (medieval) city centres. Vehicles operating in these zones require an authorisation and permits are limited to a number of 'essential users'.

2 Financial incentives include charges or taxes, subsidies and cap and trade schemes. A combination of these is also possible, particularly taxes and trading caps. In transport, subsidies have been used to encourage the purchase of energy-efficient and/or clean vehicles (to dispose of old, polluting ones).

This type of policy has been used in road transport to regulate the internalisation of various types of externalities, including $CO_2$ emissions. Although cap and trade systems have been successful in internalising a large share of environmental externalities, and in particular greenhouse gas (GHG) externalities caused by US road traffic, in Europe through the EU Emissions Trading Scheme (EU ETS), the number of companies subject to this policy has been relatively small.

Irrespective of the purpose of the policy (which could be, for example, a cap-and-trade system for road use permits), if individual hauliers also entered a cap-and-trade system, it is obvious that transaction costs would be very high, as would monitoring and compliance costs, given the very large number of users. A cap-and-trade system would therefore not be appropriate to encourage shared mobility.

Fees and taxes are often rejected by the majority of residents and it is therefore difficult to implement such a policy, especially if it proposes to charge for something that was previously free, such as the use of certain roads during peak periods. However, increasing public acceptability can be achieved when the revenue generated, or a large part of it, by imposing new taxes is returned to the transport sector.

Taxes on transport fuels can encourage fuel efficiency and more sustainable use of cars, including the use of cleaner fuels. However, the size of such taxes should reflect both environmental (e.g. carbon, lead, nitrogen, etc.) and energy details.

At present, substantial differences in fuel tax rates can be observed between EU countries. There is a general preferential treatment for diesel. Diesel is taxed less than gasoline in almost all EU countries (EC, 2018).

Fuel taxes are, in many cases, set at relatively low levels that do not reflect the externalities of fuel consumption. Low fuel taxes lead to excessive car use and can fuel urban sprawl by promoting the dispersion of jobs, housing and other key points of economic activity. This makes it even more important to set fuel taxes at levels that take full account of the environmental costs of fuel consumption. The revenues raised by imposing such a tax can be invested, for example, in the development of public transport or emergency or preventive health care. In South Africa, for example, revenues from the General Fuel Levy are used to fund the *Road Accident Fund*, which is a public health insurance for crash victims. Thus, drivers are responsible for at least some of the negative consequences of transportation.

One of the major criticisms of policies to introduce fuel taxes is that they may contribute to higher inflation and slow economic growth. If gasoline and diesel are essential for all transportation, then the cost of each economic step increases. For example, it costs more to get wheat to the mill, flour to the bakery and bread to the market, and the price of bread goes up. That is why fuel taxes need to be managed carefully.

The imposition of duties and taxes in the case of transport can have a significant effect on consumer preferences when purchasing a car. In this respect in the European Union the registration of a car is taxed in 20 countries, and in 22 road taxes are imposed on vehicles.

Currently, registration taxes depend on the $CO_2$ emissions of the car's engine in 15 EU countries. A further 12 countries take the level of pollutant emissions into account when setting the level of road taxes payable for different types of vehicles. Bulgaria, the Czech Republic, Estonia, Lithuania, Poland, Slovakia and the Czech Republic are among the countries where $CO_2$-based vehicle taxes are imposed. However, neither registration taxes nor tolls affect the marginal cost of operating a vehicle. Moreover, the lack of harmonisation of registration taxes at the EU level can create a significant administrative burden and sometimes double taxation when vehicles are transferred to another country.

The share of environmental taxes in total transport taxation can only indicate to a certain extent how the taxation system addresses transport externalities.

Favourable tax treatment of company cars is a practice that needs to be taken into account when considering the internalisation of environmental costs. Several EU countries subsidise the private use of company cars. Belgium, Ireland, Estonia and Latvia allow a partial deduction of VAT charged on the purchase of company cars for private use by employees (EC, 2018). Advantageous schemes for company cars tend to encourage ownership and often affect the choice of model and driving habits. Recent European Commission proposals seek to address these problems.

A better way to tax vehicles is through a licensing fee. This means cities or administrative regions can set context-specific rates. Rural areas where private vehicles are more needed would have lower licensing rates than large cities that suffer from congestion caused by road traffic (Speirs, 2019). However, as mentioned, such charges are difficult to implement when the number of vehicle owners (especially cars) is extremely high.

The introduction of systems of internalisation of external costs caused by transport, involving a wider application of the polluter pays principle, should be promoted and encouraged in all EU countries. In this perspective, the EC has launched a comprehensive study on transport infrastructure charging and internalisation of negative externalities of transport, which will assess infrastructure user charges, other internalisation measures and infrastructure-related expenditure. In addition, EU countries should be encouraged to make use of the possibility offered by Directive 2011/76/EU60 to collect external cost-based charges from heavy goods vehicles in addition to infrastructure user charges (EC, 2018).

3 In-kind incentives can also persuade consumers to change their behaviour. In Norway, for example, many consumers purchased EVs not only because of financial incentives, but also because EVs had access to dedicated bus lanes.

### *7.2.9 **Parking management***

Parking availability shapes mobility options. An oversupply of cheap parking stimulates driving, while a limited but very well-managed parking system encourages walking, cycling and public transport – especially in city centres. Parking should be properly priced to reflect the value of the site (site, area) and to direct its use. Parking should reflect the value of the land: a progressive charge should be introduced in city centres to encourage the free trading of parking spaces. Ideally, higher parking charges should lead to drivers choosing alternative modes of transportation (Speirs, 2019).

Every vehicle trip begins and ends in a parking space. This leads to the need to integrate special measures to promote special measures for parking areas into a green urban transport policy.

A parking policy must take account of local traffic and be the result of a combination of various other measures.

The introduction of a parking policy is, at least partly, the responsibility of local public authorities; in principle, they have effective tools at their disposal to introduce green transport policies.

It is much easier to obtain a political majority if the measures target commuters, for example, commuters living outside the city do not participate in the local vote (voting on the measures mentioned), nor are they represented in local activities.

Cooperation between the parties (public transport, central and local authorities, politicians, businessmen, residents) is the essential condition for a successful parking policy.

Parking policy applied in the centre of a city has negative repercussions on neighbouring areas. Local authorities must therefore coordinate action to introduce such a measure.

A comprehensive parking policy must take into account:

- New and existing sites.
- Public and private parking.
- Combination of charges, time, targets, users.

Measures should also take into account the quality of public transportation:

- The design of major environmental protection projects is a good opportunity to introduce new measures.
- Campaigns to foster a positive attitude towards the measures should be integrated into an effective parking policy.
- The experience of several cities has shown that restrictive measures can also have a positive influence on the local economy, especially on retail trade – if the following aspects are taken into account:
  - Using every opportunity to redesign roads and public spaces.
  - Phasing in restrictive measures.
  - Improving access for pedestrians, cyclists and public transport.
- Revenue raised from parking charges can be used to introduce environmentally friendly modes of transport.
- Blue zones (where there are major environmental restrictions) have significantly reduced commuter traffic in residential areas, especially during morning and evening rush hours. Experience has shown that:
  - Blue zones should exist in all urban residential perimeters.
  - Regular checks should be made to monitor activity.

    - Alternative solutions to commuter demands should be offered.
    - P+R schemes are suitable for introduction in blue zones.
    - Revenues from blue zones are higher than the costs of their implementation and operation.

- Paid parking, combined with reduced parking spaces, can significantly reduce traffic volumes in central areas.
- The 'P+R' system contributes to traffic volume reduction if certain principles are respected:

    - Non-centralised decisions to limit route length.
    - No competition between P+R and public transport: if new P+R systems are created on the edge of the city centre, public transport users may prefer to use their own vehicles.
    - Defining management principles for P+R (e.g. level and structure of user charges, access restrictions, etc.).

A regulation to limit the construction of new car parks should be considered, especially as this instrument is only effective in the long term.

Increased vehicle use leads to a reduction in parking times and parking spaces. Currently, most cars are parked for 95% of the day, and the number of parking spaces in the United States is estimated at about 500 million, with some speculating it could be as high as two billion. With 263 million passenger vehicles in the United States, there are at least two spaces per car. In a future of sustainable shared mobility, each vehicle could be used by multiple occupants each day, greatly reducing the amount of time the vehicle is idle and in need of parking. This would free up massive amounts of space in urban areas, and in particular, on-street parking spaces that are suitable for bicycle infrastructure.

The need for private cars is decreasing, while parking is becoming less and less of a priority. Parking is likely to prove unnecessary or be reused in surprising new ways. Looking to the future, the International Parking Institute, the British Parking Association and the European Parking Association have collaborated to develop the Alliance Standards for Parking Data. This will enable better analysis and the most effective measures to be adopted for the changing demand for parking spaces (Speirs, 2019).

## 7.3 Urban sustainable development plans and the specific transport component

Several cities stand out for their commitment to sustainable transportation, with some leading the way in promoting public transit, cycling and pedestrian-friendly infrastructure. European cities like Amsterdam, Copenhagen and Paris are often cited as examples of successful sustainable urban mobility, while other cities worldwide are also making strides in this area.

Some of the most notable cities:

- *Amsterdam, The Netherlands*: known for its extensive cycling infrastructure and high percentage of trips made by bicycle.
- Copenhagen: a global leader in cycling, with a large portion of the population commuting by bike.
- *Paris, France*: has made significant investments in public transportation, walking and cycling, with a relatively low percentage of trips made by car.
- *London, UK*: aims for a zero-emission bus fleet by 2034 and has a large number of EVs and charging stations.
- *Oslo, Norway*: a leader in environmental sustainability, including sustainable transportation, with a focus on green spaces and clean air.
- *Stockholm, Sweden*: has a goal of fossil fuel-free maritime traffic by 2030 and emissions-free traffic in the city centre.
- *Hong Kong*: has a highly efficient and well-regarded public transportation system.
- Tokyo: boasts the world's most extensive urban rail network.
- *Curitiba, Brazil*: has implemented innovative public transportation solutions, including Bus Rapid Transit.
- *Bogota, Colombia*: another city with a successful BRT system and a focus on sustainable urban planning.
- *Portland, Oregon* (USA): has a strong commitment to public transit, cycling and pedestrian infrastructure.
- *Vancouver, Canada*: a leader in sustainable urban development and transportation.
- *Mérida, Mexico*: implements innovative and sustainable transportation projects like the Va y Ven system.

### *7.3.1* ***Highlights of sustainable transport policies in some of the world's cities***

We present, briefly, some characteristic features of transport systems in some of the world's most *highly rated* cities that support sustainable mobility development.

#### *7.3.1.1 London, a city that must support sustainable transportation*

London, a city where history intertwines with modernity, is facing unprecedented transportation challenges. Firstly, London's transport system is a complex network comprising various modes of travel, including the underground, buses, taxis, private cars, bicycles and walking. Among these, the underground and bus systems serve as the backbone of city transportation, carrying a significant portion of daily commuting needs.

Secondly, London's transportation system is also grappling with serious environmental pollution and energy consumption challenges.

London is a city of transportation firsts: the world's first underground railway network; the world's first under-river tunnel; the world's first international airport and the world's first fully orbital ring road. London can be proud of its tradition of pioneering transportation and mobility infrastructure.

Yet if the city is to be as transformed by the digital revolution in the 21st century as it was by the industrial revolution in the 19th, it must invest, adapt and, above all, innovate.

To have an edge on the competition, London must enable millions of people to travel daily in, out and around the city. It must create a high-capacity, reliable, low-cost cost and integrated transport network, all without disrupting day-to-day travel.

The network must be conducive to growth and receptive to rapidly emergent technology and it must support the increasingly polycentric nature of the city.

In this context, London's transport strategy has set clear transformation goals: by 2041, 80% of all trips will be made by walking, cycling and public transport, with a commitment to achieving net-zero carbon emissions for the transport system by 2050 (Shi, 2024).

London's sustainable transport transformation policy framework is a multidimensional strategic system designed to reshape the city's transportation landscape through a series of meticulously crafted strategies.

Public transport – the strategy for ensuring public transport accessibility and service capacity focuses on enhancing the overall performance of the public transport system. By preserving and ensuring new spaces for public transport, walking and cycling, as well as ensuring necessary line connectivity and accessibility, London aims to enable the public transport system to accommodate an additional five million daily trips by 2041. The implementation of this strategy will strengthen the appeal of public transport and reduce reliance on private cars.

Non-motorised transport – the convenient bicycle transport strategy is aimed at promoting and encouraging cycling by constructing new cycling routes and high-quality related facilities, supporting the realisation of a city-wide cycling network and proposing minimum standards for bicycle parking facilities for different levels of areas, ensuring sufficient and convenient bicycle parking facilities.

Transport special zones – the refined motor vehicle parking management strategy encourages car-free areas in regions with well-developed public transport networks and follows the principle of providing the minimum necessary parking spaces in other areas. In addition, the strategy encourages each administrative district to adopt stricter enforcement standards on top of the upper limit construction standards proposed in the 'London Plan 2021' to further promote green travel. Also, the strategy supports the provision of hydrogen refuelling stations and electric vehicle fast charging points for logistics and industrial sites and encourages off-peak distribution to reduce additional freight caused by failed deliveries.

Many of the developments necessary to achieve these goals are underway: HS2 with stations in Euston and Old Oak Common will increase capacity, drive regeneration and create new quarters in the city; Thames river crossings will increase connectivity and enable freer traffic flow; Crossrail will connect new areas of the city currently blighted by their isolation and increased capacity at all five of London's airports will maintain the city's status as a global hub.

The question is, will London be reactive or boldly proactive? The long-awaited London plan must set a clear, far-sighted and integrated strategy. It needs to underwrite a vision for the future with a clear approach to mobility that catapults London into the 21st century and ensures economic productivity and sustainability.

#### *7.3.1.2 Hong Kong*

Hong Kong has long been famous for its world-class infrastructure. But as one of the planet's most densely populated cities, it faces unique challenges in providing an efficient transportation system to meet the needs of its 7.3 million people (ARCADIS, 2017).

Challenges include limited space, a growing and aging population, rapidly increasing cross-boundary traffic with China and environmental concerns.

Hong Kong International Airport (HKIA) is one of the world's best and busiest, with 70.5 million passengers in 2016. HKIA is operating at 99% capacity, with a third runway to be opened in 2024. The expansion of the airport is intended to aid Hong Kong's role in connecting mainland China with the rest of the world, a key part of the Belt and Road Initiative, Beijing's global trade development strategy.

The controversial Hong Kong–Zhuhai–Macau Bridge was opened at the end of 2017. The bridge, beset by delays and environmental concerns, is meant to bring closer economic integration between Hong Kong and the Pearl River Delta. There is no rail on the bridge, leading to concerns among some politicians and civil society groups that the number of vehicles coming from China will compromise Hong Kong's drive for better air quality.

Hong Kong's public transportation system oversees approximately 12.6 million passenger journeys every day and the Mass Transit Railway is recognised as one of the most efficient in the world.

Public transportation is relatively cheap compared to other world cities. Less than one-fifth of people in Hong Kong own a car, but pedestrianised areas are few in number.

Compared to European cities, commuting by bicycle is rare. This is partly due to Hong Kong's climate and topography, but the city has done little to encourage the uptake of cycling.

#### *7.3.1.3 Vancouver, Canada*

Vancouver is known for its ambitious goal to become the greenest city in the world by 2020 (Bron, 2023). The city has made significant progress in reducing greenhouse gas emissions, promoting green building practices and investing in renewable energy. Vancouver also boasts an extensive public transportation network and strong support for cycling and walking.

Vancouver has implemented various innovative initiatives, such as the Greenest City Action Plan, which outlines a comprehensive strategy to reduce emissions and waste, and the Neighbourhood Energy Utility, which uses waste heat from sewage to provide heating and hot water for buildings in the False Creek area.

Vancouver faces challenges in balancing rapid population growth, housing affordability and maintaining its commitment to environmental sustainability. The city continues to invest in new technologies, infrastructure and policies to promote sustainable development and maintain its status as a leading green city.

#### *7.3.1.4 Berlin, Germany*

The Smart City Strategy Berlin, submitted within the 2030 strategy, aims to find solutions to these challenges and highlights the importance of innovation and technology.

A key part of this strategy emphasises smart mobility. The main elements include: shortening distances by strengthening an intelligent and sustainable transport policy to reduce the distances of individual and commercial transport; stimulating the use of new mobility concepts in line with the 'sharing economy' (e.g. car, scooter and bike sharing) to decrease the levels of traffic and reduce adverse environmental impacts; supporting research and development in electromobility (e-mobility), to strengthen its economy by creating jobs and technical know-how, positively impacting the climate and improving quality of life in the city (ARCADIS, 2017).

To ensure these mobility measures improve, Berlin's strategic investment includes charging infrastructure for electric cars, test tracks for autonomous vehicles, new rail tracks and improvements and extensions of bicycle paths.

Berlin aims to be climate-neutral by 2050 and has made substantial progress in reducing greenhouse gas emissions and promoting renewable energy sources. The city has implemented various sustainable transportation options, such as extensive public transportation networks, bike-sharing programmes and electric vehicle infrastructure.

Berlin is home to numerous innovative projects in the field of clean technology and sustainable urban development, such as the EUREF-Campus, a hub for startups, research institutions and companies focusing on renewable energy, smart grids and electric mobility. The city also supports sustainable building practices and green spaces to enhance the quality of life for residents.

Berlin faces challenges in balancing rapid urbanisation, housing affordability and environmental sustainability. The city is committed to ongoing innovation and collaboration between the public and private sectors to address these challenges and maintain its status as a leading sustainable city.

#### *7.3.1.5 Singapore*

Singapore continues to be a top-ranked sustainable city in Asia, but the city-state is not resting on its laurels. With a population predicted to grow to more than six million people by 2030, the government has committed significant investment over the next decade to improve mobility and connectivity: two new underground lines, extensions to four MRT (metro) lines, a new terminal and runway at Changi Airport, a high-speed rail link between Singapore and Kuala Lumpur, a cross-border MRT system which will connect Singapore and Johor Baru in Malaysia and the relocation of the container port.

In this context, 66% of journeys in Singapore are via public transport, but the government wants to increase this to 75% during peak hours by 2030, and it is investing in developing a more efficient and integrated bus system (ARCADIS, 2017).

Singapore has made remarkable progress in sustainable urban development despite its limited resources. The city has implemented ambitious green building standards, comprehensive waste management programmes, and significant investments in public transportation, green spaces and water management.

Singapore's innovative projects include the development of the 'Gardens by the Bay', an iconic urban park that showcases sustainable design and advanced horticultural technologies. The city also utilises advanced water treatment technologies, such as NEWater and desalination, to ensure water security.

Singapore faces challenges in managing its limited land area, ensuring energy security and balancing economic growth with environmental sustainability. The city is committed to continuous innovation, technological advancements and international collaboration to address these challenges and maintain its position as a leading sustainable city.

A number of other initiatives are underway, which will help Singapore to evolve and remain competitive. These include the government's intention for Singapore to go 'car-lite', with a SIN$1.5 billion 15-year plan which aims to reduce the country's reliance on petrol-fuelled cars.

The government is also generally supportive of ride-sharing car and bicycle platforms, though 'bike dumping' is an increasing problem. The small city-state is also the perfect testing ground for driverless vehicles and the government has signed deals with two self-driving technology companies to develop autonomous truck fleets for use on public roads. Singapore is also ahead of the game in using big data to study commuter flows and improve planning, which will help guide future investment.

#### *7.3.1.6 Zurich, Switzerland*

Zurich is the largest city in Switzerland, providing a home to 415,000 people, significantly more than the official capital, Bern. One in 11 jobs is located in the Helvetian city, which presents challenges to the city's transportation network as people need to get to work from within and outside the city borders.

To meet these challenges, in 2012, Zurich's local government devised the Strategien Zurich 2025, which provides a framework for long-term sustainable development, including an overarching mobility plan (ARCADIS, 2017).

Although Zurich is well-connected through widely branched and modern public transportation, the use of private cars creates significant traffic stress.

The city's mobility strategy centres on the expansion of the public transportation network and improving its upkeep and access. While trams, buses, city trains and even ships and cableways already form a tight and efficient public transportation network, there is still room for further development. More tram and train lines are planned, and the entire network of the Züricher Verkehrsbetriebe will be expanded with the construction of additional lines. Other measures designed to improve the environmental sustainability of the system include replacing diesel buses with trolleybuses and supporting carless households.

In this light, developing the cycle network is an essential task. More cycle parking areas are being created, complemented by new guidance systems and efforts to reduce traffic dangers to cyclists.

A visionary part of the mobility strategy is the project Cargo *sous terrain*. A subterranean logistics network will transport goods between Swiss economic centres via a network of tunnels. Reloading hubs in the cities will ensure the distribution of goods by autonomous vehicles.

Zurich has set ambitious goals to reduce per capita $CO_2$ emissions by 50% by 2030 and increase its renewable energy usage. The city is known for its excellent public transportation system, energy-efficient buildings and strong focus on waste reduction and recycling.

Zurich's innovative projects include the '2000-Watt Society' initiative, which aims to reduce energy consumption and promote sustainable urban planning. The city also supports green building practices and has implemented advanced waste management systems to minimise its environmental impact.

Zurich faces challenges in maintaining its commitment to environmental sustainability while ensuring economic growth and affordable housing. The city is committed to ongoing innovation, collaboration and the development of new technologies to address these challenges and continue its progress towards a sustainable future.

#### *7.3.1.7 Melbourne, Australia*

Australia's second largest city and the capital of the State of Victoria, Melbourne has a reputation as a thriving city with strong business, cultural and sporting sectors.

Melbourne has become known as the world's most liveable city, one of the largest factors affecting Melbourne's mobility being its rapid recent growth.

It is projected that Melbourne's population will double to around eight million in the next generation, overtaking Sydney as Australia's largest city. Expanding fast and with high house prices in the city centre – Melbourne has the fifth least affordable housing market in the world – the city is increasingly reliant on long-range transport options.

Melbourne boasts the world's largest tram and light rail network and has recently begun the development of a cross-city Metro. This will help to free up the train network and enable more commuters to use public transport. Melbourne does not yet have a functioning Metro network, a situation that will improve upon its completion.

Further digitisation will ensure continued improvements to the mobility of the city.

Key to Melbourne rising to its challenges is the Resilient Melbourne Strategy, which sets out a clear, collaborative agenda to ensure Melbourne is in a strong position to deal with the problems of city growth. Combining this with the State Government's Plan Melbourne refresh, the city has a strong strategy in place to address its mobility and sustainability issues.

In accord with these, Melbourne aims to be a carbon-neutral city by 2040 and has made significant progress in promoting renewable energy, green building standards and sustainable transportation options. The city has implemented various urban greening initiatives, such as green roofs, walls and facades, and is recognised for its commitment to waste reduction and recycling.

Melbourne's innovative projects include the Fisherman's Bend Urban Renewal Area, which focuses on sustainable urban development, mixed-use communities and green spaces. The city also supports electric vehicle infrastructure and public transportation systems that prioritise sustainability and efficiency.

The city is dedicated to fostering innovation, collaboration and the development of new technologies to address these challenges and continue its progress towards a sustainable future.

### *7.3.2 Copenhagen, a possible model for integrated urban development*

Copenhagen is the capital, largest city and largest seaport of the Kingdom of Denmark and is also the largest city in the Nordic countries.

In the process of becoming a modern city, Copenhagen is, on the one hand, vigorously developing green transport and controlling the speed of motorisation and car use within a reasonable timeframe. On the other hand, Copenhagen pays attention to the construction of green urban spaces as well as to the protection and heritage of traditional buildings and cultural landscapes in old city areas. Innovative architectural designs are being adopted and new districts are being developed with the concomitant protection of the environment, realising energy savings and creating an overall living environment at a higher level of satisfaction for residents.

Moreover, Copenhagen enjoys the reputation of being the 'most liveable city' in the world.

Copenhagen aims to become the first carbon-neutral city by 2025. The objective is supported by the Climate Plan 2025 adopted by the City Council in

2009 (State of Green, 2017). A medium-term goal was to reduce carbon emissions by 20% by 2015; a target Copenhagen already achieved in 2011.

Since 1995, Copenhagen has reduced carbon emissions by 50%. CPH 2025 is a holistic plan with specific targets and initiatives under four focus areas:

- Energy consumption.
- Energy production.
- Green mobility.
- Municipal government initiatives.

Copenhagen's urban land-use structure and transportation corridor is known as 'Planning Fingers' (Huapu, 2020).

The principle of 'Planning with Fingers' was first presented in 1947. The structure of the hand-shaped urban skeleton is formed with the city of Copenhagen as the centre of the palm and five finger-shaped axes extending north, west and south. The wedge-shaped areas between the fingers serve as forest, farmland and open recreational space. For more than half a century, Copenhagen has been steadily improving its urban planning and building, following the principle of 'Planning with Fingers'. 'Five Planning' was formulated in 2007, clearly emphasising that the direction of urban development is to build rail transit from the centre to the periphery along the finger directions, improve public transportation facilities and build residential areas along the transportation corridor. The wedge-shaped areas between the fingers do not involve the construction and development of urban facilities, but maintain the ecological characteristics of forests and green spaces. Additional space required for urban development and the construction of urban clusters is taken from other areas.

Town centre planning and construction focuses on improving urban public transportation and non-motorised traffic service systems, such as walking and bicycling, to meet short-distance trips into the town centre and the need for ecological and environmental protection. The urban fringe areas (finger parts) provide space for urban sprawl and new city construction, while rail transit is the main mode of transportation to the city centre. Planning and construction of peripheral areas focuses on improving infrastructure construction and improving the level of public transport services.

Rail transit is the dominant mode of transportation in Copenhagen. Thanks to the 'finger planning' principle adopted by Copenhagen, traffic corridors supported by rail transit radiate from the city centre to the peripheral areas in five directions.

In the process of expanding the city to these peripheral areas, the development and construction of most of the infrastructure is concentrated near railway stations, such as residential buildings and main roads. Thus, the formation of 'Finger Planning' is formed. According to this, the rail transit network also shows a hand shape (palm and fingers) that matches the urban layout.

This shows that the implementation of 'finger planning' in urban planning and construction has also promoted the integrated development of transportation systems and land use. Copenhagen currently has almost 200 km of rail, including 85 suburban and 35 urban stations. Rail transit carries about 500,000 commuter trips daily, accounting for about one-third of all daily motorised trips. It picks up the traffic demand of the corridor between the city centre and the 'fingers' areas within a 40 km radius.

Efficient mixed-use land development was implemented along rail lines and around stations during the construction of the rail transit system so that people could quickly enter commercial, retail and residential areas without transferring to other modes of transportation after exiting the station. Some station buildings themselves are comprehensive hubs, giving people concentrated areas for traveling, working, shopping and living.

Increased mobility through integrated transportation and cycling solutions has significantly reduced traffic congestion and improved citizens' health. Since 2005, DKK 1 billion has been invested in bike lanes and super cycle highways, and 45% of Copenhageners cycle to work or school every day.

The most relevant facts about sustainable transport in the Danish capital are as follows:

- 75% of trips in Copenhagen are made on foot, by bicycle or public transport.
- 50% of all trips to work or education in Copenhagen are made by bicycle.
- 20% more passengers now use public transport than a decade ago.
- 20%–30% of all light-duty vehicles (cars) and 30%–40% of all commercial (heavy-duty) vehicles use new, clean fuels.

These are just some of the benefits of sustainable urban solutions. In Copenhagen, green urban solutions are already widely implemented and used by citizens. The urban solutions have had significant effects on reducing $CO_2$ emissions and, at the same time, have made the city more livable. These include:

- Cleaning up the harbour, which has realised attractive urban areas with better quality of life, better local business life, job creation and income generated in the area.
- Developing a highly efficient district heating system where 98% of all households are connected.
- Introducing a new district cooling system where cold is taken from harbour water. This saves 70% of energy compared to traditional air conditioning.
- The implementation of integrated transportation and cycling solutions has significantly reduced traffic congestion and improved citizens' health. Since 2005, DKK 1 billion has been invested in cycle lanes and cycle highways, with 45% of Copenhageners cycling to work or school every day.

On car policy, the Copenhagen authorities have mainly adopted a restrictive approach. Despite the low volume of car traffic, the government continues to urge people to move away from private cars and switch to environmentally friendly modes of transport such as public transport, cycling and walking. Copenhagen's car restrictions are mainly two policy elements: buying cars and securing parking spaces.

Since the Second World War, the Copenhagen government imposed a ban on car imports and maintained it long after the war. Since the late 1970s, the Copenhagen government has tripled the annual tax on cars registered in the city.

For parking, the city authorities have adopted an innovative and politically savvy policy that focuses on reducing parking spaces in central urban areas. To avoid strong opposition from car owners about reducing too many parking spaces, at one point, the city government adopted a gradual strategy over the past decades, insisting on reducing parking spaces by 2%–3% each year. This approach has led to significantly reduced parking spaces and avoided aggravating social contradictions. In addition, the price of parking is defined by zones, being directly proportional to the relationship between the supply and demand of the parking area and the level of public transport service. Local city authorities have thus considered both increasing the parking price and reducing parking spaces in order to keep the vacancy rate in central urban areas at around 10%, avoiding complaints from car owners and thus reducing resistance to policy implementation.

By contrast, and in response to growing demand for bicycle parking, the Copenhagen authorities have adopted a method of gradually increasing the construction of bicycle parking. For residential and office areas where the contradiction between demand and supply is prominent, landlords are being allowed and encouraged to convert their vehicle parking spaces into bicycle parking. For general streets, bicycle parking spaces are designed to be added at the corners of sidewalks. For commercial streets, where land is scarce and expensive, merchants are allowed to use the remaining sidewalk space to build bike racks.

Copenhagen has always been known as the 'city of bicycles', benefiting primarily from the Danish people's general promotion of green environmental and fitness concepts.

About a third of Copenhagen commuters travel by bike, including government officials, business tycoons and celebrities. Second, the Copenhagen authorities have strongly promoted bicycle traffic for a long time, increasing the bike lane infrastructure and building a huge system of bike lanes totalling 350 km. In line with the urban structure of the 'Planning of the Fingers', Copenhagen's regional bike lane system has been formed in a similar shape.

There are 40 km of green bicycle corridors in Copenhagen, which are completely separated from motor vehicle lanes. When there is a bus stop in front of the bicycle lane, the green bicycle lane is designed to turn naturally to the right, then behind the bus stop, eventually reaching back to the left direction parallel to the motor vehicle lane after the bus stop.

The standard of construction of Copenhagen's bicycle lanes is high, and the maintenance system has been perfected and well developed. The pavement of the bicycle lanes is 7–12 cm higher than that of the vehicle lanes. The separation area between the bike lane and the vehicle lane is constructed with a curb and there is an independent drainage system, which helps to maintain the bike lanes. The speed of bicycles on special roads can be comparable to that of cars. Pedestrians or other vehicles are not allowed to pass or stop in the cycle lanes, which have the same priority in traffic as the traffic lanes.

To support mass bicycle use, the link between public transport and bicycles is almost seamless in Copenhagen. Near the bus stop, bicycle parking lots are well equipped to facilitate the transfer of cyclists. Bike racks are mounted on buses so that bikes can be taken by bus and train, providing as much convenience as opportunities for cyclists.

On trips of less than 10 km, the bicycle is usually the most useful and used option. The longer the trips, the more the demand for mobility shifts towards public transport. This makes sense and is also consistent with the fact that interest in cycling drops significantly for trips longer than 8 km.

Danish experience in the use of environmentally friendly means of transport in urban areas shows that cycling is the best means of transport to get from one point to another in a city. Investing in public transport solutions that cover a larger geographical area at a lower cost is also very cost-effective in every respect. Of these, the best examples are trams or (surface) light rail.

### *7.3.3 Bucharest, between past and future, developments from a sustainable development perspective*

The structural-urbanistic evolution of Bucharest, from polycentric in 1800, to urban monocentric in 1900, went through a radial, archipelago-type distribution in 1925, concomitant with the process of urban consolidation of its initial centre (PMB, 2023). After 1950, Bucharest consolidated its archipelago character by encouraging an extensive urban development solution, which largely ignored the historical core, while developing, on the other hand, residential urban centres of collective housing, associated with industrial platforms, generating jobs.

The historical evolution of the Municipality of Bucharest defines and illustrates the path of formation and consolidation of its current configurative-spatial structure, by superimposing successive layers that either consolidate or modify and model previous stages. The lack of finalisation in many approaches imposed by the historical evolution has been the cause of the current heterogeneity of the city, of its archipelago character. The historical modules that have been the cohesive elements in the realisation of this path have been the neighbourhoods, as the only urban spaces that carry tradition, spirit and identity, always present in all stages of the city's historical evolution, and still preserving their defining virtues today.

It is also worth noting that it is the city's historical evolution that has generated the Bucharest of today, but which has also produced all the problems and challenges that will require answers and solutions in the present and the future.

In the light of past experience, a rigid and retroactive regulatory system leads to multiple inconsistencies between the provisions at the time of drafting and the dynamic context of the urban development process, as well as to a proliferation of complementary and derogatory documentation. In order to avoid the consequences of a piecemeal approach to the same system that would invariably lead to the same results, it is necessary to develop a proactive, adaptable regulatory strategy capable of effectively absorbing the contextual changes at local, regional and global levels that Bucharest has to face due to its role as European capital.

Unfortunately, the current metropolis has ended up with far more concrete than green spaces or water areas. Concrete, asphalt are materials that absorb heat and store it.

That is why the atmosphere in Bucharest often feels like an oven: it is built in such a way that it simply traps heat and amplifies it, making the city several degrees hotter. The differences are especially noticeable inside the city, near vegetation.

For example, in the middle of a park, even on a scorching day, the temperature is somewhat more bearable. But the further you move away from the green space towards the city centre, towards the pavements, the higher the temperature you feel.

Urban heat islands amplify already extreme temperatures due to climate change. These differences are not fixed, but oscillate, according to studies, between 0.5°C and more than 10°C, depending on the urban centre, season, park size, vegetation density and even the time of day (Radu and Mateş, 2023).

Unfortunately, more than 4% of summer deaths in European cities are caused by what is called the urban heat island, dense concentrations of buildings and blocks. It is a phenomenon that defines the temperature difference between urban areas and rural and surrounding areas, and between an area with green space and the urban space around it. This difference can range from 0.6°C to 16°C in some cities, depending on several criteria, the size of the green space, the density of vegetation, and even the time of day.

The same heat waves produce serious social inequalities and amplify them: for example, people on lower incomes do not have access to air conditioning or do not even live in quality housing with access to infrastructure. On the economic side, heatwaves increase hospital admissions, which means extra costs for the system, and food prices, because agriculture is no longer performing. For the energy system, it also means extra costs and the risk of power cuts due to over-utilisation.

Also, in terms of mental health, several studies have linked high temperatures to increased violence and suicide rates.

Taking into account also the above-mentioned elements, for the future development of Romania's capital, the decision-makers decided to adopt a strategic approach on the following levels (PMB, 2023):

- Permanent bucharest.
- Identity-based bucharest.
- European bucharest.
- Business bucharest.

- Public bucharest.
- Connected bucharest.

Interesting in this document is the fact that the proposed city development model overlaps with the current structures in which the population is organised.

According to this proposal, it is envisaged to maintain the characteristic archipelago-type spatial structure, which, however, is beginning to manifest a major problem, generated by spatial barriers that carry disparity. They were generated either by functional fractures (enclaves, platforms and industrial corridors) or by spatial fractures (unstructured urban fabrics or impenetrable urban corridors generated by the placations of the major boulevards, created in turn by piercing and breaking up the urban fabric of the traditional city).

This new development model for Bucharest, in line with the principle we define as an integrated sustainable urban space, therefore emphasises the possibility of establishing areas with a dominant character, resulting from morphological analysis, and thus 70 neighbourhoods are drawn, understood as urban management units, which give a clearer picture of its identity, providing an effective tool for planning urban diversity.

In order to realise a real development plan for this large urban agglomeration, which is the capital of Romania, numerous supporting projects are proposed, the most important of which, from the perspective of the 'integrated sustainable city' concept, are the projects addressing the public, connectivity and business areas.

Bucharest is already in the top ten most congested cities in the world, and we therefore urgently need to initiate actions aimed at building connected cities that offset the negative effect that traffic has on local communities.

Therefore, it is necessary to rethink city development in line with the new concept, which is tantamount to conscious urban planning, which, even if it will lead to an increase in city density, will drastically reduce the demand for long-distance mobility, preventing sprawl in the suburbs, helping to decongest traffic and, therefore, reducing pollution.

Basically, within a radius of 2–3 km there must be independent hubs that are equipped with infrastructure for mobility, health, leisure, work, education, recreation and provisioning, including grocery stores, restaurants, offices, parks and schools. These hubs must be integrated and interconnected in such a way as to ensure the totality of the living needs of the inhabitants in the entirety of the current perimeter of the capital. Of course, the large health (hospitals, polyclinics, etc.), cultural and economic establishments will be located in the same place, and will be relieved of many less important demands, which are taken over by the local centres within the hubs. Thus, many of the activities that used to require longer journeys, mainly by polluting vehicles, will be eliminated and replaced by less travel by environmentally friendly means of transport.

According to the new city development concept, we can consider the 70 neighbourhoods mentioned in the proposal of the Bucharest City Hall as the starting point for the new construction of Bucharest, where the 'vital locations' of the communities must be rethought so that, with a minimum travel effort, the vital requirements (social, economic, cultural, etc.) can be met.

The transport network should be rethought with a view to giving priority to public transport and non-polluting alternatives (walking or non-polluting vehicles). The overall connections between neighbourhoods (communities) must be efficient and sufficient to support the harmonious and balanced development of the city from all points of view, which will be permanently on an upward trend (a permanent pole of socio-economic attraction)

In response to the negative effects of urban activities, numerous solutions can be adopted to mitigate or even eliminate these effects.

As presented in the first chapter of the current study, the development of a *green city* implies, among other things, the presence of a green (possibly agricultural) belt around the city (to reduce the levels of air pollutants entering the area). Moreover, at the current juncture of the development of the binomial 'transport infrastructure-sustainable city', it is desirable that the green belt be built between the city's outskirts and its transport belt.

Strictly in terms of transport, in addition to supporting the development of efficient public transport and reducing the demand for transport by private vehicles, a viable solution is also to develop as many green spaces as possible, particularly areas with trees.

Taking Bucharest as an example, the future development of Romania's capital should also include two green belts, one before each of the transport bypasses.

The existence of green spaces in the urban environment is an essential prerequisite for sustainable development, which puts care for nature, but also care for the inhabitants of a city, first and foremost. Moreover, urban forests have numerous benefits, the most important of which are (Smile Media, 2023):

- Trees take up the largest amount of $CO_2$ and through photosynthesis convert it into oxygen, thus becoming the main tool to protect against air pollution mainly caused by automobile traffic.
- Trees help regulate the local climate. Trees, especially mature trees with crowns, lower air and surface temperatures by providing shade and evapotranspiration.[6] In summer, when temperatures rise to 40°C, the base of the tree
- Trees protect against flooding by trapping water from rainfall and sending it to the water table, thus regulating the flow and velocity of runoff. Moreover, trees maintain moisture and reduce dry periods.
- Forests store most of the carbon dioxide in their crowns, trunks and roots, as well as at ground level, and in doing so filter the air and give us oxygen in return. Trees also act as sponges that absorb pollutants and prevent them from reaching the air and water.
- Urban forests provide living conditions for many species of plants and animals that are necessary to maintain the food chain in optimal conditions.

## Notes

1 According to the Shared-Used Mobility Centre (USA), sharing mobility is defined as those transportation services and resources that are shared among users, either concurrently or one after the other. This includes public transportation; micro-mobility (bike sharing, scooter sharing); automobile use (carsharing, ridesharing and micro transit); and commute-based or sharing modes (carpooling and vanpooling) (Shared-Used Mobility Centre, 2022).
2 Crossrail is a project to build a rail line mainly in central London. Its aim is to provide a high-frequency hybrid rail and rapid transit system that crosses the capital from the suburbs west to east, by connecting two major rail lines terminating in London: The Great Western Main Line and the Great Eastern Main Line. The project was approved in 2007, and construction began in 2009 on the central section and connections to existing lines that will become part of the route, to be named the Elizabeth Line in honour of Queen Elizabeth II, who opened the line on 17 May 2022 during her Platinum Jubilee. The central section of the line between Paddington and Abbey Wood opened on 24 May 2022, with 12 trains per hour.
3 The Autostrada Soarelui later became the name given to the A1 high-speed road linking the cities of Bucharest and Constanţa.
4 According to statements from that period, 'We were only 8 people with seats that we brought from an airplane. We were wearing seatbelts' (Gadget Report, 2022).
5 The technology has been used in supermarkets to transport money from homes to cashiers, for example.
6 Evapotranspiration is a complex process of converting water into vapour through a series of physical (evaporation in the case of the liquid phase and sublimation in the case of snow and ice) and biological (transpiration) processes.

## References


Andriescu V., *Hyperloop versus Henri Coandă. The projects that preceded Elon Musk's revolution*, 2013, https://adevarul.ro/stil-de-viata/stiinta/video-hyperloop-versus-henri-coanda-proiectele-1464582.html

ARCADIS, *Sustainable cities mobility index 2017*, 2017, https://www.arcadis.com/campaigns/scmi/images/sustainable-cities-mobility-index_singles.pdf

Băltărereţu R., *How could the transportation of the future work? The hyperloop is the dream of one of the biggest entrepreneurs*, 2013, https://adevarul.ro/stil-de-life/science/photo-how-could-the-means-of-transportation-1457404.html

Bawangaonwala M., Wadhwa D., Nandeshwar U. V., Dhurate S., Ramteke S. P. and Bante P. N., Ansari S., A review on development of intelligent transportation system to compare with Nagpur transport system, *International Journal of Computer Science and Mobile Computing (IJCSMC)*, vol. 7, Issue. 4, 2018, https://pdfs.semanticscholar.org/65a0/bee457082f1de036c79aae570b8a88d6bbb3.pdf

Bron D., The Top 10 Most Sustainable Cities in the World, 2023, https://medium.com/chain-reaction/the-top-10-most-sustainable-cities-in-the-world-407e1ee93e37

Casiello B., et al., *Increasing mobility in Dubuque: developing alternative mode-sharing opportunities*, 2012, https://iisc.uiowa.edu/sites/iisc.uiowa.edu/files/2022-09/increasing_mobility_in_dubuque.pdf

Chisăliţă D., *Urban electric mobility on the rise in Romania. Our buses, getting greener*, 2021, https://asociatiaenergiainteligenta.ro/mobilitatea-electrica-urbana-in-crestere-in-romania-autobuzele-noastre-tot-mai-verzi/

David R., *NXP semiconductors România dezvoltă soluţii software pentru maşinile autonome (Semiconductors Romania develops software solutions for autonomous cars)*, 2022, https://economie.hotnews.ro/stiri-companii-25861770-nxp-semiconductors-romania-dezvolta-solutii-software-pentru-masinile-autonome.htm

EC, *Annual accident report 2018*, 2018, https://road-safety.transport.ec.europa.eu/document/download/680f0863-aa0a-44f5-b3b8-39fe139924b0_en?filename=asr2018.pdf

Fistung D., *Teorie economică, ecologie, legislatie* (Transportation. Economic theory, ecology, legislation), ALL BECK Publishing House, Bucharest, ISBN 973-94435-03-3, 1999.

Gadget Report, *Coanda's electric train! Another invention that could redefine Romania*, 2022, https://gadgetreport.ro/trenul-electric-al-lui-coanda-inca-o-inventie-care-putea-redefini-romania/

Huapu Lu, *Eco-cities and green transport*, Elsevier, 2020, doi:10.1016/B978-0-12-821516-6.00001-1

Leuenberger D. Z., Bartle J. R. and Chen C., Sustainability and transportation, *Public Works Management & Policy*, vol. 19, issue 4, 2014, https://www.researchgate.net/publication/319102281_Sustainability_and_Transportation

Măciucă A., *Germania a inaugurat prima flotă de trenuri cu hidrogen din lume* (Germany has inaugurated the world's first fleet of hydrogen trains), 2022, https://stirileprotv.ro/stiri/international/germania-a-inaugurat-miercuri-prima-flota-de-trenuri-cu-hidrogen-din-lume-video-si-galerie-foto.html#

Martin V., Anholon R. and Quelhas O. L. G., *Sustainable transportation methods*, 2019, https://www.researchgate.net/publication/329359932, doi:10.1007/978-3-319-63951-2_192-1

PMB (Bucharest City Hall), *Despre PUG București* (About the PUG (General Urban Plan) Bucharest), 2023, https://pug-bucuresti.uauim.ro/despre

Radu C. and Mateş I., *Tree-free cities*, 2023, https://www.libertatea.ro/stiri/orasele-fara-copaci-marile-centre-cuptoare-4676036

Santos G., Sustainability and shared mobility models, MDPI, *Sustainability* Revue no. 10/2018, Cardiff, 2018, link: www.mdpi.com/journal/sustainability

Shared-Used Mobility Centre, 2022, https://sharedusemobilitycenter.org/what-is-shared-mobility/

Shi J., Research on London's sustainable transport transformation from a micro perspective, *Industrial Engineering and Innovation Management Journal*, doi:10.23977/ieim.2024.070313, 2024

SLOCAT (Partnership on Sustainable, Low Carbon Transport), *Transport and demand*, 2022, https://tcc-gsr.com/transport-demand/transport-demand/

Smile Media, 2023, https://www.hotnews.ro/stiri-esential-26404660-video-sunt-importante-padurile-urbane-dan-barbulescu-directorul-asociatiei-parcul-natural-vacaresti-arborii-sunt-prietenii-nostri.htm

Speirs J., *Safe, attractive, and sustainable cities: A mobility-oriented approach*, Ecole Polytechnigue Federale de Laussane, 2019, https://iglus.org/wp-content/uploads/2019/05/Safe-Attractive-and-Sustainable-Cities-A-mobility-oriented-approach.pdf

VIRTA, *The global electric vehicle market overview in 2022: Statistics & forecasts*, 2022, https://www.virta.global/en/global-electric-vehicle-market

# 8 Prospects for adapting markets for sustainable transport in line with society's development requirements

As presented above, from a transport perspective, there are mainly several solutions that can be adopted for a seamless integration in future urban areas developed on the principle of integrated sustainability:

1 Reducing mobility demand with polluting vehicles.
2 Increasing the share of public transport in the overall mobility plan of congested urban areas.
3 Adopting green technical solutions for all types of vehicle fleets.
4 Redesigning urban transport infrastructure in line with the new development concept. At the same time, the creation of green areas (nature protection areas) for the absorption of pollutant emissions caused by traffic in cities (presented in detail above) should be considered.

In the following, we will try to sketch some of the most important elements characterising each type of solution.

## 8.1 Adapting mobility demand

In terms of reducing the demand for mobility, the most important thing becomes the urban redevelopment based on integrated hubs, as detailed above. In this context, traffic intensity will decrease significantly, especially over medium and long distances. The accelerated development of remote work, education, etc., should also be considered, which would significantly reduce the need for mobility.

But there are also solutions that can help reduce the demand for mobility, especially for non-essential trips.

The Institute for Transport Policy and Development (ITDP), based in New York (USA), has developed a scenario called Sustainable Shared Mobility that calls for the efficient deployment of new technologies and the full use of the benefits of automated and electric vehicles, focusing on supporting walking, cycling and public transport. In this scenario, the cost reductions apply to public transit and other higher occupancy vehicles, rather than to individual driving and other low-occupancy modes (Mason, 2018).

DOI: 10.4324/9781003743064-9

Sustainable shared mobility could be achieved in two ways: by discouraging inefficient travel (low-occupancy vehicles) and by encouraging efficient travel (e.g. walking, cycling and public transport). There are a variety of ways to discourage inefficient travel, including by internalising external costs as effectively as possible and increasing specific taxes such as vehicle registration taxes. These are vital to counteract the savings generated by new mobility technologies.

Efficient travel is being encouraged in a variety of ways, including through the planning and zoning of compact, mixed-use cities centred around public transport, as well as by building infrastructure that supports and prioritises walking, cycling and public transport. Para-transit (partial transit) users would similarly benefit from sustainable shared mobility. Door-to-door transport, provided and subsidised by public authorities, is essential for people with mobility problems that prevent them from using public transport. In many places, government agencies are ill-equipped to manage such services but are legally mandated to do so. The result is often incredibly high costs and poor service.

At the same time, in a future scenario of sustainable shared mobility, increased vehicle use will reduce parking times and parking spaces. Currently, most cars are parked for 95% of the day, and in the United States, the number of parking spaces is estimated at around 500 million, with some speculating that it could reach as many as two billion. With 263 million passenger vehicles in the United States, this is at least two spaces per car. In a future of sustainable shared mobility, each vehicle could be used by multiple occupants each day, greatly reducing the amount of time the vehicle is idle and in need of parking. This would free up massive amounts of space in urban areas, and in particular on-street parking spaces that are suitable for bicycle infrastructure.

Urban mobility also has societal and environmental implications. How we move or do not move is also a class issue. While the wealthy have access to more modes of transport and can live in more accessible parts of cities, poorer communities often face mobility problems in marginalised areas of cities. Sometimes, these areas have no access to public transport. Other times, they lack decent sidewalks or pedestrian zones. This causes more deaths and injuries among those living in poorer communities.

Shifting urban mobility from personal car use to safer and more sustainable options requires techniques that simultaneously attract and 'push'. Alternatives to private car use must be offered. They must be attractive enough to persuade people to choose them. Underground or surface trains must run to a respected timetable, buses must have good frequency, cycling must be safe and walking must be pleasant. When alternatives are attractive, they attract users to them, making cities more sustainable, adaptable and efficient.

However, cities are changing rapidly, and technology is disrupting mobility options and land-use patterns. What was embraced a decade ago (before smartphones were prevalent) is almost no longer the case today. Cities need to plan their infrastructure for an uncertain future. How cities develop legacy infrastructure in this climate is difficult to manage.

Many cities have invested heavily in self-oriented infrastructure only to redesign it in the next few years.

The 'Cheonggyecheon' and 'Seoullo 7017' projects in Seoul (South Korea) are two illustrative examples of inner-city roads that have been reoriented at great expense. Their reconfiguration was necessary to improve quality of life and, counterintuitively, to reduce congestion. Similarly, the removal of San Francisco's 'Embarcadero' freeway has helped the city become one of the most livable in the United States, while bringing thousands of tourists each year. Cities should be discouraged from building costly roads that may prove to be a solution only in the short term, but which increase demand and congestion in the medium term, prove obsolete in the long term and eventually become costly to redesign.

The need for private cars is falling, while parking is becoming less and less of a priority. Shawn Conrad, CEO of the International Parking Institute, explains that 'airport pick-up and drop-off points have led to declining revenues for cities in the US as more drivers opt for ride-hailing services' (Speirs, 2019). In addition, autonomous vehicles are developing at a rapid pace. They will change the way people have access to cars and the necessary infrastructure. Parking is likely to be rendered unnecessary or reused in new and surprising ways.

Looking to the future, the International Parking Institute, the British Parking Association and the European Parking Association have collaborated to develop the Parking Data Alliance standards. This will enable better analysis and the most effective measures to be taken for the changing demand for parking spaces.

Cities around the world are ageing rapidly. People are living longer than was foreseen for urban planning and design should reflect this (cities are in many cases no longer suited to the requirements of a population that lives by different rules, has a different structure and a different mentality). In many such cases, the solution lies in supporting the development of good, safe and attractive public transport.

Increasing road capacity is subject to diminishing returns. Fighting congestion by building more roads is like fighting obesity by loosening the beltway. It fails to address the cause of the problem while giving more room for the problem to get worse. If cities aim to increase mobility without exacerbating private car problems, more innovative solutions are needed. However, innovation does not have to be just technological.

In the United States, for example, the transportation community has shown a growing interest in sustainable transportation and its links to land use and urban development patterns, economic growth, environmental impact and social equity. Many US transportation agencies are re-examining their policies, planning approaches and evaluation methods, and are considering changes in every aspect of practice, from the materials and designs used in construction to the alternatives considered for implementation. Federal, state and local agencies, as well as private organisations, are working to achieve the broad goals of sustainability in specific transportation policies, goals and programmes.

By 2050, EU leaders have endorsed a goal of reducing Europe's greenhouse gas emissions by 80% below 1990 levels, as part of efforts by developed countries to cut their emissions by a similar amount. The European Commission has published a roadmap for building the European low-carbon economy that will require this (EUROSTAT, 2019).

Key EU initiatives to reduce greenhouse gas emissions included the following:

- Development and implementation of the EU Emissions Trading Scheme, with the ultimate goal of building an international carbon trading market, including aviation.
- Monitoring the implementation of Member States' emission reduction targets by non-EU ETS actors ('Effort Sharing Decision').
- Implement legislation to increase the share of energy consumption from renewable energy sources such as wind, solar and biomass to 20% by 2020.
- Increasing Europe's energy efficiency by 20% by 2020 by improving the energy efficiency of buildings and a wide range of household equipment and appliances.
- Binding targets to reduce $CO_2$emissions from new cars and vans.
- Support the development of Carbon Capture and Storage (CCS) technologies to capture and store $CO_2$ emitted by power plants and other major industrial installations.

Of course, the emergence of the coronavirus pandemic and the Russian–Ukrainian military conflict have led to some of these objectives being reformulated and adapted to the new situations.

## 8.2 Urban transport in the sustainable mobility equation

Urban transport is a key factor in underpinning any strategy for the sustainable development of localities.

Meeting increased mobility needs, while reducing environmental pollution, accidents and improving the living environment are the challenges facing all major conurbations. It is city dwellers who bear the brunt of the negative effects of their own mobility more than anyone else, and therefore need to be open to innovative solutions to create sustainable mobility.

Today, around 80% of European citizens live in urban areas. In this context, European cities are increasingly facing transportation problems. At the same time, the migration of city dwellers to the suburbs is leading to the emergence of residential patterns that require people to commute longer distances. This phenomenon of urban sprawl has developed in parallel with a significant increase in car ownership and commuter traffic.

Satisfying the travel needs of people in large cities by personal car use is a form of transport that is detrimental to both those involved and those not involved in the transport system.

Traffic and air pollution are among the main problems facing the world's biggest cities. That is why local authorities are trying to get people to choose public transport, restricting car access to city centres and increasing parking charges.

Public transport, cars, trucks, cyclists and pedestrians all use the same infrastructure at the same time. Urban transport generates 40% of $CO_2$ emissions from road transport and approximately 70% of other transport pollutants. One-third of road fatalities are caused by urban accidents. Traffic congestion problems also occur mainly in and around cities.

The production and purchase of cars are constantly increasing, while road infrastructure has exhausted its possibilities for expansion, leading to incoherent use of space, its degradation with economic (congestion alone consumes about 2–5% of GDP), environmental and social implications.

An effective plan to mitigate these negative effects certainly includes the development of public passenger transport, especially in large urban areas. One of the measures strongly promoted by transport specialists in this context is the development of high-capacity transport networks so that they can become attractive to users. In this respect, metro networks in major cities should become the main axes of travel and be preferred by users in their modal choices.

Moving towards sustainable mobility, at the urban or regional level, implies increasing the share of public transport in the modal distribution of travel, to the detriment of private car transport.

Enhancing the attractiveness of the public transport system is a key element of the urban development vision, and there are a number of priority measures that can contribute to improving the quality of the services offered, that is, the local public transport service. These include the following:

- Increasing the level of coverage with this type of service.
- Connection to an integrated passenger information system, with details of activities (timetable, destination, connections with other routes, prices, etc.).
- The creation of multimodal transfer nodes (between different types of surfaces and underground public transport, between personal means of transport and public transport, etc.).
- The use of environmentally friendly and socially adapted means of transport (disabled access, bicycle compartments, etc.).

A city's territorial, social and economic development strategy depends to a large extent on the level of its public transport network.

Public transport has a major influence on the economic and social development of a city and can be a factor in speeding up or slowing down its development.

Urban public transport, in this context, has certain characteristics such as accessibility, mass character, periodicity and uniformity.

*Accessibility* means that anyone can use public transport services. The person's main obligation is to have a travel permit (ticket, identity card, etc.), which in fact constitutes the conclusion of the transport contract.

The second specific feature concerns the *mass character of* the transport. From this point of view, the activity is designed to transport individuals or groups. The number of persons present in the vehicle is irrelevant, as the carrier is obliged to carry out the journeys in accordance with the timetable approved by the local public administration. As a rule, the flow of passengers is not uniform throughout the day, depending on the time of day.

The third peculiarity is *periodicity*, which is manifested in the fact that the traffic is always running at certain intervals according to a timetable approved in advance.

Finally, the *uniform character of* public transport is expressed both in the specific arrangements for organising the activity and in the contractual terms, which are the same for all passengers.

There are discussions in policy circles about the meaning of the term 'urban public transport'. Neither legislation nor the legal literature defines this concept, which is why, in practice, there are difficulties of interpretation as regards the essence of the relationship between the movement of persons by means of urban public transport, the means of interaction, the rights and obligations of the parties, etc.

There is also a lack of clarity as to the criteria according to which transport operators can be included in the category of 'urban public transport' and the types of transport that fall within that category.

DEMAND: It is clear, however, that urban passenger transport is intended to take passenger flows from the transport network in both directions.

SUPPLY: In this respect, it is necessary to establish the hourly variation in passenger flows for each line (route) and then, depending on the length of the route and the capacity and speed of the vehicles available, to determine the number of vehicles needed at different times of the day and to draw up the corresponding traffic timetable.

As in any economic field, if supply and demand are not in balance, distortions arise, sometimes negligible, sometimes serious.

Urban public transport is not just a social service provided by local authorities to ensure a certain degree of mobility for the whole community. It is, in fact, the most efficient mode of transport in terms of space consumption and air pollution (calculated per passenger transported). Therefore, several aspects should be taken into account when implementing such a system (Bălăşoiu and Bălăşoiu, 2009):

- *Financing*. The erosion of public funds, rising fuel prices and competition from private cars mean that public mobility lacks the resources to invest in quality. Funding can be provided through taxation or 'polluter pays' schemes. Car competition reduces revenues from public transport and leads to inefficiencies in public services where cars and public vehicles use the same infrastructure.
- *Accessibility*. People with reduced mobility, including disabled and elderly people, as well as people with temporary injuries or children (who account for around 35–40% of citizens who frequently use public transport), need to

be taken into account in plans for the development of a public transport system. Moreover, as the population ages, accessibility becomes an increasingly serious issue.

- *Decentralisation*. Public transport systems are not designed to serve the low-density or sparsely populated urban areas that increasingly dominate the landscape (uncontrolled urban sprawl). The larger the area of urban activities, the more difficult and costly it is to serve them, especially specific infrastructure such as metro and rail (in the case of trams).

The development of passenger transport in general, and of urban and suburban transport in particular, has been driven by the economic, political and socio-cultural development of society at different historical stages.

Throughout the world, different urban public transport solutions have been developed over time, based on the technological developments of the respective eras, the increasingly varied and complex infrastructure supply and transport demand. In this context, five characteristic periods can be identified:

1 The period between 1820 and 1870, characterised by animal traction of vehicles. Thus, in 1820, so-called omnibuses were introduced in Paris, London and New York, which were larger horse-drawn covered carriages that could carry up to ten people at a time.

   After George Stephenson had pioneered the first public railroad in 1825, the first horse-drawn tramcar (carriage on rails) was put into service in New York in 1832 (MOOVIT, 2018).

   Later, this type of transportation was introduced in other cities around the world: in 1855 in Paris (France), in 1858 in Santiago (Chile) and in 1860 in Sydney (Australia).

2 The period 1870–1920, characterised by the use of electric-powered transport vehicles. The development of the electric motor revolutionised urban public passenger transport and beyond.

The first electric tramway was tested in 1873 in San Francisco (USA), where the specific characteristics of the city (hills) required special engineering solutions.

However, the first public electric tramway was put into service in Berlin (Germany) in 1881.

In 1890, the UK's Metropolitan Railway built the first underground electric tramway in London.

Since then, this type of transportation has developed rapidly in many parts of the world.

In the USA, for example, such systems were developed in 1892 in Chicago and in 1897 in Boston and Massachusetts.

1 The period between 1920 and 1950, characterised by the expansion of urban public transport by means of vehicles powered by internal combustion engines (buses in general). Even though the first step in the development of

this type of transportation was experienced at the end of the nineteenth century, the high costs and special conditions of construction of these means of transportation did not support the large-scale development of this type of vehicles until after 1920 (Rodrique, 2022). Subsequently, with the decreasing costs of manufacturing and operating these vehicles, their use increased rapidly.

2 The period from 1950 to 1990, during which all types of urban public transport (surface and underground) were developed simultaneously, according to the concept of integrated transport, designed to meet demand by offering the best and safest conditions of travel. The rapid development of travel by private car within large conurbations has led to the adoption of solutions that are constantly being adapted to make public passenger transport more attractive. However, the trend has been in favour of private transport to the detriment of public transport.

3 The period after 1990 and up to the present day has been and is characterised by the need to reduce the impact of private car transport in urban areas. For this reason, the concept of sustainable urban mobility was adopted. In this context, the implementation of environmentally friendly transport alternatives, such as the use of electric vehicles or increasing the efficiency of the management system of the whole public transport activity, has been supported with the aim of reducing the external costs caused by urban public transport activity. The first city in which electric buses were widely tested was in China (Shanghai) in 2000. Currently, China has almost 99% of the world's existing electric buses in operation (385,000 buses), and the fleet in China is growing by about 1,900 electric buses every week (MOOVIT, 2018).

In fact, electric buses have grown in popularity since 2020. In 2022, nearly 66,000 electric buses were sold globally. China dominated the market, accounting for over 80% of global electric bus sales. The country also excels in manufacturing electric buses and is a major exporter to countries in Latin America, North America and Europe (VIRTA, 2023).

In EU countries, the Clean Vehicles Directive sets targets for public procurement of electric buses. France, Germany and Spain are just a few EU countries that are seeing an increase in electric bus sales.

Finland had the highest share of electric bus sales in Europe in 2022. Electric bus sales accounted for over 65% there.

In this context, the importance of public passenger transport should be emphasised, highlighting that in the EU alone:

- Some 60 billion people travel by this service every year. In comparison, commercial air or long-distance rail transport carries, on average, only one billion passengers annually.
- The role of this activity in relieving traffic congestion in major conurbations is extremely important, estimated at around €100 billion a year, equivalent to 1% of EU GDP.

- The sector's contribution to EU GDP is 1–1.2%, that is, around €130–150 billion annually. At the same time, the level of investment in the sector is estimated at almost €40 billion annually.
- The economic benefits from public passenger transport activities are five times higher than the investments in the sector (UITP, 2020).
- This type of activity creates many local jobs, providing equal development opportunities for all citizens.
- The sector employs the largest number of people locally, around two million, accounting for 20% of all employees in the transport sector.
- Thanks to public passenger transport in large urban agglomerations, the environmental quality is better. In this context, public transport is a key tool in the development of a sustainable way of life, in line with the new requirements of The European Green Deal, presented at the end of 2019.

## 8.3 Green alternatives to current unsustainable transport systems

Analysing the energy sources required for vehicle propulsion, a tentative framework could include the following categories:

- ABU: this type of fuel is used in 'external combustion' engines being produced outside the engine itself. This type of engine has been in use since the nineteenth century and has been superseded by 'internal combustion' engines, but new concepts for modern high-power steam engines for vehicles appear to be available.
- HYDROCARB FUELS (GAS FUELS and GASOLINES): steam engines have gradually been replaced by those using this type of fuel (diesel or gasoline). Over the last few decades in particular, this type of engine fuel has become increasingly undesirable because of both the negative effects on the environment (noxious emissions) and the consumption of non-renewable energy products. In this context, at least on the European continent, there are plans to reduce or even eliminate the sale of vehicles powered by petroleum-based engines by 2030–2035. Discussions are heated, and major car manufacturers, in particular, are expected to strongly resist these plans.
- LPG: Liquefied Petroleum Gas (LPG) has been a fairly good alternative to hydrocarbons, in terms of both emissions and its lower cost. However, the more intensive use of LPG has not been put into practice because of both technical problems (faster degradation of engines) and reduced storage possibilities (small tanks for the consumption realised and, thus, reduced driving range).
- BIOFUELS: biofuels, such as bio-ethanol (which can be used instead of gasoline), are made from corn and sugarcane, while bio-diesel is made from vegetable oils and animal fats. Many see them as the best medium-term solution for sustainable fuels, but the land used to grow the crops from which these fuels are made reduces the land that can be used for other purposes, mainly food.

- ELECTRICITY: electric vehicles have increasingly 'caught up speed' in recent years amid intensifying demands to reduce pollutant emissions, fluctuating hydrocarbon prices and a desire to promote new sustainable transport models. Despite the real 'zero' pollution advantages of the vehicle in operation, the production of electricity and the batteries used to store it cause significant environmental damage, even greater than the hydrocarbon-based but energy-efficient versions of engines (EURO 6 and later models).
- Many electric vehicles (and a growing number of internal combustion engine cars) have regenerative braking energy systems, which convert the energy normally wasted during braking into electrical energy. The use of such systems is expected to increase in the future in order to better harness the energy in motion held by a car and thus use less fuel overall.
- HEATING: two-thirds of the energy generated by gasoline or diesel is wasted as heat. Thermoelectric technology, which converts heat into electricity, can help reduce this and is already being developed by several car manufacturers. One solution is the use of thermoelectric panels to convert waste heat from the tailpipe into electricity, which can reduce fuel consumption by 5%.
- Hydrogen can be used instead of fossil fuels in internal combustion engines. Hydrogen vehicles have no harmful emissions – they only remove water. Hydrogen can also be used to power a fuel cell and produce electricity. This is what many consider one of the best long-term energy sources for cars: it produces zero emissions and overcomes the limitations of electric vehicle batteries. Currently, however, fuel cell technology is too expensive, and critics point out that it shifts energy consumption and pollutant emissions to the plant that produces the hydrogen, and there is currently no hydrogen fuelling infrastructure.
- AZOT: liquid nitrogen, stored in a pressurised tank, can be heated to produce high-pressure gas. This can be used to power a piston or rotary engine. Liquid nitrogen is, however, a less efficient energy carrier than fossil fuels and requires electricity to produce it.
- AIR: several concepts have been discussed over the years, and some automakers, such as Tata, have even proposed cars for series production with air-powered engines. Stored in 4500 psi tanks, air as a power source is much less energy-dense, but produces no exhaust emissions. Compressed air can replace gasoline in a combustion engine to drive pistons and produce power. Improved storage and propulsion systems will, in our view, be the best solution for sustainable mobility in the not-too-distant future.

As mentioned earlier, by 2022, there will be 1.49 billion cars in use worldwide, of which 1.11 billion will be passenger cars and 380 million freight/commercial vehicles. To put this in perspective, the number of cars in the world in 2009 was one billion (which means that the world car fleet has grown in 13 years by more than 67%), equivalent to an average annual growth of about 5.2%, only 1.1% less than the total population growth (6.88 billion in 2009 and 8 billion in December 2022).

Unfortunately, despite all the efforts in recent years to promote environmentally friendly vehicles such as battery-electric, plug-in hybrid and fuel cell electrics, more than 98% of the total number of cars in use in 2023 were still powered by gasoline or diesel.

The 1.49 billion cars emit the equivalent of 6.9 billion tonnes of $CO_2$ about $17 trillion) into the atmosphere each year. In comparison, this is the equivalent of burning all the coal in a train that stretches some 5,58,000 kilometres, long enough to wrap around the Earth 13.9 times around the equator.[1]

Of all cars in use worldwide, 390 million were in China (a 56.7-fold increase since 1990), 294 million in the USA and 283 million in Europe, with the three areas accounting for 62.9% of the world's vehicle fleet (ACEA, 2022).

In the EU, despite a strong sales growth in recent years, green passenger cars still account for only 5.3% of the total fleet. Electric cars and plug-in hybrids[2] represent only 0.5% and 0.6% of the fleet respectively.

Light-duty, diesel-powered commercial vehicles are still dominant in all EU countries except Greece: 91.2% of the EU van fleet is diesel-powered, and only 0.4% of EU vans are battery electric.

Some 96.3% of all goods vehicles in the EU run on diesel, while gasoline powers less than 1% of the fleet; 0.24% of trucks on EU roads have a zero-emission powertrain, up from 0.04% in 2019.

Diesel buses account for 93.5% of the EU fleet, with only 0.9% electric and 1.4% hybrid electric. Significant fleets of electric buses can be found in the Netherlands (12.4%) and Luxembourg (6.6%).

It is clear that the future is forcing mankind to put increasingly accessible, efficient and cheap energy solutions into practice. In this context, clearly, powering vehicles with fuels from petroleum processing is becoming less and less desirable and will soon remain an extremely limited segment of the specific market. But viable alternative solutions have to be found with which to replace the energy sources of the current world fleet, which is mainly based on diesel and gasoline consumption.

### *8.3.1 Hydrogen vehicles market*

Hydrogen plays a vital role in reducing/eliminating carbon emissions from transportation activities, enabling cities and governments to meet their goals for a sustainable society. Hydrogen is also the most abundant chemical element in nature. As noted by the International Energy Agency (IEA), the global demand for hydrogen for use as a fuel has tripled since 1975 and reached 70 million tonnes per year in 2018 (Iberdrola, 2020).

For hydrogen-powered vehicles, compared to those powered by electric batteries, hydrogen's fast refuelling time and higher density give it an advantage over battery-only solutions in many situations – including for freight and passenger transport in heavy-duty vehicles (trains, ships, planes, coaches, etc.). In addition, green hydrogen is a clean energy source that emits only water vapour and leaves no residues in the air, unlike hydrocarbons, coal and oils.

Hydrogen is essential for decarbonising industries and businesses around the world and supporting the energy future. Hydrogen can unlock the decarbonisation of industries and businesses worldwide, enabling an efficient energy transition.

Hydrogen can be produced from a range of resources, including fossil fuels, nuclear, biomass and renewable energy sources. This can be done through a number of processes.

Hydrogen is classified into the following categories, based on how it is produced:

- ALB: a naturally occurring version that can occasionally be found underground, but there are few viable ways to extract this – so experts are looking to generate it artificially.
- GRI: the most common form which is generated from natural gas or methane through a process called steam reforming.
- BLACK (BROWN): the most damaging to the environment because both $CO_2$ and carbon monoxide, generated during the production process, are not retained (captured). Black (bituminous) or brown (lignite) coal is used for production.
- BLUE: is sometimes also referred to as 'carbon neutral' because the emissions are not dispersed into the atmosphere but captured and stored underground.
- GREEN: also called 'clean hydrogen' is produced by using clean energy from surplus renewable energy sources, such as solar or wind power, to split water into two hydrogen atoms and one oxygen atom through a process called electrolysis. It currently accounts for about 0.1% of total hydrogen production, but is expected to grow as the cost of renewable energy continues to fall.
- TURCOAZ: produced by the pyrolysis of methane, which generates solid carbon. Its production is still at the experimental stage.
- RED: like green hydrogen, it is created by electrolysis of water, but the latter is powered by nuclear rather than renewable energy.
- YELLOW: representing hydrogen made by the electrolysis of water using solar energy, although some use it to designate hydrogen generated by the electrolysis of water using mixed sources, depending on what is available.

Hydrogen can also be generated from biomass and, depending on the type of biomass and production technologies, can have lower net carbon emissions than black/brown or grey hydrogen.

Hydrogen's high versatility allows it to be used in those consumption niches that are very difficult to *decarbonise*, such as heavy transport, aviation and shipping. There are already several projects in this area, such as Hycarus and Cryoplane, which are being promoted by the European Union (EU) and aim to introduce it into passenger aircraft.

The hydrogen we currently have access to, known as grey hydrogen, is produced from fossil methane and emits 14% more $CO_2$ than fossil diesel. However, a switch to green hydrogen, produced by renewable electrolysis powered by electricity, could reduce $CO_2$ emissions by up to 62%.

This transition will take place slowly and at the cost of significant investment, and road transport will still be in competition with other sectors that have no other way to decarbonise. Its widespread use cannot therefore be expected before the next decade.

Hydrogen fuel cells are today's most advanced technology. These cells convert hydrogen and oxygen into electricity. An alternative use for hydrogen is in combustion engines, which can use a less pure form and require less cooling. However, a disadvantage is that such systems emit a small quantity of noxious emissions, making their use in cities unlikely, and consume slightly more hydrogen than fuel cells.

For a functional transition to hydrogen energy, several actions can be taken. First, create a natural demand in the market by coalescing economic actors convinced of the benefits of hydrogen technologies. Secondly, there is a need for regulations applied to the field, for example, establishing an index on the pollution degree of the various hydrogen technologies and thus encouraging green hydrogen among the existing options. The next steps are to couple production potential with storage and transportation potential by converting existing infrastructures. Last but not least, collaboration is needed: innovation does not come under regulatory pressure, but through common will and systemic approaches. The solution to integrate these measures is the so-called 'hydrogen hubs', that is, geographical areas that have accessible production resources, investment potential and market-stimulating regulations, and research shows that Europe is best placed to materialise such projects by 2030 (Deloitte, 2023). With its significant potential in the area of renewable energies, Romania can also play an important role.

Unlike a few years ago, the exploitation of hydrogen as an energy carrier is now supported by the increasing availability of renewable energy, which makes hydrogen a 'green fuel'. Storing energy as hydrogen is a valid option to balance the electricity grid, making large volumes of green hydrogen available when produced in an electrolysis plant. To make hydrogen attractive, the cost of production would come down in a way that supports the hydrogen economy in the transportation sector.

The current increased interest in hydrogen as an energy carrier has the potential to have a significant positive impact on the decarbonisation of the transport sector, in particular where the limitations of electric (battery) solutions represent relevant economic and/or technical burdens, such as in the case of heavy-duty long-distance vehicles. Hydrogen will therefore be able to contribute to the decarbonisation of transport, working together with electrification and not in competition with it, bridging the gap that could not be technically or cost-effectively bridged by batteries.

The size of the global fuel cell vehicle (FCV) market has been valued at \$1.14 billion in 2021 and is anticipated to grow at an average annual growth rate of 49.9% from 2022 to 2030 (Grand View Research, 2023).

Unfortunately, fuel cell vehicles are very expensive, and the cost of hydrogen, especially green hydrogen, is also very high. These drawbacks are some of the biggest impediments to bringing more types of such vehicles to market. However, many car manufacturers are testing fuel cells with platinum-free metal components to reduce some of the costs.

The market for fuel cell vehicles is still in the early stages of development and growth. Many technologies are in the testing phase, with several cars from Toyota Motor Corporation, Honda Motor Corp. and Hyundai Motor Group running in specific regions. European manufacturers such as BMW AG and Audi are developing fuel cell vehicles based on their earlier prototypes, which will be launched in the coming years. Many other companies are awaiting the achievements of BMW and Audi. For example, Renault has produced its prototype hydrogen fuel cell-powered SUV. The SUV called the Scenic Vision will be equipped with a 16-kW fuel cell, which will help extend the car's driving range to 800 kilometres without the need for charging. The battery will be lightweight, and the carbon footprint will be reduced by 75% compared to the traditional electric vehicle.

Hydrogen fuel cell vehicles are expanding their applications to heavy commercial vehicles to increase efficiency. For example, a major global player in zero-emission hydrogen fuel cell-based commercial vehicles – Hyzon Motors Inc. – has announced the development of hydrogen storage system technology, which can lower the production cost and weight of the commercial vehicle. The new technology has not yet been patented and has the potential to reduce the number of manufacturing components by 75%, system weight by 43% and system storage cost by 52%. This technology utilises lightweight composite material for the metal stamping of the system, which is based on a single rack system capable of storing five hydrogen cylinders simultaneously. The development of this on-board storage system technology is a collaboration between the European and American divisions of Hyzon Motors Inc.

As stated earlier, fuel cell vehicles are expensive, and the cost of hydrogen (per kilogram) in some places is high, limiting the growth of the fuel cell vehicle market. In addition, the number of gas stations is limited or inadequate in regions around the world. The progress of the industry is limited by a poor distribution network, fear of electric shock and flammability due to the chemical attributes of hydrogen fuel cells, which acts as a restraint to the market growth. The manufacturing cost of hydrogen fuel cell vehicles is higher than the production cost of conventional vehicles. The cost of production, coupled with the cost of maintenance services, significantly increases the total purchase cost compared to a standard vehicle. In addition, it faces competition from battery electric vehicles and plug-in hybrid electric vehicles,

The fuel cell vehicle market can be described as a competitive market due to increased adoption and investment in the market by private and government players. The market has a few established players such as Toyota Motor Corporation, Audi AG and Volvo AB, among others. These companies invest heavily in research and development for fuel cell vehicles. More mergers and acquisitions, partnerships and collaborations are expected to be some of the growth strategies these companies are pursuing to expand their business by leveraging each other's expertise. For example, Plug Power Inc. and Renault Group have announced a HYVIA *joint venture* to develop hydrogen fuel cells and low-carbon mobility vehicles.

The vehicle type segment is divided into passenger cars, LCVs (low calorific power) and HCVs (high calorific power). Passenger cars dominate the market, accounting for the largest share of 84.3% of the total industry in 2021 and registering the highest growth of 50.6% over the forecast period (2021–2030). Rising passenger car purchases in countries such as Japan and South Korea are driving the growth of the fuel cell vehicle market segment. The LCV segment is expected to grow steadily over the forecast period. Light commercial vehicles are instrumental transportation vehicles used for supply chain purposes. The use of a fuel cell instead of a battery ensures electrochemical conversion of energy into torque and high efficiency. Thus, the use of fuel cells in long-distance travel vehicles such as LCVs is an ideal choice from the perspective of efficient and economical travel (Grand View Research, 2023).

### *8.3.2 Electric vehicle market*

On average in Europe, vehicles using electricity produced from renewable or nuclear energy sources currently save 55% of $CO_2$emissions compared to fossil diesel. In some countries, such as France, this saving is over 80%.

Electric motor technology is within reach, and battery workload is increasing.

Free from the direct effects of emissions, electric cars and trucks currently appear to be the most effective solution to urban pollution problems. They can be recharged at night – during off-peak hours – either at a three-phase industrial connection or at a base-load charging terminal, operate silently and offer great convenience for drivers.

Although electric vehicles continue to be more expensive than those that run on diesel, the gap is narrowing all the time.

From 2025, the total cost of owning and operating a battery-electric truck for urban use will be less than that of a diesel-powered truck. By that date, all urban uses will be covered by electric trucks.

Despite these trends, calculations of the level of environmental emissions produced in the production of electric vehicles do not seem to support the widespread use of this solution. Thus, in one such study, researchers at the Fraunhofer Institute (Germany) calculated that an electric vehicle equipped with only a 40kWh battery that is charged at a wall socket needs to travel 72,000 km to create a $CO_2$ emissions advantage over a gasoline vehicle (Fraunhofer ISI, 2022). For an electric vehicle equipped with a 58kWh battery, the distance needed to be travelled to create a $CO_2$ emissions advantage over a similar gasoline-powered vehicle increases to 100,000 km, and for an electric vehicle powered by a 95kWh battery, the distance increases to 166,000 km.

This means twice as much environmental pollution goes into making an electric car as a similar car running on conventional fuel (internal combustion engine). For example, a fully electric car uses about 80 km of copper in the composition of batteries, winding cables and rotors (Deloitte, 2018). It is well known that in cars powered by petrol and diesel, several steel components are used, the production of which already causes environmental damage in the manufacturing areas. In electric cars, however, the raw materials and processes used in battery manufacturing cause much more environmental damage. The larger the battery produced, the higher the level of environmental damage.

At the moment, the cost of filling up with electric batteries is also only 50% of the cost of filling up a car of similar capacity with a electric battery.

More specifically, for a Logan 1.5 dCI with a 55-litre tank, the cost of a full tank is about 413 lei (taking into account the 7.5 lei/l of diesel). At an average fuel consumption of 7L/100 km, the diesel vehicle has a range of about 800 km. In the case of a Dacia Spring model, full-electric with a 25.6 kv battery, the cost of a total supply is about 61 lei for a range of about 230 km (the standard cost of supplying a 110 kv battery was, in August 2023, 263 lei at a parking lot at Kaufland, Bucurestii-Noi). To reach the autonomy of the Dacia Logan, the Dacia Spring will need 3.5 charges, which means a cost of 212 lei, about half the cost of diesel for the Logan. But small batteries, numerous charges, low range and shifting pollution to the source of production do not offer a viable prospect for electric cars.

The realisation of electric vehicle designs with increasingly powerful batteries, made from raw materials and using polluting technologies to support longer and longer driving ranges, is akin to significantly increasing levels of environmental damage.

According to the same researchers at the German Fraunhofer Institute, current Li-ion battery technology used in vehicle propulsion is expected to slowly reach its limits over the next decade (Fraunhofer ISI, 2022).

For every 1% growth in the electric vehicle market, demand for lithium is estimated to increase by 70 kilotonnes per year, even though supply constraints could drive up prices and prompt supply from new markets (Deloitte, 2018).

In contrast, solid-state batteries (SSBs) using solid-state electrolytes that are under development could reach the market in the next few years, improving several key performance indicators such as energy density, safety, lifetime, cost and fast-charge capability. At the beginning of their market introduction, the cost of solid-state batteries is likely to be significantly higher than that of current batteries, partly due to lower production volumes. The fast-charging capacity of solid-state batteries is currently limited by the generally low ionic conductivity of solid electrolytes, but their design could be specifically adapted for this purpose. In terms of projected market developments, it is anticipated that solid-state battery production, which is currently below 2 GWh globally and based on SSB polymer, will increase significantly between 2025 and 2030 – when solid-state batteries based on oxide and sulphide electrolytes will reach the market. Production capacity is estimated to be between 15 and 55 GWh in 2030 and between 40 and 120 GWh in 2035.

The automotive industry's transition to electric vehicles is accelerating. The year 2026 has emerged as a starting point for an acceleration in the adoption of electric vehicles, which will drive automotive electrification trends forward. By 2030, one in four new cars sold will be an electric vehicle. Many major vehicle manufacturers around the world have signalled the end of an era of internal combustion engines (ICEs) as the transition to zero-emission vehicles (ZEVs) intensifies.

Leading automakers are expected to account for more than 70% of global EV production by 2030, up from 2022, when they accounted for just 10% of all EV manufacturers (S&P GLOBAL, 2023).

But despite the rapidly modification in the preferences that electric vehicle consumers are shifting and the unprecedented loyalty rates among electric vehicle buyers, the industry as a whole needs to address some of the issues that still characterise consumer restraint, particularly for those without garages or those who travel long distances.

The solution must be a joint effort among multiple industries: auto, utilities, government and private property owners such as shopping malls and apartment complexes. As these pathways converge, the trends for increasing the share of electric vehicles could increase significantly.

At the moment, however, the price of an electric vehicle and the limited driving range due to battery power are hurdles that have not been overcome.

For example, in 2022, a Centre for Automotive Research calculation showed that in just a few months, prices for electric cars rose by an average of 15%, due to higher raw material, wage and energy costs. The shortage of components also played a role here, as did the fact that batteries did not become cheaper at the expected rate (Barza, 2023).

In the EU, 819,000 new electric cars were sold in the first seven months of 2022, giving them a record 13% share of the total car market. The increase was 54% compared to a year ago. In Romania, 9,000 new electric cars were registered, more than 10% of the total market.

There is currently no perfect electric car and no cheap electric car. Even the most affordable full-electric models on the market are not cheap, at €20,000–25,000, and the battery remains the point where there is most room for improvement, in terms of both cost and range. But the problem is that when the capacity increases, the size and weight of the battery also increase, so if we want more autonomy, the production costs are also higher. Technology in recent years has allowed batteries to have higher capacity without huge increases in size. But batteries remain very expensive.

Prices per 1 kWh for lithium-ion batteries have been falling for the last 15 years, but by 2022, they are up about 7%. It came to a cost of about $155 per 1 kWh at the industry level. From 2008 to 2022, the price per 1 kWh dropped by nearly 90%, according to estimates from the U.S. Department of Energy's Office of Vehicle Technology.

Thus, while in 2008 it cost around $1,350 to generate 1 kWh, by 2022, the cost for the same amount of power has gone down to just $150–155. The main reasons for the price decrease are the increase in production volumes and the improvement of lithium-ion battery manufacturing technology. It is expected that the $100 per kWh per kWh threshold will be reached in 2026, two years later than previously estimated.

At the automotive industry level, batteries with a capacity of 100 kWh are seen as a limit beyond which electric cars become very competitive with thermal cars (in terms of range).

By way of example, the 26.8 kWh (gross) battery of a Dacia Spring costs around $4,000, the 60-kWh battery of a Tesla Model 3 costs around $9,000, the 82-kWh battery of a Volkswagen ID Buzz costs more than $12,000 and the 105-kWh battery of a BMW i7 is around $16,000.

For now, electric cars are very expensive, with only a few small models costing less than €30,000 before subsidies.

Research firm JATO did a study last year showing that the average price of an electric car in Europe has risen from €49,000 to €58,000 in the past seven years, and that is also because larger (SUVs, sedans) and more expensive models (Barza, 2023) have emerged.

In China, the average price of an electric car has dropped from €66,000 to €32,000 in seven years, and that is also because a lot of smaller models have appeared, many more than in Europe. And when there are €20,000 electric cars, we should not expect high ranges of 400–500 km, because it is not possible with those batteries. Real ranges will be clearly below 250–300 km.

A negative point is that electricity has become more expensive across Europe, and charging a car is not as cheap as it was 3–4 years ago. But it is clearly cheaper than gasoline cars.

For the future, many European, South Korean, Japanese and, of course, Chinese car companies have set their sights on city models costing around €20,000–23,000 to make them even more attractive.

The most striking example is Germany's Volkswagen, Europe's leading carmaker. Volkswagen has a range of electric IDs that includes several models, some costing upwards of €60,000, but no small city models yet, with the exception of the e-Up! (which will be out of production in 2024).

Those at the German group have reported that the Volkswagen and Skoda brands are working on developing full-electric models costing from €20,000 and are convinced that the launch of such cars could be possible, but no sooner than 2026–2027.

Against this global backdrop, the EU has the strongest regulatory strategy and is backing the original equipment manufacturers' ambition to phase out vehicles with internal combustion engines. Some 64% of cars sold in 2030 are expected to be battery-electric, compared with 55% in the previous forecast in Q1 2021 (S&P GLOBAL, 2023).

Despite the recession, sales of electric vehicles across the EU continue to grow steadily, reflecting the recent implementation of stricter $CO_2$ emission standards, such as a 100% $CO_2$ reduction for new cars and vans from 2035 (VIRTA, 2023).

Among other reasons, the increase in sales of electric vehicles can be attributed to incentives introduced by many European governments and various tax benefits and subsidies implemented in major markets. In this context, Norway, Sweden, the Netherlands and Germany remain the largest European markets.

The S&P GLOBAL forecast for global passenger car production is 41 million cars in 2030 – representing about 42% of the total global production of 97 million cars.

Regarding energy consumption, an important element of policies to support the purchase of electric vehicles, in 2022, the total fleet of such vehicles consumed about 110 terawatt-hours of electricity, doubling from the previous year. By 2030, electric vehicles are expected to account for less than 4% of global electricity consumption (VIRTA, 2023).

As the demand for electricity to charge electric vehicles increases, so does the need to protect the electricity grid. Careful planning of the electricity infrastructure, widespread use of smart charging and implementation of smart load management solutions will be key to ensuring healthy and balanced energy systems.

By the year 2022, the use of electric vehicles had reduced more than 80 million tonnes of GHG globally. In practice, all the emissions from electric vehicles arise from the manufacturing process of the vehicles and especially from the production of electricity, while we cannot apply a similar logic to internal combustion engine (ICE) cars. However, the shift of pollution centres from large urban agglomerations to power generation areas and especially battery factories are issues that need to be considered at a global level to benefit from real GHG emission reductions.

Future projections, based on the state policy scenario, show that by 2030, the use of electric vehicles could contribute to a reduction of 700 million tonnes of carbon dioxide equivalent ($CO_2$-eq) in GHG emissions (VIRTA, 2023).

### 8.3.3 Other types of cleaner vehicle solutions

#### 8.3.3.1 Bio-diesel

A fatty acid ester produced from vegetable oils such as rapeseed oil provides a 65% saving in $CO_2$ emissions compared to fossil diesel and could be a good supplement for decarbonising road freight transport. At present, it is clearly one of the cheapest solutions. However, the use of bio-diesel for road transportation is constrained by the limited quantities available, and its production may also risk deforestation and reduce food production, thus increasing the likelihood of global food shortages.

However, it is estimated that by 2040, it is very likely that no more than 10% of trucks will use this type of fuel (RENAUL TRUCKS, 2023).

Synthetic fuels (XTL, HVO), second-generation biofuels, offer further $CO_2$ emission reductions. Made from animal fats, waste oils or forest residues, they do not compete with food. However, the quantities currently available are very small and are likely to remain so for decades to come, as the amount of usable raw materials is limited.

E-diesel, produced using renewable electricity, water and $CO_2$ emissions from the atmosphere, also offers a 65% reduction in emissions compared to fossil diesel. However, to be available, it requires massive investments that are not yet a sure bet.

#### *8.3.3.2 Gas*

Fossil natural gas provides only 5% $CO_2$ savings compared to diesel. It is therefore not a viable option for decarbonisation.

Bio-methane or biogas, produced by fermentation or gasification of organic matter, emits 75% less $CO_2$ than diesel. However, the quantities of biogas available are and will remain limited. Thus, the transport sector will have to compete with other sectors for its use, and its price (four times that of natural gas in 2019) will increase due to the likely decrease in public funds available for its production.

Biogas emits oxides of nitrogen (NOx), which will disqualify it for city use. Finally, its potential greenhouse gas effect is 86 times greater than that of $CO_2$ over an 80-year period and therefore requires very careful monitoring to minimise the risks of leakage during its production and transport.

However, around 10% of trucks, mostly heavy-duty, heavy-duty and construction vehicles, will run on biogas from 2040.

#### *8.3.3.3 Electric vehicles and hydrogen fuel cells*

Some *electric vehicles* contain a battery, and others contain a hydrogen fuel cell and a battery. All electric vehicles have electric motors.

In battery-only electric vehicles, the electricity directly charges the battery. In hydrogen fuel cell vehicles, hydrogen is stored as fuel in a tank. The hydrogen stores energy, flows into a fuel cell, reacts with oxygen in the air and creates electricity that powers the electric motor.

Advantages of vehicles that use the hydrogen fuel cell–electric battery couple as an energy source:

- Zero pollutant emissions: the tailpipes on hydrogen fuel cell vehicles produce only heat and clean water, with no pollutants.
- Fuel cells convert the fuel's chemical energy into electricity and can be two to three times more efficient than internal combustion engines.
- Fuel cells make the vehicle more efficient and quieter because there is less vibration from moving parts. Hydrogen-based fuel allows vehicles to travel longer distances with fewer refuelling, so it is ideal for powering heavy

tractor-trailers and public transport buses, which travel hundreds of kilometres at a time.
- Increased driving range and fast hydrogen refuelling.

Hydrogen is sold on the basis of kilograms. The energy in a kilogram of hydrogen is equal to a gallon of gasoline. Hydrogen can fill a vehicle's tank in minutes, like gasoline at the pump. Currently, about 50 *refuelling stations in the US provide hydrogen* to support more than 12,000 hydrogen fuel cell vehicles and nearly 70 buses. The cost of hydrogen ranges around $13 per kilogram (kg), but as technologies advance and costs come down, hydrogen fuel will become more affordable.

Vehicles powered by hydrogen fuel cells travel longer distances using less energy. One kilogram of hydrogen contains about the same energy as a gallon of gasoline. Today, a fuel cell electric vehicle with one kilogram of hydrogen can travel about 60 miles, compared with conventional vehicles, which get about 25 miles on a gallon of gasoline.

#### *8.3.3.4 Hydrogen as a fuel for internal combustion engines*

Hydrogen is an energy carrier capable of powering on-road vehicles in two ways: as a fuel to be injected (ICE) or as a fuel to power a fuel cell (PC) system.

ICE can be modified or redesigned to run on 100% $H_2$, exploiting the properties of hydrogen as a fuel: wide flammability range, high flame speed, low ignition energy and high diffusivity (Genovese, 2022).

ICE is a mature technology and could be transformed to run on hydrogen ($H_2$-ICE), given the benefits of already existing production facilities.

$H_2$ -ICEs may be cheaper, less dependent on scarce and expensive materials and more tolerant to hydrogen impurities compared to PC technology.

Although $H_2$ -ICE emissions are water vapour, traces of carbon-based emissions (CO, HC) might be present in the exhaust gas due to petroleum lubricants and NOx resulting from the oxidation of nitrogen ($N_2$) in air by high combustion temperature.

Alternatively, stoichiometric combustion can be combined with a three-way catalytic converter to reduce NOx. Fuel injection into the intake manifold is the solution used in many demonstration prototypes, but this technology suffers from low power density, as a large volume of $H_2$ leads to air displacement in the cylinders.

This disadvantage can be solved by direct injection, a solution capable of increasing volumetric efficiency with an injection pressure above 100 bar, which overcomes the high cylinder pressure. In 2009, BMW equipped an ICE $H_2$with a high-pressure direct injection system (300 bar) and achieved a maximum efficiency of 42%, similar to a diesel engine.

In the HyFLEET: CUTE (2006–09) and CHIC (2010–16) projects, naturally aspirated injection-based ICE $H_2$ buses have been put into regular service, indicating a service availability of more than 89%.

#### *8.3.3.5 Fuel cells (FC) for road transport*

As is well known, a fuel cell is an electrochemical device in which hydrogen and oxygen react, converting chemical energy into electrical energy and yielding water as the reaction product. The reactants are individually introduced into two electrodes separated by an electrolyte that avoids direct mixing of the reactants but allows the transfer of ions.

The most common technologies adopted in the field of transportation are based on the use of a polymeric membrane (PEM) as an electrolyte. This type of electrolyte makes this PC-type technology extremely attractive due to its low operating temperature (70–90 °C), compactness and power density. Usually, a platinum catalyst is used to facilitate the reduction and oxidation reactions.

The performance of PEM PCs depends not only on the different components but also on operating parameters such as humidity and temperature. The high efficiency that characterises this system makes it energy competitive, while its rather short lifetime is an impediment for the time being.

In transportation applications, the benchmark lifetime of a PC system (defined as the maximum lifetime of a PC with an efficiency loss of no more than 10%) is 6,000 hours for cars and 20,000 hours for buses, respectively.

For these reasons, at least for buses, it is desirable to implement solutions that increase the lifetime of a PC towards 50,000 hours.

Currently, PC vehicles are expensive, although mass production could reduce costs by bringing them closer to other powertrains.

One very important problem that needs to be solved to increase the market share of PC vehicles is hydrogen storage.

Pressurised vessels up to 350 or 700 bar are the adopted option for passenger cars, but other solutions, such as cryogenic liquid hydrogen, are being investigated to increase the energy density of storage.

PC vehicles have a longer range compared to electric vehicles and a refuelling time comparable to ICE vehicles.

There are a small number of PC cars on the market, such as Hyundai Nexo, Honda Clarity or Toyota Mirai.

#### *8.3.3.6 Using hydrogen in rail transport*

Using hydrogen as a fuel is proving to be a viable alternative to diesel-powered trains on non-electrified lines in Europe. Hydrogen trains are considered competitive for non-electrified, low-frequency and long-distance rail lines.

These conditions are common in rail transport, making hydrogen rail economically attractive and an excellent opportunity to reduce greenhouse gas emissions in public transport.

Fuel cell technology, unlike other 'clean' technologies such as batteries, is able to offer high autonomy in terms of range and power rating.

The train 'Coradia iLint' of the Canadian company Alstom, for example, has $H_2$ tanks that provide a range of 600–800 km, depending on route conditions and service offered.

The Coradia iLint first entered commercial service in Germany in 2018 and has covered more than 220,000 kilometres in eight European countries. The train is powered by a hydrogen fuel cell that emits only water vapour during operation, ensuring optimal transport conditions.

On 15 September 2022, the Coradia iLint covered the record distance of 1,175 kilometres without refuelling (ALSTOM, 2023). The Coradia iLint has a top speed of 140 km/h and an acceleration and braking performance comparable to a standard regional diesel train – but without noise and emissions.

The Coradia iLint stands out with its combination of innovative features: clean energy conversion, flexible battery energy storage, intelligent traction and energy management. Designed specifically for non-electrified lines, it enables safe, clean and sustainable operations. So far, 41 linings have been ordered by customers in Europe.

The national plan for the development of hydrogen uses for rail transport in Italy, launched by the IT $H_2$ consortium, underlines that, in Italy, there are favourable conditions for the consistent introduction of hydrogen trains (Genovese, 2022).

The refuelling of hydrogen trains is programmable, so that the refuelling station can also be used for other users, such as buses or private cars.

The $H_2$ Iseo project developed by the Italian railway company FNM involves the purchase of 6 hydrogen trains produced by Alstom, with the option of supplying 8 additional trains, which from 2023 will serve the non-electrified Brescia-Iseo-Edolo line instead of the current 14 diesel trains; this project foresees the construction of hydrogen production plants, initially for the new clean energy trains and, by 2025, including local public transport and freight logistics.

#### *8.3.3.7 Hydrogen uses in the maritime sector*

The maritime sector is a major consumer of petroleum products, accounting for about 5% of world demand. Because it uses heavy oils as fuel, it produces major negative impacts on air quality, especially around ports.

In 2018, the International Maritime Organization introduced and adopted a clear strategy to reduce total annual greenhouse gas emissions from international shipping by at least 50% by 2050 compared to 2008 (Genovese, 2022).

The current role of the maritime sector in the deployment of hydrogen propulsion is limited to demonstration projects involving inland waterway vessels, coastal craft and ferries powered by hydrogen produced from renewable sources.

For example, the first experience of a hydrogen-powered vessel in Italy was realised in 2009 in Venice with the construction of the 'Accadue' water bus, a fuel cell-powered prototype. Subsequently, a new, improved version of the above-mentioned prototype, the ferry called HEPIC, was also built in the Venetian lagoon.

As far as large ships are concerned (several tens of MW of installed power on board), it currently seems difficult to make hydrogen fuel cell propulsion, especially given the large volumes that would be needed for storage. It therefore seems more likely that for systems requiring high power, hydrogen could be used more for powering auxiliary services rather than propulsion.

Consequently, hydrogen offers a real potential use for passenger ferries operating short distances with generally limited power demand and range, so that the space required for on-board hydrogen storage is not excessive.

Turning industrial ports into strategic hubs for the use of hydrogen for different types of activities would strongly stimulate the hydrogen market in closely related activities, including, for example, freight movement, material handling and heavy transport.

#### *8.3.3.8 Hybrid vehicles*

As mentioned earlier, *Plug-in Hybrid Electric Vehicle* (P*HEV*) is a highly regarded automotive technology for two very clear reasons (Grosoiu, 2023):

- The vehicle is equipped with two propulsion engines, one thermal (gasoline, diesel) and one electric (powered by a battery).
- The battery has a large capacity (perhaps even more than 10 kWh), but no more than in a 100% electric vehicle, that is, it can be charged both while braking and individually at a household socket or charging station.

There are therefore two important advantages. Firstly, it consumes less fuel than a conventional car with a combustion engine. Secondly, there is a greater range than that of the electric car, of course, and with the internal combustion engine.

There are several types of such cars:

- *Mild Hybrid*, which has a starter-generator that assists the internal combustion engine when the car starts up and when the car is moving off, thus contributing to lower fuel consumption and emissions. A mild hybrid means that the engine is switched off when the vehicle is stationary, so that the additional 48V battery resources can be used for starting. The battery is automatically recharged when braking by recovering electrical energy.
- *Full Hybrid*, which has an electric motor that can start the car on its own or in combination with a combustion engine. As the battery is charged by the internal combustion engine, full hybrid cars do not need to be charged from a plug or an electrical station. In electric mode, the car travels short distances and up to a certain speed limit. The simplest example of this is the Toyota Prius.
- *The Plug-in Hybrid* is similar to the full hybrid, except that it can also be charged at a plug socket or electric charging station (hence the name 'Plug-in'), and the battery capacity is larger, as is the range, which can average more than 50 km, depending on the manufacturer. What is more, energy recovery during braking is more efficient.

Basically, the plug-in hybrid vehicle, or a rechargeable hybrid vehicle,[3] can be considered as a bridge, combining two different propulsion technologies, towards one of mankind's most important missions: a green future.

## Notes

1 In making this assessment, the following were taken into account: a car using conventional fuel (gasoline or diesel) produces about 4.6 tonnes of carbon dioxide per year; burning one tonne of coal produces 1.9 tonnes of carbon dioxide; each 15-m-long coal car holds about 100 tonnes of coal; and the circumference of the earth at the equator is 40,075 km, i.e. 40,081 km (RFIDTIRES, 2023).

2 The Plug-in Hybrid Electric Vehicle, abbreviated 'PHEV', is a highly appreciated automotive technology for two very clear reasons: 1) the car is equipped with two propulsion engines, one thermal (gasoline, diesel) and one electric (battery powered); 2) the battery is of large capacity (perhaps even more than 10 kWh), but no more than a 100% electric vehicle, that is, it is charged both while braking and individually at a household outlet or charging station. There are therefore two major advantages. First, you have lower fuel consumption than a conventional car with a combustion engine. Second, you get a longer range than an electric car, of course, also with the combustion engine (Grosoiu, 2023).

3 According to Gabriel Grosoiu, 'The first hybrid was built in the early 1900s, under the signature of the young electric enthusiast Ferdinand Porsche. The same man who, many years later, founded the famous Porsche automobile company. The Lohner-Porsche "Mixte" was a carriage-type vehicle equipped with both an electric battery and an internal combustion engine. This was followed after the two world wars by the plug-in Hybrid XP-883 concept developed by General Motors, and the Audi 100 Avant Quattro "Duo" wagon, produced in 1989, was the first plug-in Hybrid produced by a European car company. After the 2000s, car manufacturers became increasingly interested in designing a hybrid car. Initially, the Chinese and American markets enjoyed the most of the new car technologies, with mass-produced hybrid vehicles. By 2020, the Mitsubishi Outlander PHEV was the best-selling plug-in Hybrid SUV, and the Toyota Prius continues to carry the tradition forward into a sustainable future' (Grosoiu, 2023).

## References

ACEA, *Vehicle in use in Europe, report 2022*, 2022, https://www.acea.auto/publication/report-vehicles-in-use-europe-2022/

ALSTOM, *Rolling stock solutions*, 2023, https://www.alstom.com/solutions/rolling-stock

Bălăşoiu D. and Bălăşoiu T., Transportul public urban – de la exemple de bună practică spre strategii europene (Urban public transport – from examples of good practice to European strategies), Buletinul AGIR nr. 4/2009 (AGIR Bulletin No. 4/2009), Bucureşti, ISSN 2247-3548, 2009.

Barza V., 2023, https://www.hotnews.ro/stiri-mediu-26499571-nu-vom-avea-prea-curand-masini-electrice-oras-sub-20-000-euro.htm

Deloitte, *Making the future of mobility*, 2018, https://www2.deloitte.com/content/dam/Deloitte/br/Documents/energy-resources/DI_Making-FoM.pdf

Deloitte, Hydrogen. Making it happen, https://ro.register-deloittece.com/forms/registration.html?docid=354&utm_source=CP&utm_medium=RO&utm_campaign=deloitte-romania-hydrogen-making-happen, 2023

EUROSTAT, *Greenhouse gas emission statistics – Emission inventories*, 2019, https://www.sipotra.it/wp-content/uploads/2019/06/Greenhouse-gas-emission-statistics-emission-inventories.pdf

Fraunhofer ISI, *What will solid-state batteries be like in the future?*, 2022, https://www.isi.fraunhofer.de/en/presse/2022/presseinfo-12-Roadmap-Feststoffbatterien.html

Genovese A., *Hydrogen and "green transport"*, 2022, https://www.eai.enea.it/archivio/pianeta-idrogeno/hydrogen-and-green-transport.html

Grand View Research, *Fuel cell vehicle market size report*, 2023, https://www.grandviewresearch.com/industry-analysis/fuel-cell-vehicle-market

Grosoiu G., *What plug-in hybrid means and what models are on the market*, https://www.autovit.ro/blog/ce-inseamna-plug-in-hybrid-si-ce-modele-sunt-pe-piata/#Ce%20%C3%AEnseamn%C4%83%20plug-in%20HybridRFIDTIRES, 2023, How many cars are in the World, https://www.rfidtires.com/how-many-cars-world.html

Iberdrola, Green hydrogen: an alternative that reduces emissions and cares for our planet 2022, 2020, https://www.iberdrola.com/sustainability/green-hydrogen

Mason J., *The future of transport is sustainable shared mobility*, IITDP, 2018, https://itdpdotorg.wpengine.com/wp-content/uploads/2018/02/The-Future-of-Transport-Is-Sustainable-Shared-Mobility.pdf

MOOVIT, *Commuter kate: The history of public transit*, 2018, https://moovit.com/blog/history-public-transport/

RENAUL TRUCKS, 2023, https://www.renault-trucks.ro/static/ce-tipuri-de-surse-de-combustibil-trebuie-utilizate-pentru-decarbonizare

RFIDTIRES, *How many cars are in the world?*, 2023, https://www.rfidtires.com/how-many-cars-world.html

Rodrique J.P., Transportation and energy, 2022, https://transportgeography.org/contents/chapter4/transportation-and-energy/

S&P GLOBAL, *Electric vehicle trends*, 2023, https://www.spglobal.com/mobility/en/topic/electric-vehicle-trends.html?utm_source=google&utm_medium=ppc&utm_campaign=PC023665&utm_term=ev%20sales&utm_network=g&device=c&matchtype=p&gclid=Cj0KJ0KCQCQjwgNanBhDUhARIsAAeIcAsWOViNfwgFBKK3hO487SILXAJdKc198x-nXethH2O5zSY-p6KIFLUYaApycEALw_wcB

Speirs J., *Safe, attractive, and sustainable cities: A mobility-oriented approach*, Ecole Polytechnigue Federale de Laussane, 2019, https://iglus.org/wp-content/uploads/2019/05/Safe-Attractive-and-Sustainable-Cities-A-mobility-oriented-approach.pdf

UITP (The International Association of Public Transport), *Bus tender structure report 2020*, 2020, https://www.uitp.org/publications/bus-tender-structure-report-2020/

VIRTA, *The global electric vehicle market overview in 2023: Statistics & forecasts*, 2023, https://www.virta.global/en/global-electric-vehicle-market

# Conclusions

Sustainable development has been interpreted in different ways. Sometimes it has been used simply to emphasise the importance of continued and steady economic growth. In other circumstances, however, it has been limited to an environmental objective. The operationalisation of the concept, therefore, depends on how it is defined. Because of all the existing interpretations of the concept, it is necessary to be quite explicit whenever it is used.

In a world in which environmental degradation is a major problem and possibly a critical issue for the long-term existence of our society, solutions should be sought from different starting points and from different fields.

Recently, global concerns about climate change, environmental impacts and limited financial resources illustrate the need for a different approach in selecting transport solutions. There is, therefore, a growing desire to ensure environmental sustainability in the transport infrastructure system. A sustainable transport system needs to be safe, healthy, available and renewable, operating in an equitable way and limiting pollutant emissions and the use of non-renewable resources.

The need to remedy some of the serious distortions and negative effects of transport on the environment and human health has forced mankind to take the issue of sustainable transport development increasingly seriously. New elements are constantly being added to what were initially timid achievements, so that there is now a worldwide commitment to implementing sustainable transport systems as quickly and effectively as possible.

The increase in the negative effects of transport activities and infrastructure (chemical and noise pollution, occupation of land which could be used for other purposes, accidents, reduced health standards, congestion of transport networks, constantly increasing energy consumption, etc.) has made it not only necessary but also mandatory to implement numerous methods and techniques to support sustainable transport.

Given the speed of their implementation, as well as the introduction of new technologies and products that are much more environmentally friendly, we can say that sustainable transport will become a major segment of this economic sector in the near future.

DOI: 10.4324/9781003743064-10

Unfortunately, the degree to which sustainable transport levels can be achieved depends both on the level of development of the area in which it can be developed and on the policies of the authorities, which means that there are currently major differences not only at the global level, but also at the zonal, regional or even local level. This is an aspect that needs to be reviewed and resolved.

The future, from a transport perspective, will look good if we, the people of today, put the needs of our children and grandchildren before our own desires. Everything else will have a favourable resolution.

Considering the increased need for mobility and the low degree of modernisation of the infrastructural networks, and for the favourable evolution of the socio-economic environment, several priority measures are necessary, in our opinion, among which we emphasise:

- Developing a coherent, sustainable transport system. Permanent encouragement of those transport subsystems that support sustainable development processes. The concrete measure is the internalisation of external costs.
- Redirect budgetary financial allocations away from highway construction towards modernisation and increasing the capacity of European, national and county roads. Taking into account a minimum value of about two million Euro for the construction of a single kilometre of highway and 300,000 Euro for the modernisation/rehabilitation of one kilometre of national/European road, we can conclude that the total restoration of the current national road network to the highest standards and European parameters would cost on average the same as the construction of 2,000 km of highways, and would create many more jobs, due to the scale and geographical distribution of the works.
- The development of high-capacity and environmentally friendly transport networks in regions where this is possible.
- Creating economic facilities for business incubators to provide financial support for the development/modernisation of regional transport infrastructure.

As the economy grows and output, sales and incomes increase, it is generally fair to assume that demand for transportation would increase. At the same time, it is clear that there are a number of market failures due to the unaccounted-for impacts of transport (*non-internalised* external costs) that can be corrected by appropriate pricing and could therefore help to internalise negative externalities and reduce inefficiencies and negative impacts of transport on society in general.

However, the internalisation of the various external costs of transport affects the level of transport activities and the choice of travel modes, thus further reducing the levels of externalities.

Negative externalities caused by transport activity are an important factor in the overall process of environmental degradation. In Europe, for example, tens of thousands of people die every year due to air pollution (mainly particulate matter). Transport-related accidents are another major cause of death. In Europe, where tens of thousands of people die each year, mainly from road accidents, this phenomenon is the leading cause of death in people under 40.

Moreover, in all OECD countries, road transport is responsible for most of the negative impact of the transport sector on human health and the environment. It is responsible for more than 80% of the sector's total energy consumption, most accidents, most emissions of chemical and noise pollutants and habitat degradation. Maritime transport, although generally associated with lower environmental impacts, continues to raise concerns about oil pollution caused by major accidents and spills of residual fuels from tanks. More recently, there has been growing concern about the impact of air traffic on the global environment, particularly in relation to tourism and leisure travel. Air transport currently accounts for about 11% of transport-related energy consumption.

From this point of view, it is necessary to compare trends in transport activity, the level of external costs and, of course, general economic development (GDP) over a given period of time.

Most modern societies value the independence of movement. The desire for individualised, unrestricted mobility is another desideratum that influences and will continue to influence transport demand. The ability to travel where we want, whenever we want and without being tied to fixed timetables has been and still remains a difficult step to take. As a result, some transport companies are already rushing to make sure their businesses are ready for the future, whatever shape it takes. One thing is certain: the next generation of transit will not be characterised by a smooth ride in terms of mobility policies and strategies.

Mobility will increase, but the transportation sector is in a dilemma, facing major challenges such as:

- The situation is where oil reserves will decline in the coming decades, and more and more will have to be imported from unstable regions of the world. Oil prices are expected to rise at least two to three times between 2005 and 2050. Current events demonstrate the extreme volatility of oil prices.
- The reality is that although transport has become more energy efficient, 96% of the fuels used are oil-based.
- Evidence that transport produces massive pollution with greenhouse gas emissions, which are responsible to a significant extent for climate change (accelerating the destructive phenomena of planetary protection and disrupting the natural cycles of planetary evolution).
- There are big differences between the eastern and western regions of the EU in the level of infrastructure development. In the new Member States, there are, moreover, less than 5,000 km of motorways and no special high-speed railways, while conventional railways are often in poor condition.
- The fact that Community transport is facing increasingly fierce competition in specific, rapidly expanding world markets.

Multimodal transport, because of its many advantages in both economic and environmental terms, is, in our view, a viable alternative to the current situation, which is characterised mainly by a profound imbalance between road and other modes of transport. However, the economic and social implications are enormous. If the efficiency of the multimodal transport network is directly linked to the investment resources that can be attracted for the modernisation and development of infrastructure, equipment and multimodal transfer points, the diversification of funding sources is extremely important. Attracting private capital, the Structural and Cohesion Funds, and the existence in the national budget of funds earmarked for this purpose are mandatory conditions for achieving the proposed objectives.

Sustainable mobility must have two major elements addressed simultaneously: full coverage of mobility needs and a decrease in transport demand.

Clearly, looking also at the many examples presented, technical solutions exist. The only condition, without which it will not be possible to achieve a real sustainable transport system, is the political will to do so.

Apart from this, there are several other influencing factors in the process of implementing such a transport concept.

The use of information technology (IT) can influence the development of transport in two essentially different ways. One is due to the introduction of intelligent transport systems (ITS), that is, information technology used in traffic, including public transport information systems and dynamic route choice systems. The other is due to the effects on transport demand, mainly through the 'relocation' of activities, which may lead to changes in the organisation of society.

In another vein, as outlined in the current study, we have to accept that transport remains almost entirely dependent on fossil fuels and that this would have to change completely in order to make transport comply with the new sustainability status. This is a requirement that is seen by many as perhaps the most difficult task for future transport.

To strive for sustainability, investments in transport infrastructure should be evaluated in terms of the service they can provide, not in terms of the money that has historically been spent on it. Thus, the principle opens the way for solutions to transport problems outside the traditional transport domain; for example, it demonstrates that in some cases it may be possible to trade the value of physical mobility for access through information technology.

Important work also needs to be done to reduce the demand for transport without affecting the natural needs of human society (work, education, recreation, etc.). Increasing opportunities for a person to travel provides greater accessibility to services if nothing else changes. However, where this increased mobility reflects higher car use, it also includes social costs in terms of, for example, car dependency. At an aggregate level, increased mobility for some is likely to reduce accessibility for others. This happens when services requiring access to cars drive other competing businesses out of the market.

Unfortunately, transport inequalities between regions are only a reflection of global inequalities and cannot be addressed by transport strategies alone. They can only be addressed as part of a more general redistribution of wealth.

Therefore, the implementation of general policy measures, such as an increase in the $CO_2$ tax, would, in the short term, probably lead to a greater reduction in leisure travel than business travel. If it is true that leisure travel is generally preferred to business travel, then alternative measures should be addressed to prevent a situation where high transport prices force people to reduce leisure travel.

Reducing net pollution, encouraging modal diversity and improving the infrastructure network will create a safe environment on city streets, which will further increase the attractiveness of non-motorised travel. More citizens using non-motorised transport could lead to less sedentary lifestyles and an overall healthier community. In addition to the health and safety benefits, a transportation network with a more diverse set of options will also increase its attractiveness for housing.

Unfortunately, as can be seen from the examples presented, the measures taken so far have not been concerted and complementary, and as a result, stand-alone actions have been taken in different regions of the planet without a relevant impact at the global level. For this reason, but not only, it is also necessary to address activities to support the development of a sustainable transport system at the global level.

However, this can only be achieved through the adoption of a common agreement recognised by all states. Perhaps a strategy to support a sustainable transport system adopted under the aegis of the UN would be one of the best solutions.

In addition, the need for a sustainable mobility plan in the field of transport requires the implementation of appropriate legislative instruments. The overall objective of these measures must be to ensure that each mode of transport enters the specific market under the same competitive conditions. In other words, each mode of transport must also be analysed in terms of the level of total social costs, eliminating (reducing) as far as possible external costs that cause competitive distortions and even 'hidden subsidies'. One way, often used recently, to solve this problem is to internalise external costs.

In order to implement a strategy for internalising external costs, naturally, the first step is to assess the extent of all these types of costs.

In this paper, an attempt has been made to present the complexity of approaching this type of approach, the most important techniques and methodologies of analysis and evaluation and their relevance for each external transport cost.

The analysis has shown that there is no single 'state-of-the-art' approach to estimating external costs. For example, for infrastructure user costs, there may be a preference for econometric estimation, but it is rarely possible to do so at a level of detail that directly lends itself to pricing decisions; hence, a mix of econometric research and various engineering and cost accounting approaches is needed. In terms of user costs, road congestion can be extensively investigated using either direct (single) link models or network models based on speed relationships and formulas to estimate the delays (congestion) caused by congestion.

In the case of accidents, progress has been made in estimating the key parameters that explain how the risk of accidents varies with traffic growth, but this is still a controversial issue.

In terms of environmental costs, detailed estimates have been obtained for a variety of circumstances using the methodological approach of estimating the impacts produced. An important aspect of this approach is the uncertainty created by simpler approaches, such as regression, which certainly does not provide an acceptable level of precision, is not clearer and is not cheaper in any given situation.

The analysis carried out in this study, in terms of the size of external costs, has shown that the highest external costs are in the area of road transport. In this area, as expected, congestion costs predominate, followed, only in urban areas, by accidents and noise, with global warming and air pollution being much lower. Overall, the differences between areas are quite large. Outside urban areas, noise and accident costs are, of course, much lower.

For other transport modes, external costs are generally lower, with congestion being much less important, but we have no quantification of pure deficiency costs, which may be important for rail and air transport.

In the end, we can conclude from all the analyses carried out that it is difficult, if not impossible, to establish clear links between the evolution of the economy in general and the level of external costs specific to transport. However, without the adoption of policies and strategies to support sustainable mobility (use of less polluting modes of transport, internalisation of external costs, etc.), it is certain that the development of transport activities (freight and passenger transport) against the background of growing travel demand, in tandem with the general level of economic development, will invariably lead to an increase in pollution, congestion, accidents, etc. In this context, sustainable economic growth can only be underpinned by a policy to support sustainable mobility.

The 'technological revolution' is currently continuing in the field of road transport, with the aim of replacing vehicles powered by polluting fuels with vehicles with lower $CO_2$ emissions, with electrification being the main solution for the time being. Battery-powered cars and motorcycles have a high potential to reduce $CO_2$ and pollutant emissions with zero emissions 'on the road', while overall emissions depend on the energy mix used to generate electricity. The range of battery electric vehicles (BEVs) and recharging times are now more responsive to users' needs and make BEVs more attractive to meet short and medium distance mobility requirements. Current car batteries have an energy capacity of up to 100 kWh, offering a real range of 300 km and more.

All the examples analysed converge towards the idea of making energy consumption in transport more efficient, while at the same time changing the types of engine propulsion sources, towards solutions without air and noise pollution, and reducing the demand for mobility without affecting the independence of movement of people and goods.

The integrated sustainable city, in which transport plays an extremely important role, is a natural necessity in view of the current trend for large cities to 'expand' at the expense of areas that could be used for afforestation or agriculture. The new urban concept should not, however, be understood as a restriction of rights or freedoms, but simply as a model that offers more by reconfiguring services and logistic centres with various uses (economic, social, cultural, ecological, etc.). In short, an actual city reconfigured on the basis of the concept of integrated sustainability will have all the current functions, but it will be made up of several hubs or neighbourhoods that can offer the inhabitants of the area access to all kinds of essential services while significantly reducing the amount of travel to reach the different locations. The most illustrative way in which we have imagined this new model of (integrated sustainable) urban development is that of a molecule (the city as a whole) made up of many atoms (the neighbourhoods or hubs). Each molecule in turn has all the elements necessary for survival and multiplication, and together with the other molecules, they form and support the development of the cell, in our case, the integrated sustainable city.

A requirement for the foreseeable future, linked to the sustainability of transport and thus of specific infrastructures, is to reduce the total demand for travel (goods and passengers), while at the same time fully meeting mobility needs without harming the environment and the health of living beings. This sounds like a definition of sustainable transport, but, in our opinion, this should be the immediate aim of the transport development strategy in Romania.

In order to develop such a type of infrastructure, in the present study, we have defined both the concept and the main instruments necessary for its practical realisation.

Thus, we considered that sustainable transport infrastructure represents the totality of networks supporting the mobility of passengers and goods, designed, built and maintained in accordance with the principles of the circular economy and which must ensure a high degree of adaptability to the mobility demand of future generations, without damaging environmental factors and the health of living creatures.

Structurally, for this new concept, the four pillars (different from only three in the definition of sustainable development) are the premises (social, economic and environmental) and a guiding mechanism (resilience). The mechanism is the binder of the three premises, and the result is the continuous process of sustainability (sustainability).

The result of this ongoing resilience process of sustainability has been defined as neo-sustainability (or SUSTAINABILITY 2.0)

To realise such a system in practice, however, it is necessary to:

- Much of the current infrastructure network should be reoriented to support sustainable transportation.
- Invest, as a priority, in the construction of new infrastructure to support sustainable development in general and cleaner transport in particular.
- Give priority to the use of recycled, recyclable and easily maintainable materials for the construction of new infrastructure, and adapt them to future changes in transport demand. Moreover, the circular system must be implemented in the life cycle of transport infrastructure.

Throughout the entire transport infrastructure lifecycle chain (planning, design, construction, maintenance, upgrading and recycling), consideration should be given to the implementation of effective analysis and control measures that can confirm the 'correct life trajectory' of the transport infrastructure under analysis or, failing this, the most effective and rapid corrective measures to correct any deviations/non-compliance.

In conclusion, we must restate the obvious truth: *we are bound to nature and we cannot live without it*. It is, however, equally obvious that we humans must fundamentally change some of our habits and some of our concepts of the evolution of systems on the planet, with the simple idea that we cannot control nature, but that it can protect us or not. It is up to us. So, only in understanding with nature can we move forward and coexist on this planet (*Integri procedamus!*).

This is why we cannot let transportation evolve as it has in the last decades, and we have to make a major, strictly necessary shift towards a sustainable mobility system. And in this approach, in our opinion, the 'key' element is to reduce the overall demand for transport for non-essential trips, supported by the development of the right solutions to ensure optimal living conditions and economic and social development in balance with the environment.

# Index

Pages in *italics* refer to figures and pages in **bold** refer to tables.

For Product Safety Concerns and Information please contact our EU
representative GPSR@taylorandfrancis.com
Taylor & Francis Verlag GmbH, Kaufingerstraße 24, 80331 München, Germany

www.ingramcontent.com/pod-product-compliance
Ingram Content Group UK Ltd.
Pitfield, Milton Keynes, MK11 3LW, UK
UKHW022030061126
477531UK00003B/138

* 9 7 8 1 0 4 1 2 4 6 6 9 5 *